DRIVING WITH MUSIC: COGNITIVE-BEHAVIOURAL IMPLICATIONS

From listening to music and singing in the car ('car-aoke') to the sound tracks that accompany movie car chases and computer car games, music is essential to human beings' engagement with the automobile. Warren Brodsky tells the history of cars-&-music, explains the different functions performed by music-listening in everyday life, and discusses drivers' perceptions of the effects of music in the autosphere on them, their passengers and (particularly in the case of "boom cars") the public at large. Informed by the critical review of a wealth of research, and reports of recent studies by the author intended to develop the use of music as a strategy for safe driving, Driving with Music *not only provides a highly readable and enjoyable account of a fascinating phenomenon, familiar to us all, but also demonstrates the potential impacts, positive as well as negative, of drivers' in-car musical behaviours.*

Jane Ginsborg, Royal Northern College of Music, UK

This timely and unique volume brings an encyclopaedic knowledge of everything to do with music and cars to bear on the critical issue of road safety. Music in cars may have benefits, but too often can also cause accidents. Informed by a masterful review of existing studies, Warren Brodsky and his collaborators have developed and trialled music specifically designed to enhance driver safety. In doing so, a much needed gap between research and effective practice has begun to be filled.

John Sloboda, Keele University, UK

In this book, Brodsky's deep passion for music, respect for car culture and forensic scientific analyses emerge in a symphony of high quality writing. This book will be of interest to readers wanting to understand how the accompaniment of music fundamentally shapes the driving experience.

Nick Reed, TRL Academy Director, UK

Human Factors in Road and Rail Transport

Series Editors

Dr Lisa Dorn
Director of the Driving Research Group, Department of Human Factors, Cranfield University

Dr Gerald Matthews
Associate Research Professor, Institute for Simulation and Training, University of Central Florida

Dr Ian Glendon
Associate Professor, School of Psychology, Griffith University

Today's society confronts major land transport problems. Human and financial costs of road vehicle crashes and rail incidents are increasing, with road vehicle crashes predicted to become the third largest cause of death and injury globally by 2020. Several social trends pose threats to safety, including increasing vehicle ownership and traffic congestion, advancing technological complexity at the human-vehicle interface, population ageing in the developed world, and ever greater numbers of younger vehicle drivers in the developing world.

Ashgate's Human Factors in Road and Rail Transport series makes a timely contribution to these issues by focusing on human and organisational aspects of road and rail safety. The series responds to increasing demands for safe, efficient, economical and environmentally-friendly land-based transport. It does this by reporting on state-of-the-art science that may be applied to reduce vehicle collisions and improve vehicle usability as well as enhancing driver wellbeing and satisfaction. It achieves this by disseminating new theoretical and empirical research generated by specialists in the behavioural and allied disciplines, including traffic and transportation psychology, human factors and ergonomics.

The series addresses such topics as driver behaviour and training, in-vehicle technology, driver health and driver assessment. Specially commissioned works from internationally recognised experts provide authoritative accounts of leading approaches to real-world problems in this important field.

Driving With Music: Cognitive-Behavioural Implications

WARREN BRODSKY
Ben-Gurion University of the Negev, Israel

CRC Press is an imprint of the
Taylor & Francis Group, an **informa** business

CRC Press
Taylor & Francis Group
6000 Broken Sound Parkway NW, Suite 300
Boca Raton, FL 33487-2742

First issued in paperback 2017

ISBN-13: 978-1-4724-1146-4 (hbk)
ISBN-13: 978-1-138-74888-0 (pbk)

Visit the Taylor & Francis Web site at
http://www.taylorandfrancis.com

and the CRC Press Web site at
http://www.crcpress.com

Contents

List of Figures

List of Tables

About the Author

Professor Warren Brodsky is Director of Music Psychology in the Department of the Arts at Ben-Gurion University of the Negev (Beer-Sheva, Israel). After a short-lived career as a music performer, he completed an artist-degree (BMus) in Orchestral Percussion, and majored in early childhood musical development leading to a Diploma and Certificate of Teaching (K-6) from the Rubin Academy of Music (Jerusalem, 1982). Warren trained as a music therapist at Hahnemann Medical University (Philadelphia, USA, 1984); during his 10 year clinical career he was registered (RMT), certified (CMT), and board-certified (MT-BC) in the USA, registered (RMTh) in the UK, and licensed as a Creative and Expressive Therapist in Israel. Warren completed his PhD in Psychology at Keele University (Staffordshire, UK, 1995) under the eminent music psychologist John A. Sloboda. At Ben-Gurion University of the Negev, Dr Brodsky served two 2-year Post-Doctorial Fellowships (1996–2000). Among his research initiatives and projects are: Mental Representations of Musical Notation; Handclapping Songs as a Spontaneous Platform for Cognitive Development; Positive Aging among Symphony Orchestra Musicians; Functional Application of Music in Automobile Branding; and The Effects of Music on Driver Behaviour and Vehicular Control. Professor Brodsky can be reached at: wbrodsky@bgu.ac.il.

Source: Photograph by Elisha Brodsky.

Preface

A long time ago, in fact what now actually seems to have been *as if* in a previous lifetime, I trained as a music therapist at Hahnemann Medical College in Philadelphia (USA). I was registered (*RMT*) with the National Association for Music Therapy, certified (*CMT*) by the American Association for Music Therapy, board certified (*MT-BC*) by the Certification Board for Music Therapy, and registered (*RMTh*) with the Association of British Professional Music Therapists. Subsequently, I was licensed as a Creative and Expressive Therapist by the Ministry of Health in Israel. I practiced music therapy in special education schools, vocational centres, hospitals, and mental health settings. I engaged in a Jerusalem-based private practice and trained music therapists at a college in Tel Aviv. These experiences offered me a deep respect and understanding of human behaviour – how to manage and advance adaptive functioning through the use of music.

Then, in the late 1990s, the use of music in everyday life had finally reached the centre stage of Music Science research. Before that, only a few investigations had documented how real people employ music in social spaces. But in the final years leading up to the new millennium, several initiatives capitalized on a conceptual underpinning demonstrating that the effects of music were not at all disassociated from the specific contexts in which we use music. These studies pointed out that not only do we do things *to* music, but also most of the time we do things *with* music playing in the background. The first large-scale study to document the usage of music in everyday life was implemented by John A. Sloboda (my PhD supervisor and mentor) who surveyed 500 representative correspondents in a British National Survey about their lifestyle and listening habits (Sheridan, 1998). The findings unequivocally demonstrated that activities that were accompanied by music were predominantly domestic or solitary; most frequently these were reported to involve housework or driving (Sloboda, 1999). At the time, I remember thinking: *Driving? That seems a bit odd.* The study also stressed that while many situations containing everyday activities offer little room for music involvement, other types are more open to background music – and these more likely occur when people are alone and have the opportunity for personal choice over the music. It appeared to me to be somewhat absurd that the popular location where individuals seem to be found when listening to music was not in the comfort of their living room, nor was it shared with social agents such as intimate partners, extended family, or friends. But rather, the circumstance most frequently reported while listening to music involved unaccompanied vehicular driving.

The more I looked into the phenomenon, the more I found that *driving with music* did not interest music psychologists or traffic researchers. Yet, so many people all over the world drive automobiles! I wondered:

- Does music have any effect on how people drive?
- Does music enhance or detract vehicular performance?
- What kind of music do drivers listen to?
- Do drivers have the same preferences for in-cabin music as when they listen to music in their home?
- Do drivers pre-plan the albums and/or music tracks they take along for the ride?
- Do drivers differentiate songs by type of trip depending on where they are going, the distance dictating how long the trip will last, or how many passengers may be seated in the vehicle?

What seemed to me as evident was that driving with background music might be *the* most prevalent music behaviour ever taken on by the human species. That is, the frequency of driving hours worldwide per day seems unfathomable, and perhaps even more massive than the number of those who attend religious services where music behaviour such as communal singing is an integral part of the experience. And yet, my questions about music effects on driver behaviour fell on deaf ears. Foremost, Music Science (an umbrella term for music psychology, music perception and cognition, and music education) has all but redeployed to the 'next frontier' of neuroscience; most researchers involved in music and brain seem to view low-tech everyday behavioural manifestations as an agenda of yesteryear. Second, even traffic scientists who explore driver behaviours saw little value in investigating music (beyond the possible distraction effects caused by the structural mechanical actions of manipulating a car radio). That is, the *music* – the actual sounds drivers hear – did not in the least seem to be of concern to anyone, including legislatures, law enforcement agencies, traffic safety authorities, traffic accident examiners, crash prevention investigators, driving behaviour psychologists, driving instructors, parents of young novice drivers, the actual drivers themselves – and certainly not those with the greatest ability to enhance our understanding of how music may have causal effects on driver affect and vehicular performance – music scientists.

This book addresses these issues and explores the everyday use of music listening while driving a car. It presents the relationship between cars and music in an effort to understand how music behaviour in the car can either enhance driver safety or place the driver at risk for incidents, near-crashes, accidents, and fatal crashes. Initially, the text outlines the automobile, its relationship to society, and the juxtaposition of music within the automobile as a complete package. Then, the text highlights concepts from the field of Music Perception and Cognition; these will enable the reader to gain a better understanding about the functional use of background music in our everyday lives. Subsequently, music behaviours of drivers – both adaptive and maladaptive – are outlined focussing on benefits, contraindications, and the ill-effects of in-car music listening. Finally, implications, countermeasures, and applications for driving with music are suggested, and an experimental music background designed to increase driver safety is outlined.

The text has been written for several professions and interest groups. Foremost, the text accommodates research scientists related to transportation psychology, driver behaviour, and music psychology; these professional are those capable of studying human behaviours linked to the engagement of music activity in the car. The text presents a highly critical review exposing the methodological blunders employed by researchers who demonstrate an almost total lack of insight, as well as less than acceptable level of rigour, towards employing music in empirical settings. The book also offers much information to transportation-related specialists such as law enforcement practitioners, safety and accident investigators, legislators, insurance companies, professional driver training instructors and organizations, and high school educators; these readers need to gain new insights about music as a factor component of driving performance. Finally, the book has been tailored for the actual drivers themselves; namely, the text has taken on an overall friendlier language with many explanations and examples to boost interest for non-academic readers. The book is clearly intended as a resource sourcebook with appendixes of materials for all peripheral fields of interests, including: psychology and human development, sociology, safety and traffic studies, media studies and communications, advertising and marketing, human factors, automotive sound design, and music composition studies.

The subject of *cars-&-music* is actually one of great importance. Over the last decade many studies that have targeted our everyday music experiences point to the car as the most prevalent location where we listen to music. Yet, not much has been written about this social behaviour, and very limited evidence indicates if drivers actually choose music to hear in the car. Some drivers may have specific preferences for driving with music in a car, and these may be quite different than their general preferences and musical taste. But, there are also contraindications when it may be best not to play music as well as specific *ill-effects* of music on driver behaviour and vehicular performance. These are virtually unknown to the public, and therefore knowledge that music may place drivers more at risk is simply not passed on. The automobile is perhaps the only everyday circumstance where specific features in the background music can increase the possibility of human fatality. Transportation researchers have begun to look at multi-sensory input in vehicles and how auditory information (i.e., in-cabin sounds) effect the human-machine interface. Finally, with the car becoming a mobile telecommunications terminal including online entertainment and information retrieval, several investigations have contemplated if drivers should be warned about conflicting sensory inputs and the costs of these on their road safety. But yet, little has been published about these subjects, nor have scientists made the public aware about the (in)congruency of temporal inputs such as between music and the visual-optic flow that comes across the windshield.

The literature points to many driving tasks that interfere with one another. These demonstrate how the decrement of vehicular performance mostly occurs from *structural interference* subsequent to poor mechanical configurations. That is, for the most part, music is seen by road safety researchers and legislatures as: 'turning on the radio', 'toggling a channel knob', 'adjusting the volume', 'fine-tuning the tone button', 'flipping-over a cassette tape', 'swapping between compact disks', or 'scrolling MP3 playlists'. Hence, the music – the actual tones heard – have never been considered thoroughly. Certainly, *capacity interference* to central attention subsequent to overtaxed cognitive faculties is just as important to investigate as is structural interference. Yet, only few studies have even considered that in-cabin aural backgrounds are a contributing factor to human driver error and traffic violations. The current book attempts to fill this gap.

The book opens with two background chapters that set the foundation for the specialised topic under discussion. Chapter 1 highlights the association between music and cars; the chapter looks at the culture of automobiles and its effect on society (including a sociological and historical overview), the development of car-audio, and the associated links between popular culture and automobility. Chapter 2 outlines the effects of background music as experienced in everyday life; the structural features of music, music complexity, and genres of Western music are described as well as interlinked to human sensation, perception, cognition, and emotion. After these, the book presents four topic-specific chapters. Chapter 3 outlines in-car music behaviours; the extent of employing music in the car is described through survey data from naturalistic driving studies in North America and Europe, and trends such as specialty compilations of driving songs are discussed. The chapter also looks at benefits and the emotional aspects of driving with music; surveys illustrating widespread popular beliefs about in-car music listening are outlined, and research findings showing positive outcomes to listening to music while driving are presented. Chapter 4 delineates the contraindications to in-car music listening; structural interference related to HMI ergonomics and mechanical configurations, perceptual masking, capacity interference to central attention subsequent to overtaxed cognitive faculties, and social diversion are all presented and discussed. Chapter 5 explores specific *ill-effects* of in-car music listening. The three variants of music effects on driver behaviour and vehicular control outlined, are: music-evoked arousal, music-generated distraction, and music-induced aggression. Chapter 6 presents implications, countermeasures, and applications towards driver well-being; efforts to promote safe driving through music choice, volume adjustment and control, and the dynamic temporal flow are presented. Finally, the chapter outlines a unique original music background designed for increased driver safety. Chapter 7 brings the book to a close with a more personal Postscript that reviews the overall text, summarizes the main purpose of each chapter, and reveals the overriding conceptual message that has been threaded into the fabric throughout the entire book.

Acknowledgements

The author wishes to acknowledge the following people, without which this book could not have materialized. Erez Ein Dor for preparing the *Car Movies* list, the *Car Games for Video & PC* list, and the *Music CDs for Driving* list (Appendixes B, C, F); David Louis Harter (California Technology) for permission to reproduce the *Ultimate List of Car Songs* (Appendix D); Lewis Carlock (Digital Dream Door) for permission to reproduce the *100 Greatest Car Songs* list as compiled by Motor DJ 'Golden Gup' (Appendix E); Richard F. Weingroff (Office of Infrastructure, Federal Highway Administration, United States Department of Transportation) for permission to reproduce the *Road Movies* listing (Appendix A); Alistair McKillop (Group Communications Manager, Market Management, Allianz Australia Insurance Ltd) for permission to reproduce *Allianz Car Insurance Car and Music Infographic* (Figure 1.2), and Mark Bishop (Allianz Corporate) for sharing the *Allianz Music Survey Data Set* (Table 5.5); Leah Griffiths, Rosie Sheridan (SAGE Publications Ltd UK) and David J. Hargreaves for permission to reproduce the *Reciprocal Feedback Model of Musical Response* (Figure 2.1); Carin Oblad for permission to reproduce the *In-Car Listening Model* (Figure 3.1); Elsevier Publishers for permission to reproduce materials from Oron-Gilad et al. (2008) *Alertness Maintaining Tasks* (Figure 4.2), Elsevier Publishers for permission to reproduce materials from Just et al. (2008) *Brain Activation Associated with Driving* (Figure 4.5); Michael Austin (Vice President, BYD America) for permission to reproduce the *Shakespeare 14-Speaker Configuration System* (Figure 4.4); Michael Regan (Chairman, *2nd International Conference on Driver Distraction and Inattention*, 5–7 September 2011, Gothenberg, Sweden) for permission to reproduce materials from Reed and Diels (2011) *Effects of Everyday Driving Conditions* (Figure 5.1); Natasha Carr (Public Relations Associate, Kanetix Insurance) for sharing the *Kanetix Music Survey Data Set* (Table 5.6); Tal Oron-Gilad (Ben-Gurion University of the Negev) for permission to reproduce *Application of countermeasures for underload and overload driving conditions* (Figure 6.3); and Dani Machlis (Ben-Gurion University of the Negev) for the photography on the back cover.

To all those at Ashgate Publishers who assisted me in bringing this book from a proposed idea to the final publication. Special thanks to Guy Loft, Charlotte Edwards, Carolyn Court, and Emily Pace.

To my family Lila and Seymour Brodsky, Tanyah and Asaf Murkes, and Myra Pludwinski for reading through early manuscript versions, for their comments, and for their overall support. To Elisha Brodsky for the photography on the front cover, in the 'About the Author' section, and the *In-Car Music* CD (Figure 6.5). Finally, to my wife Ada, who was very sensitive to my needs for the time, space, and distance that were required to complete this journey.

Chapter 1
Automobility: Car Culture and Popular Music

This chapter is the first of two background chapters. As an introduction to the subject matter of the book, the text explores the culture of automobiles within a historical framework, illustrating the social embracement of the automobile as an integral component of everyday life. Then, the chronological development of car-audio from the 1920s is outlined. Finally, the chapter targets the interface between music and the automobile as seen through popular culture. Cars-&-music have truly become a highly widespread predominant and overbearing constituent of our modern times.

1–1. Automobility and Car Culture

We live in an *autocentric culture*, a society preoccupied with automobility. Many sociological studies (Gartman, 2004; Gilroy, 2001; Paterson, 2007; Thrift, 2004) point out that the automobile, introduced to Euro-American societies over a century ago, has long become the most common feature of our everyday lives. Gartman declares that automobility exerts a highly specified form of domination over almost every society across the globe. As mass motorization quickly spread throughout the first half of the twentieth century, the motorcar became the overriding form of daily movement throughout the world, and an autocentric culture could already be recognized (Sheller, 2004). By the 1950s the car could no longer be seen as a simple object that was constructed and/or purchased by consumers but rather was grasped as part of a wide-ranging system relating to developing technologies and communal habituation. Social scientists such as Wright and Curtis (2005) observe that by the 1960s, humans were living a 'car-based lifestyle' in which the automobile reflected social and cultural forces that transformed our everyday life and environment. Moreover, automobility appears to emulate a social structure that greatly reconfigured human civilization, as we know it today. Wollen (2002) explored the extent to which the automobile had affected human behaviour and concluded that the invasiveness of the motorcar has been so profound that it is seen to have inspired the food we eat, the music we listen to, the risks we take, the places we visit, the errands we run, the emotions we feel, the movies we watch, the money we spend, the stress we endure, and the air we breath.

Certainly, car travel imparts benefits that go beyond the functionality of mobility. More than a means to get from one place to another, the automobile embodies symbolic and affective values (Sheller, 2004; Steg, 2005; Steg & Gillford, 2005; Steg, Vlek, & Slotegraaf, 2001). When we imagine our favourite car, sign a contract of purchase, or actually go for a ride, a host of feelings flood our minds and heart. We might envision comfort, freedom, individualism, speed, power, and privacy. Thereafter, we might experience the seductive desire of ownership, the hedonistic pleasure of driving, an aggressive control over a powerful engine, the exhilarating thrill of speed, the belligerent outburst of road rage, and the comforting sanctuary of a safely designed automobile. Moreover, the car is also a platform for self-expression on both conscious and unconscious levels. Particular brands serve us as icons of fashion and style, while trademarks symbolize our professional status, and models are seen to mirror our financial achievements. For some people even the vehicle types themselves (such as the sedan, sport, convertible, coupe, hatchback, SUV, pickup, van, hybrid, and crossover) link to social affiliations, subcultures, gender roles, and sexual orientation (Dittmar, 2004). A *Roper* survey from 2004 found that 46% of Americans reported their car as the object that most reflects their personality and individuality (Lutz & Fernandez, 2010). In short, the automobile has truly evolved as a 'looking glass' to our being.

1–1.1. Production, Consumption, and Usage of the Car

We can only but feel at awe when considering the momentous proportions of car production, ownership, and usage since its promotion. During the twentieth century one billion cars were manufactured, and by the turn of the millennium there were over 500 million cars on roads worldwide (Urry, 1999, 2004, 2006). Between 1970–1990, the number of kilometres travelled by private cars per capita increased by 13% in the United States, and by 90% in Western Europe (Steg & Gillford, 2005). The 1990 USA average number of passenger miles was 11,589 miles, which at the time was more than double the Western European average of 5,420 miles. Vanderbilt (2008) presents figures indicating this rise in incidence: In 1951 there were about 49 million cars in America, and by 1960 the average American drove 21 miles a day, but in 1999 there were already 200 million cars on US roads and in 2001 the distance increased fourfold to 32 daily miles. Although the American population increased by 50% between 1950–1980, automobile ownership for the same period increased un-proportionally by 200% (Joo, 2007). By 1990 the number of vehicles in the world grew to 600 million (Steg & Gilford, 2005), and by 2004 there were 700 million cars on highways, freeways, and thoroughfares (Urry, 2004, 2006). In the USA alone, between 1999–2005 over 800,000 new vehicles were sold. A similar picture is seen for Britain where car use more than doubled after the 1960s, when as few as 29% of households had the regular use of a car, but by 1999 72% of households owned an automobile (Dant, 2004). Then, in 2004 70% of the UK population held driving licences, and there were nearly 23 million cars on British motorways. More recent appraisals reveal that automobile use has continued to increase during the last decade. For example, Gatersleben (2012) details the fact that by 2008 the average UK resident spent an annual ten hours cycling, 70 hours walking, and 70 hours on public transportation – all the while spending 230 hours per year in a car.

A survey by the *Arbitron/Edison Media Group* from 2003 found that Americans were spending about 11 hours in the car per week on weekdays and 4 hours per weekend day – that is roughly 15 hours per week or 14% of all waking hours. Consequently, commute times in 2003 averaged 51 minutes (a 14% increase from 45 minutes during 1990–2000) and covered about 307 miles per week. *Arbitron/Edison* report that the time spent in a car (roughly 2.25 hours per day) was already triple the time people spent reading a newspaper (41 minutes), but significantly less than the time they watched television (3.25 hours). A 2007 survey reported by Vanderbilt (2008) found that Americans spent more time in traffic than they actually did on any other daily activity – including eating meals with families, going on vacation, or having sex. Accordingly, the average American spent 18.5 hours a week in the car, which is 962 hours per year or 1.5 months per year. Looking at these figures, it would seem that Americans also devoted more of their monthly earned income on driving than they did on nutrition, commodities, leisure, pleasure, or their own well-being. Considering the above, one might simply summarize that Americans spend roughly one out of six hours of their waking lives engaged in car travel (Lutz & Fernandez, 2010).

In a review on vehicle consumption Lees-Maffei (2002) delineates the expansion of car use over time. In the early 1950s heyday of American private car use, only 13% of households owned a car (1% owned a second car); by 1986 17% owned two cars (3% owned a third car); and by 1996 23% owned two cars (5% owned a third car). Another way to view this evolvement was presented by Joo (2007): There was one car per every five people in the 1930s, then one per every three people in the 1960s, and then one car per every two people in the 1970s. Lutz and Fernandez (2010) point out that by 2006 the average US household owned 2.3 vehicles, while in 2009 nine out of ten US households owned a car, with 65% of American families reporting to be multi-vehicle households (owning more than one car).

An increase in automobile ownership not only reflects the fact that there are more vehicles on motorways, but also indicates that there are simply more drivers of both sexes on the road. It may be hard to comprehend, but the social norms in 1933 only afforded 12% of women to obtain a driver's licence. While it may be true that some women had been driving from the inception of widespread use of cars in America and England, before the 1950s it was assumed that males had an almost exclusive privilege and entitlement over the automobile. Only in the late 1950s did the proportion of American women engaged in driving a car change: In 1964 56% of males and 13% of women were licensed drivers, by the late 1970s 68% of men and 30% of women were licensed drivers, and then in 1993 the proportion was 50% to 50% (Lees-Maffei, 2002). It is surely a sign of modern times, and equality between the sexes, that figures from 2002 indicate that 42% of cars sold in the USA were bought by women.

Finally, Urry (1999, 2004, 2006) projects that by 2015 there will be roughly 1.4 million automobiles in transit worldwide, and by 2030 our children and grandchildren will live on a planet with over 1 billion cars cruising the streets. Albeit, Lutz and Fernandez (2010) claimed that already in 2008 there were 1 billion vehicles globally, and hence by 2030 there will be no less than 2 billion cars on Earth's roadways. When looking at the bigger picture, we can see that by the year 2050 car travel will have more than tripled during the first 50 years since the turn of the millennium.

1–1.2. Historical Insights of Automobility

The car itself, and the philosophy it represents, is the basis of autonomy and self-determinism. The 'freedom of the road' not only allows one to travel in any direction they choose, at any time of day or night, at nearly any speed one commands but is also a vehicle of social liberty. On the one hand, automobility links together our cities and neighbourhoods, our homes, offices, factories, shops, and leisure places. Yet, on the other hand, automobility is an allegory for the complex road along which Western societies have travelled over the last two centuries in a quest for development, modernization, liberty, and equality between the classes and sexes. The car echoes where we have gone and what as humans we were able to do (Urry, 1999, 2006). Nonetheless, this road was not without potholes, alternative routes, and dead ends. Gartman (2004) quotes the 1906 *New York Times* that cited the worrisome attitude of US President Wilson, who regarded the class-divisive effects of the automobile as the most destructive agent that ever existed against America's future. The development of the automobile is indeed parallel to the evolvement of our culture; the car served as the 'vehicle of change' as well as the platform for social reform. In Gartman's view, there have been three *ages of the automobile*, each defined by a significant unique cultural logic and identity. Perhaps now, more than one hundred years after the birth of automobility, the experience of driving is so much part of our everyday experience that it has sunken into our collective unconsciousness, and we take it for granted – as if it had existed since the dawn of evolution. But, in fact, the motor vehicle is historically novel (Thrift, 2004), and should be seen as a telltale sign of who we have become.

1–1.2.1. The first stage of automobility (1900–1925)

This time period centres on the automobile as it entered American society in the late nineteenth century. During this era large specialist crafted luxury cars functioned as upper class status symbols in an elaborate game that furthered the division of the social classes (Featherstone, 2004). There was economic crisis and shared conflict in which the vehicle was inescapably linked to the struggle. Because of a very high purchase ticket, motorcar ownership was far beyond the means of anyone except the extremely affluent. The price of early vehicles was based on the skilled labour of coach-building artisans who were less attentive to the mechanical functions of the engine and steering than to the aesthetic appearance of the body; these early motorcars were created in highly elaborate styles to match the tastes of the upper classes. Wollen (2002) outlines the development of the first vehicles, which had been assembled on the technologies of the late nineteenth century: In 1869 a Belgian mechanic invented the battery-powered spark plug; in 1876–87 the four-cylinder engine was discovered and constructed using a fuel mixture compressed within the cylinder; and in 1893 the German mechanic Benz produced a commercial four-wheel car with a redesigned engine and electric ignition. At this very beginning of what was later to become the automobile age, France was the world leader in car design and production. According to Inglis (2004), although the first motorized vehicles had already been developed in Germany, by the late 1880s French industrialists (such as the bicycle manufacturer Peugeot) became the leading designers that made cars commercially viable. Yet, while the number of automotive vehicles in France rose from 300 to 14,000, by 1900, there were already more vehicles in the USA and UK than in all other European countries together.

In the early days of the American twentieth century (ca. 1906), the wealthy classes who owned motorcars more often employed them for leisure activities than for daily transport. Subsequently, the automobile became an indispensable fixture for touring, racing, and exhibiting class hierarchies by cruising down Main or Market streets. For most people the early motorcar symbolized leisure and wealth. Clearly, during the first age of the automobile entire classes of working people were denied access of ownership. Then in the 1920s, American middle-class professionals and managers began to acquire income that afforded them the ability to purchase a motorcar (albeit, most were often second-hand vehicles). These consumers also saw the motorcar as an icon for their newfound prosperity, and automobility now served them also as an outlet to publically display their social standing. Such a trend had stimulated automakers to add less expensive models to their production lines, and in doing so revolutionized methods for mass production. For example, by 1908 the Ford Motor Company introduced a new a manufacturing process based on specialized machines, and unveiled the *Model T*. In 1913 industrialized moving assembly lines successfully contributed to lowering prices with Ford targeting the income of upper-middle working class families. Joo (2007) explored the impact of the automobile and its culture in America. Accordingly, before the introduction of the assembly line, Ford assembled one automobile every 12.5 hours, then with mass production each vehicle required only 93 minutes, but by 1927 Ford constructed one car every minute. Such a manufacturing process reduced prices across Ford's automobile brand. Between 1908–1924 the cost of the *Model T* was reduced by 60% (from $950 to $290), and a lower sticker price made ownership all the more viable to the general public. In the 1930s car sales increased by 3.5 million vehicles. At that time 80% of automobiles in the world were found in the USA. Although mass-produced American cars were still highly distinguishable from the luxury brands exclusively owned by the wealthy upper classes, such differences were of no consequence to working class consumers, for whom just having an automobile epitomized status in itself (Gartman, 2004).

Certainly the pace at which machinery and gadgets develop makes us contemplate the wonders of new technology. But, at the same time, we must question *if* and *how* these advances and devices also influence human nature and social behaviour. While it is clear that as a species we create technologies to advance our abilities, there is a possibility that the technological advancements we create also shape our species. This dualism seems to have been the way of our evolution since fire was first discovered; breakthroughs and inventive endeavours seem to have primed communal adaptation, behavioural modification, and social reform. Along these lines, we should note that when cars first arrived roads were not built for motorized transportation, and the motorcar simply joined the other more chaotic movement found in the street. Vanderbilt (2008) claims that initially, the only rule was to 'keep to the right', but quickly automobiles established an almighty priority over vendors, pedestrians, and most other street activity. It seems that casualty rates were unfortunately disheartening, and as Vanderbilt points out, the inhabitants of metropolitan centres had for some time been accustomed to an average four people per week killed by horses. In fact, the daily newsprints of the late nineteenth century were full of columns referring to spooked runaways who trampled pedestrians, reckless carriage drivers who paid little attention to the 5 mph speed limit, and to the overall lack of acceptance for the 'right-of-way'. When the bicycle boomed between 1880–1900, even more fatalities occurred in public streets; these machines were fast, spooked horses to an even greater extent, and subsequently were banned altogether from main streets and pedestrian footpaths. Hence, the public eventually met the dawn of the automobile on main streets and boulevards with amazement, aloofness, and resistance. Unfortunately, many citizens (especially children) did not grasp the dangers of the motorcar till too late, and if they were not especially content to walk within narrow footways, they could have fatally been mentioned in one of the frequent early twentieth-century daily papers headlines 'Death by Automobile' (Wright & Curtis, 2005).

In this initial stage of automobility, qualitative differences in cars not only mirrored disparity between social classes, but also illuminated socially accepted inequalities between the sexes. Gartman (2004) points out that unlike the women who were to confine themselves to the more private domestic domain, men were expected to coexist between the domestic world and the public sphere. To further such socially accepted family roles (and sex biases), ownership and mechanical operation of a motorcar were considered befitting only for men. As automobiles were defined as masculine, during this first period of automobility, if a woman did have access to a car it was usually an electric model. Electric vehicles were highly limited in both travel distance (between battery charges) and cruising speed, and thus these models were believed more suitable for women who were not to wander too far from the proximity of their home duties.

1–1.2.2. The second stage of automobility (1925–1960)

This time period was an era of mass consumption. Although the appearance of the simple functional mass-produced car symbolized the newly standardized mass culture of working-class people, it also represented the dominant icon for trendsetters across all social classes and sexes (Urry, 1999). The automobile became the ultimate emblem of all things that could establish the 'good life', and in such a capacity emulated all of the images that were to be seen or heard in popular culture. Representations of motorcars surfaced in literary novels, the visual and plastic arts, Hollywood's motion pictures, and the popular tunes of Tin Pan Alley. This second period began as North American families adopted the car as a means for touring and camping in the countryside among the wildlife. As if almost overnight, motor camps sprouted to accommodate the touring motorists, and car travel was seen as a voyage – a journey through the life and history of the land (Urry, 2006). As the real trend was to take pleasure in the process rather than go to a specific destination, the touring mode mandated one to drive slowly, to take the longer route, to stop, and to sightsee. This was also the demeanour during inter-war Britain, where car use was associated with exploring the countryside. As early as the 1920s, large numbers of campsites were established near American national parks, and patches of wilderness that had previously been frequented solely by the privileged were now available to the masses. By 1926, 400,000 vehicles per year were parking in nature grounds for extended vacations.

By the mid-1920s the three largest mass-production automakers in the USA (Chrysler, Ford, and General Motors) accounted for 72% of total automobile output worldwide. Yet, General Motors head Alfred Sloan sensed that consumers wanted more than cheap cars; he claimed that buyers would pay more for a car that went beyond basic transportation. Hence, in 1927 the *La Salle* became GM's most successful cheap mass-produced car, but with a style and look of a handcrafted luxury classic (Gartman, 2004). Between 1929–1939, as sales of the extravagant automobiles slumped, all luxury models followed suit and were incorporated in the mass-production lineups.

During the 1930s' despair, travelling by car was the way for many who had been evicted from their homes – usually seen as having been robbed of their hard-earned income. The road itself became a connotation for hope of one finding a better life. The car was a way out of the old community that held one back, a journey down uncharted avenues, a highway boosting the odds for new employment, and a means to fulfil the legendary 'American Dream'. Gartman (2004) claims that to accommodate consumer demand for individuality, automobile manufacturers produced many different types of cars and models – all created on the same vehicle platform but yet variegated by a structured pricing strategy to attract buyers at any income level. Based on a graded price tag, the lineup at General Motors was (from lowest price to highest): Chevrolet, Pontiac, Oldsmobile, Buick, and Cadillac. For the same reasons, Ford formed the Mercury and Lincoln division, while Chrysler bought out Dodge in the late 1920s (Joo, 2007). Although very few attributes actually differentiated between them, the brands and model types were all seen as distinctive simply because of superficial differences in exterior designs and in-cabin features; the higher the price the greater the number of alterations and accessories.

It is interesting to note that amongst the widespread financial scarcity of the 1930s, two major automotive developments surfaced: the actual design of the vehicle body evolved, and 'hot-rodding' became popular. Gartman (2004) relates to the fact that in the 1920s car bodies were still linked to the nineteenth-century rectangle buggy since size and weight were equated to both quality and sturdiness, and the 1930s' car bodies were constructed as large rectangular heavyweight frames. But, then when chassis became a strongbox of sheet-steel moulded by hydraulic presses and welded at the seams, a new elongated oval reshaped body of the motorcar replaced the former (Wright & Curtis, 2005). Accordingly, this shape not only offered improved durability during vehicle impact but also invigorated an evolving trend for streamlining. Automobile designers of this period (like Harley Earl of GM) believed that chrome trim and stainless steel accessories suggested superiority and gave the impression of a higher price. Both consumers and the pedestrian onlookers saw these associations too; they valued chrome and stainless steel as they did silver coins and jewellery. Unlike the angular linear designs of carriages that were perceived as passé remnants of a previous disenchanted world, the new design craze involving subtle curves and horizontal lines emphasized swiftness and virility. Then, hot-rodding surfaced. This trend began with groups of young racers that met in parking lots or at previously arranged abandoned sections of straight flat road. By 1957 hot-rodding was seen to be an official 'sport' with over 100 drag strips serving more than 100,000 racers to the cheering shouts of more than 2,500,000 spectators (Wollen, 2002). In many cities (especially on the American West coast) hot-rodding led to automotive customization and ornamentation – later in the 1950s this was referred to as 'pimping' or 'souping-up' a car (Lutz & Fernandez, 2010).

By the mid-1930s, there were fairly large numbers of car dealers in most large urban areas, and while the excessive prices of private cars meant that purchases and ownership were still somewhat restricted to middle class families who were financially comfortable, the period between the wars saw a rapid growth of motoring and automobile clubs. As mass production afforded automobiles to eventually spread to lower socio-economic classes, mere ownership was no longer able to express status. Gradually, consumers recognized that car type – the actual brand and the model – was an emblem of prestige, prosperity, and elegance (Gartman, 2004). As communities overseas were still adjusting to the ruins seen everywhere throughout Europe, post-war Americans who owned 75% of all the cars in the world, pampered themselves with tailfins, power steering, power brakes, automatic transmissions, electric windows, vacuum ashtrays, retractable roof tops, and wraparound windshields (Marling, 2002). Americans felt that they deserved nothing less after having experienced the Great Depression and the hardships of the war. Marling claims that as millions of American GIs came home, each and every one of them was in search of a girl, a car, and a house – but not necessarily in that order.

The landscape of 1940s America would have been simply impossible to imagine without the advent of the automobile. The increase in car production meant an upsurge in car consumption; the more cars that reached the streets, the more roads and highways had to be built. The more motorways were jammed with cars packed with passengers going somewhere, the more commercial traders had to move elsewhere. Retailers and service providers customarily situated in city centres were now moving to large open plots in the suburbs. Joo (2007) lists five constituent elements of automobility that make up the *drive-in culture* as popularized in the 1940s:

1. *Retail*. The first suburban mall opened in 1923, and then in 1956 the first enclosed climate-controlled out-of-town precinct was established. By 1960 there were more than 4,000 American shopping malls.
2. *Travel*. By 1929, 33% of Americans (i.e., 45 million people) were taking family holiday vacations by car or in mobile homes. By 1960, 27 million tourists per year were visiting nature and amusement parks (such as Disneyland, which opened in 1955). To accommodate travellers, motor hotels (i.e., motels) created by major chains such as The Alamo Plaza Hotel Courts were established on main motorways. All of these had a predictable room size and decor, a standardized price and service package, as well as a car park just outside the guestroom door. By 1960 there were roughly 60,000 motels in operation for the motoring public.
3. *Cuisine*. Drive-in fast-food restaurants with carhop waitresses became highly popular hangouts for youths, and motor-bound families patronized on-the-road rest-stop eateries such as McDonald's (opened in 1948, by 1953 grew to 228 shops). These offered a standardized menu with guaranteed food quality and free parking.
4. *Entertainment*. Drive-in theatres began operation in 1933 with space to accommodate roughly 50 cars, and flourished throughout the 1950s. By 1958 there were 4,063 car cinemas in operation. Yet, these mostly closed down by 1970.
5. *Commerce*. Businesses and homes that were located far from city centres required a host of services. Among the drive-in amenities, were: automats, banks, beer distributors, car washes, gas stations, laundromats, and drive-through theme parks. Car culture had put a premium on speed and encouraged the idea of convenience whereby the customer never left the confines of the car. Urry (2006) also notes that among the advancement of numerous services in which the provider never leaves the automobile was drive-up mail delivery services and drive-by shootings.

Between 1940 and the mid-1950s, US workers experienced a 25% rise in income. Home ownership increased from 43% to 62%, and by 1954 conveniences such as indoor plumbing rose by 15% to 80%, telephones by 44% to 80%, refrigerators by 47% to 91%, and the TV from 0% to 61% (Joo, 2007). Similarly, the more money people had, the more cars they wanted to own. Between 1929–1955 car sales more than doubled reaching 7.9 million; car registrations increased by 35.9 million cars to 61.7 million in 1960. Further, automobile models embodied a temperament of their own; in the 1940s the American male identified 'convertibles' as if a mistress, while the 'saloon' was seen as a stable and loyal wife. By the 1950s, the automobile was more or less adopted from aircraft design, meant to fulfil every driver's dream of an imagined future somewhere off in a motorway utopia (Wright & Curtis, 2005). Even the World's Fair Auto Show – *The Motorama* (Las Vegas) – was inaugurated simply as a spectacle performance to offer a preview of the latest enhancements. Marling (2002) attests to the fact that auto shows were fashioned as a way for customers to anticipate the annual model change that had become entrenched in American minds since the mid-1930s. For over 20 years every brand and model had been somewhat altered each year by transforming a bit of the external body or featured fittings. Although beneath the hood all mass-produced mechanical parts remained stable for more than a decade, new looks and appearances allured consumers into dreaming about trading-up for next year's model (Gartman, 2004; Lutz & Fernandez, 2010). Yet, most car buyers were also aware that all features would one day trickle down from the most lavish vehicles to those within the income range of average families.

Throughout the 1940s and 1950s there was a massive expansion of low-density family housing – each with a standardized lawn, garden, and parking space. Figures cited by Joo (2007) outline this earth-shattering growth: 114,000 single-family suburban houses were built in 1944, increasing to 937,000 in 1946, rising to 1,183,000 in 1948, and exploding to 1,692,000 in 1950. Automobile ownership greatly influenced the designs of these homes. Accordingly, houses of the previous Victorian style had a front entranceway and parlour that were typically used as a barrier (to keep the public contact and social exchange from entering the secluded domicile life). However, as driveways and garages replaced these safeguards, a link was formed between the house and the outside through a kitchen side door. A car parked in the driveway was certainly *the* symbol of suburban lifestyle. Domestic activities required a car to be used by the wife, while commuting lengthy distances to get to work mandated car use for the husband (Urry, 1999, 2006). Undeniably, both the need and the consumption of automobiles by suburbanite families was greater than any other locale. For example, Joo points to the fact that in 1950 suburban car ownership covered 90% of households (with 20% owning second car) compared to an overall single car ownership for 13% of inner-city households (with a second car registered to only 1%).

By the late 1950s the number of private cars, both in the US and Europe, far exceeded the total number of automobiles produced before the war (Inglis, 2004). Accordingly, in post-war France the three larger car manufacturers (Citroen, Peugeot, and Renault) targeted different sectors of the market in order to stimulate development and sales; Citroen was oriented for the luxury market, Peugeot aimed at the middle-market range, and Renault focussed on the economy market. By the mid-1960s France was as motorized as any other Western European country. Having said that, Joo (2007) points out that car-based suburbanization occurred in Britain and Europe only several years later, while only in the 1980s did Europe actually attain the 1950 level of US automobile ownership.

1–1.2.3. The third stage of automobility (1960–present)

In this final time period the car is seen as part of a disintegrated series of subcultures. Featherstone (2004) claims that an overall focus on distinction between car brands led to the initiation of a full range of new vehicle types. Each vehicle was aimed at an identified market in such a way that every car served a highly specified function. Accordingly, there were: subcompacts, compacts, intermediate-size cars, muscle cars (powerful performance cars), pony cars (sporty youth-orientated cars), sports cars, and personal luxury cars. Gartman (2004) views this era as a time in which automobile branding reflects lifestyle choices. Namely, car consumers were in search of a vehicle that expressed their unique individuality and temperament. From the 1970s there were many proposals for turning the car into a kind of mobile living room (Wright & Curtis, 2005); the idea of having an additional vehicle to spend large portions of the year on holiday became highly popular. There were towable caravans, motor-caravans, camper-vans, and auto-sleepers; it was here that the urban legends of the *Winnebago* in the US and *Marquis* in the UK appeared. By the mid-1970s, there was even a small movement of retirees and middle-aged couples who had sold their houses for a motor home, with a goal to live out the rest of their lives on the road.

Although between 1970–1980 the American market for cars was stagnant (Gartman, 2004), by the mid-1980s the income of young professionals (referred to as 'yuppies') grew, and these newly enriched consumers packed the auto market. Young professionals not only sought to display their wealth but viewed the car as an object that could be packaged through a combination of individual styling and mechanical specifications to mirror their cultural life styling (Wright & Curtis, 2005). The resulting boost in car sales revived the automobile market. Prior to this trend, the overriding automobile design was structured for everyday domestic use. The typical configurations were: family-oriented sedans, hatchbacks, estatecars, station wagons, and people-carriers (often referred to as a multi-purpose vehicle or MPV). Other designs, aimed at consumers who perceived that they were more vigorous, were sports cars, sport hatchbacks (and the sport utility vehicle or SUV), and pickup trucks. But now, the auto industry responded to demanding young professionals with a burst of distinct car types, each echoing a stereotyped lifestyle, specific leisure interest, or some other identity. Truly, none were considered to be a better vehicle than any other, but rather they were just different. These included eco-cars, hybrid-cars, micro-cars, minivans, and retro-cars.

By the 1990s car manufacturers no longer sold automobiles, but rather a brand that reflected a meaningful image of life. Gartman (2004) points out that while the rise of diversified non-hierarchical automotive products injected a much-needed financial boost to the auto industry, it also fragmented the market and threatened its core foundation. When every driver stressed the need for a car to articulate some idiosyncratic temperament, then 'context' became the problem. For example, "a Yuppie software executive may need to express his high-tech corporate persona by driving a BMW to work, but on the weekends wants an off-road vehicle to express his back-to-nature leisure persona" (p. 192). Consequently, as the number of vehicle types and car models grew, with an unlimited number of options available for the same car within the same product line, dedicated manufacturing plants had to be established. These market trends went against the traditional auto industry that had previously been founded on a mass market, with the production expectation of standardization through a consistent and uniform assembly line process. Subsequently, the industry entered a disastrous free-fall with devastating repercussions felt in 2008 that led to a US government buyout in 2009.

After the millennium, buyers were more than willing to pay higher prices for comfort at every vehicle level, and the industry responded by adding more size, power, and the number of accessories – until the gap between the low-priced cars and expensive cars was minimal. Even the most humble automobiles could be ordered with a variety of options including leather seats, satellite-linked navigation systems, and a CD player with an auxiliary input for external MP3 or *iPod* playback. It is interesting to note that at the beginning of the twenty-first century, even the personal PC computer had not replaced the automobile as the centrepiece of the US economy (Lutz & Fernandez, 2010). Accordingly, in 2002 the sales of GM automobiles were still seven times the dollar value of Microsoft products.

For some time drivers have required their cars to be fully loaded with the most updated accessories. With each new model year, technological improvements motivate drivers to trade-in their present model for one that is more enhanced by the next generation of inventiveness and development in vehicular performance, ergonomic design, luxury, style, and security. Looking back at past consumer trends Lutz and Fernandez (2010) point out that in the 1950s automobile consumers sought power windows, in the 1980s the fashion was anti-lock brakes, and in the early 2000s the rage was for voice activated cell phones and GPS navigation systems. By 2005 features that were originally an optional choice with an added expenditure, were all but listed as manufacture's standard stock features, including: air-conditioning, AM/FM radio, CD player, cruise control, and power door locks; by 2008 some auto manufacturers were also offering wireless Internet connectivity and a premium sound as buying incentives. Horrey (2009) observes that by 2008 new technologies of navigation systems were offered as standard in at least 80% of American vehicles, with Bluetooth interfaces in 70% of cars, and touch screen displays in more than 55% of automobiles. Moreover, cars could come with optional extras, including: comfort technology (such as heated leather seats), smart headlights (that angle themselves as the car turns the corner), mud flaps, bicycle holders, personal communications (such as in-cabin Bluetooth technology), and state of the art entertainment systems (such as 300w audio and DVD players). Finally, the various safety features that came as standard to increase crash avoidance (and in the event of a crash to improve survival), were: seatbelts, antilock brakes, electronically controlled vehicle stability and traction, front and side mounted airbags, steel-beam reinforced door panels, crumple zones, roll bars, active cruise control (that can sense the traffic ahead and make computer-controlled adjustments), onboard computerized sensors that provide drivers with real time vehicle diagnostics, support systems that warn of impending lane departure or collision, and vision enhancement.

1–1.3. Automotive Emotions and Attachment

This historical survey illustrates that the motorcar, which was initially a mechanism for rich men's leisure and then a much valued and safeguarded means to identify and measure the wealthy upper-middle classes, eventually came to occupy a central position in the life of all social classes (Inglis, 2004). Indeed, automobile use has reconfigured our lives, and some sociologists (such as Urry, 1999) even feel that for roughly 50 years people have been living their lives *in* and *through* their car(s). That is, driving a car is not simply a way of getting from one place to another, but rather car use allows the driver to *feel* and *be* an active participant in contemporary society. Most certainly, driving a car is a goal in itself – there are skills that must first be learned and polished, an operating licence that must be obtained, and then great efforts are exerted in maintaining a 'no-claims' insurance policy over long periods of time. Nevertheless, driving a car is also an endeavour that offers great enjoyment, and there may be both gratification and pride in possessing an automobile – an unconscious bond between the owner and a particular brand (or model). Further, the car provides an abundance of memories, and perhaps is even an object of sentimentality (Featherstone, 2004; Fraine et al., 2007; Lutz & Fernandez, 2010; Sheller, 2004). Most drivers have recollections of a family car, the first car they ever bought, going for a Sunday drive, packing bags and suitcases into the trunk for a vacation, loading up the car with contents of one apartment when moving to another, and their first car-date. Moreover, drivers often remember doing something irrational in the car, such as turning-up the CD player volume and singing as loud as possible, swerving from side to side while tipsy, and having sex in the backseat.[1]

On some level, those who do not drive or do not own a car are kept at a distance from participating fully in Westernized automobile-centred societies. The inability to drive may lead to feelings of social exclusion and disempowerment. Certainly, the ever-widening range of daily activity that is part of our modern lifestyle makes it extremely difficult to contemplate abandoning the automobile (Sheller, 2004; Wright & Curtis, 2005). On this subject, Lutz and Fernandez (2010) claim that non-drivers or car-less individuals are not only restricted in personal mobility, but are also limited as far as job options are concerned. Without access to a car one pays more for goods and food (as inner-city marketplaces have little fresh food available), and many services are much more restrictive in city centres than in peripheral areas (such as medical clinics, grooming, home and car repair).

Steg (2005) outlined the use of a car as compelled by three overriding incentives (and stereotyped kinds of driver behaviour):

---

1 Even dating patterns changed with the advent of the car, as backseats offered sexual comfort and privacy unavailable in family living rooms (Joo, 2007).

1. *Instrumental motives* that reflect utility, suitability, and accessibility. These relate to the ease and function of car use including speed, flexibility, and safety. Drivers with incentives of instrumental value indicate that cars are solely intended for travel from one location to another. Therefore, the characteristic different utilities of the vehicle are far more important than any brand name or model type. As far as these drivers are concerned, the car can easily be discarded should it not be required. Finally, these drivers would not be extremely concerned over theft of their automobile as they see the car as a functional object that can be replaced.
2. *Symbolic motives* that reflect expression of self and social position. These relate to the identity of either the car or the driver in comparison to other cars and drivers. Drivers with incentives of symbolic value indicate that cars represent who they are and what they do. Because they believe that people judge others via the cars they own, their car to some extent *is* their public image, and therefore a brand name is far more important than the actual features or utilities of the vehicle. These drivers feel it is the car's image that triggers the esteem and position that is due them. As they identify strongly with their cars, they are highly irritated when a passerby touches the car. In the event that their car is stolen, they would feel distraught and overwhelmed as a result of losing both independence and lifestyle.
3. *Affective motives* that reflect needs and desires. These relate to emotions that surface from driving a car that cue overall moods and feelings. Drivers with incentives of affective value have specific desires about possessing certain cars, because they have an intense passion that can only be gratified by driving, and they find themselves at times on the road just for pleasure. Drivers who are more emotionally attached to their car drive more frequently because they like to drive, and driving is perceived as enriching the quality of their lifestyle rather than simply because there may be a concrete useful reason to drive (Gatersleben, 2012). Car theft can bring on responses of anger and feelings of abuse (as if someone had violated the sacredness of their home or even their own self), as well as grief that would not necessarily subside when a replacement was obtained. Finally, these drivers do not feel the loss of *a* car, but the loss of *that* car – the inanimate object itself.

Indeed people develop an emotional attachment towards the automobile. Sheller (2004) astutely mentions that we not only *feel* the car, but also we *feel through* the car, as well as *feel with* the car:

> [Some drivers] physically respond to the thrum of an engine, the gentle glide through a gearbox, or the whoosh of effortless acceleration, and in some cases the driver becomes 'one' with the car... Some feel content with a smooth and silent ride (historically aligned with ideas of luxury, privilege and wealth), others prefer an all-wheel drive that shakes the bones and fills the nostrils with diesel and engine oil (historically aligned with ideas of adventure, masculinity and challenge)... It not only appeals to an apparently 'instinctual' aesthetic and kinaesthetic sense, but it transforms the way we sense the world... (p. 228).

1–1.3.1. The meaning and attractiveness of car use

There are many different ways that drivers see their perceived motivations for using a car. Such attitudes have been referred to as the 'attractiveness' of car usage (Maxwell, 2001; Steg, 2005; Steg & Gillford, 2005; Steg, Vlek & Slotegraaf, 2001). For example, some drivers may *hate* cars and driving, while there are others who *do not like* driving but do so occasionally, as well as some who absolutely *adore* driving but could do without it if they had to, and then there are those who *love* spending their time in the automobile as an essential component of their lifestyle. The attractiveness of the car can be found in: adaptability (used for home, work, or socialization); availability (used any time day or night); comfort (many cabin features and accessories); convenience (parked right outside the front door, nearby for use on weekdays and holidays); capacity (much storage space for transporting goods, a number of passengers can accompany the journey); durability (needs little upkeep, it holds the road well); efficiency (decreases journey time); freedom (used to go anywhere and with anyone); flexibility (choose any route, can stop and start repeatedly); independence (solo unaccompanied driving not dependent on others); privacy (a place to be alone, a home-away-from-home, a place of anonymity); protectiveness (a shell of armour to withstand a crash, used in all types of weather conditions); safety (a means to overcome fear of being alone, shielded in the nighttime); and speed (thrill of acceleration, sensation seeking). In short, the car provides a host of feelings: adventure, anxiety, boredom, envy, euphoria, ecstasy, fear, frustration, hatred, joy, leisure, nostalgia, pain, passion, pleasure, power, prestige, rage, self-expression, self-identity, and status. Nonetheless, we must also accept the fact that some drivers see car usage as unattractive. Among the motives these drivers list, are: danger (possibility of road breakdown, a flat tyre, uncontrollable road conditions, an accident, getting lost); environment (air pollution and quality, ozone depletion, noise, use of natural resources, non-recyclable hazardous waste, littering); finance (costs of annual licence, car-wash, maintenance upkeep and repairs, parking, petrol); personality (not able to drink alcohol, car use is addictive and habitual, feeling trapped if car is inaccessible, causes stress-related anxiety and requires intense attention); and traffic (delays from road construction, highway congestion, aggressive behaviours of other drivers, speed limitations, restrictions entering city centres). Yet, considering all the above, for the most part we still see the car as more attractive than not. To illustrate the point, in 1975–1991 two American surveys found that more than 70% of respondents reported the car as the most essential factor for having a good life (Lutz & Fernandez, 2010). Note this figure was significantly higher than how Americans rated owning a house, a happy marriage, or having children.

In a series of interviews, Lutz and Fernandez (2010) found that 70–80% of commutes are solo (i.e., unaccompanied driving) as drivers like the fact that their car provides privacy. Accordingly, the car is especially appreciated by those who find no other time to be alone, to think, or to listen to music for themselves – an opportunity for escapism, an unplanned spontaneous road trip, and a way out from the everyday banality of working in our modern society. Further, car use can also serve as an essential sanctuary for personal recovery and restoration. That is, the car may just be the only place in which we are far from the needs of our employers, colleagues, clients, partners, relatives, kids, and pets. But, quite to the opposite we can also choose the car to be an effective place for social engagement, to chat with friends, spouses, or children. Most specifically, the car delivers the optimal conditions of a secluded sound booth or karaoke club stage for drivers to belt out their favourite tunes without fear of humiliation.

1–1.3.2. Inhabiting the car

If the period before the 1960s was characterized by 'inhabiting the motorway' (i.e., dwelling on the road), then surely since the 1960s most drivers have 'inhabited the automobile' (i.e., dwelling within the car). The car has become a place of its own (Inglis, 2004; Urry, 1999, 2006). It is somewhat of a paradox that the more homely aspects of the car also allow it to take on a very distinctive level of individualized ownership. That is, we personalize our vehicle, both the external body shell and the interior cabin surface in such a way as to emulate a cosier place. Ironically, the car can simultaneously make the driver feel at home, while at the same time distance the driver farther away from home. Inglis points out that such an ambiguity sanctions the car to be both a mobile place, while at the same time functioning as a stationary dwelling. Those who inhabit the car are certainly able to create a personalized ambiance by filtering-out sights, sounds, and odours from entering the cabin from the street – replacing these with ones that are more sympathetic. Moreover, we can control the atmosphere of sociability (i.e., the social mix) as we do at home by regulating who may or may not enter. Finally, Urry highlights the fact that the driver is surrounded by in-cabin control systems, a never-ending array of technological developments that allow for simulating domestic environments. We control heating and air-conditioning, lighting textures, surround-sound audio timbres, high-definition visual screening, telecommunications, terminal networking, and refreshment amenities. Subsequently, the car has become a place for anything and everything – from corporate business, to intimate romance, with domesticity somewhere in between. It is no wonder that within their attempt to understand the relationship between drivers and their automobile, researchers have explored the concept of the car as a form of 'territory' (Fraine et al., 2007; Gatersleben, 2012). From a theoretical vantage, defining the automobile as a territorial object refers to five dimensions:

1. *Importance*. The functional and psychological significance of the car to the driver, the driver's level of control over access to the car and the driver's ability to regulate activities in the car.
2. *Invasion*. The use of alarms and immobilizers against potential thieves, stickers that warn intruders as a measure to protect the car, how the driver responds to infiltration of the zone or the car being stolen.
3. *Marking*. The driver's motivations and objectives of customizing the car through modification of parts, placing decorative knickknacks in the car cabin, and employing stickers on bumpers, windows, and doors.

4. *Ownership*. The driver's permanence over car use.
5. *Time*. The duration of time spent in the car.

Essentially, we have all been primed to be inhabitants of the automobile – from birth. It is a well-known fact that already in infancy we take pleasure in the kinaesthetic vibratory experience of the car ride (Featherstone, 2004). So much are babies' positive reactions to the automobile that parents often take them for a ride to relax and fall asleep. Within our automobile-centred society, parents motivate toddlers' leaning towards four-wheeled mobility by buying them toy cars to play with, and later child-sized vehicles for them to ride (Sheller, 2004). By age two, children have been exposed to the names of transportation by vehicle types ('car', 'truck', 'motorcycle'), as well as brand label ('Dodge', 'Mac', 'Harley Davidson'), and perhaps even model type ('Viper', 'Titan', 'SuperLow'). Lutz and Fernandez (2010) claim that automobile companies target kids as young as pre-school. As early as elementary school, children have already formed automotive inclinations. By second grade, youngsters have clearly developed preferences for specific brands and models. Primarily, these may be stereotyped images about people as seen on television or web-based media (that are essentially consequences of 'business arrangements' formed between car manufacturers and agencies of children's media). Some examples delineated by Lutz and Fernandez are General Motors who contracted MTV to advertise and promote the 2005 Chevrolet *Uplander SUV Minivan* on children's channels, web sites, and magazines, including: *SpongeBob SquarePants* (Nickelodeon's most watched animated televised series) and *Blue's Clues* (an early child development television show produced by the cable network to capture pre-school children's attention). Another example is the Disney Corporation whose film production *Cars* included a cross-promotion marketing scheme of food, as well as *Matchbox* and *Tonka* toy cars and trucks – all for the purpose of encouraging younger toddlers and older children to experience pleasure and develop a passion for automobiles. And why not? Cars are often presented as friendly animate living creatures that have personalities of their own – for example the loveable *Herbie* or the almighty *Optimus Prime*. In the visual media, cars have hip names; some demonstrate a slightly rebellious character, and like their humanoid youngster relatives, they also go through developmental life-stages such as aging. Even before cartoon animation and digitized special effects, there were TV shows that depicted automobiles *as if* possessing a human personality; two examples are the early 1950s TV series *My Mother the Car* (a dilapidated 1928 Porter touring car), and the late 1980s *Knight Rider* (a heavily modified Pontiac *Trans Am* called KITT). Nowadays, vehicles are transformed to people and back again, by way of a process referred to as 'anthropomorphics' (Beckmann, 2004; Urry, 1999). An anthropomorphic entity is an object made by humans, replacing the actions of humans and permanently occupying the persona of a human, shaping human action by advancing human ability. Albeit, perhaps in regards to cars, a more appropriate catchphrase should be *automorphology* –

which although may have associations to the morphological qualities of the 1990s famed TV series *Mighty Morphin Power Rangers*, clearly describes the associated behaviour of the vehicles in the 1990s action series *Viper* (*Dodge Viper RT/10 Roadster Coupe* and *Viper GTS*). By middle school, children have already been exposed to tens of thousands of hours of car ads, video games, and TV/Hollywood car adventures. Namely, by age ten youngsters have ultimately developed an internalized impression of what kind of people own what kind of car make and model. For example, they believe that the owners of larger cars are wealthier, luckier, and more successful people than owner of smaller cars. Considering all of the above, it would seem that all of us – our children and our grandchildren – have undergone a form of psychological indoctrination to automobility that has become the undercurrent of our society.

Then comes the 'rite of passage', which is one's certification as a driver. "It's like you go through puberty, and then you go through cars!" (Lutz & Fernandez, 2010, p. 20). Accordingly, somewhere between the ages 12 and 14, teens fantasize about cars, looking ahead at a future in which they will occupy the driver's seat. Moreover, dispositions for specific car brands and/or models have been seen to be a potent predictor of later-to-come purchases. It may be that cars are a focal point for boys and serve them in a developmental manner as fashion clothes function for girls; while girls seem to develop a devotion to specific styles and labels, boys do the same by cultivating interests in different automobile brands and models. For most teens, getting a driver's licence signifies freedom from the family. A driver's licence is independence for them to go where they want, when they want, and to exert a degree of control over their life. Albeit, many teens may simply view a driver's licence as the ultimate permit for taking control of the radio. Fraine et al. (2007) point out that when young drivers first get their own vehicle, the car is one of the few 'territories' that leads to high levels of attachment. Is it any wonder that teens view the onset of driving also as a licence to experiment – with new experiences including sex, drugs, and alcohol.

Finally, the emotions of automobility have much to do with gender stereotypes. Lutz and Fernandez (2010) are very explicit about the romantic feelings that many teens and adults have towards the car. Apparently, these originate from socially accepted ideas that have become entrenched in our minds since toddlerhood. Myths about masculinity and the car insinuate that the boy with the biggest flashiest wheels always gets the girl, or that the car enables one to display his 'manhood'. It is interesting to know that research has indeed shown that everyday people (even children) can match photos of individuals to the cars they actually own with more than 66% success. It seems that our decisions are somewhat biased on our perceptions of age, income, facial expressions, grooming, and sexuality. For example, in a study described by Gaterslaben (2012) women who saw photo images of men in a high status cars rated the men as twice as attractive than when they saw the same men in more common cars.

1–1.3.3. Inhabiting a smart car

At the beginning of the twenty-first century, a shift occurred towards inhabiting the intelligent car (Urry, 2006). The car does not simply allow driver mobility, but empowers a range of actions available only to the *driver-car* (Dant, 2004). Many social scientists (Featherstone, 2004; Thrift, 2004) point out that such a hybrid serves as both an extension of the human body, as well as an extension of technology. Actually, most elements of the modern automobile are monitored and coordinated by pre-calibrated settings. Proprietary software operates the complex feedback system to control engine management, braking, suspension, wipers, lights, cruising, acceleration, parking manoeuvres, speech recognition systems, communication and entertainment, sound systems, security, climate control, in-car navigation, and crash protection. As a result, automobiles have increasingly become command centres for a host of activities other than driving (Fraine et al., 2007). The automobile is an enclosed dwelling space in which at the touch of a finger, it can become a place of work (employing communications components via cell phone and Internet, doing paperwork, and reading), or a place of refuge and entertainment (enjoying television, video and DVD movies, or music tracks via a surround sound audio system). Featherstone states that subsequently, drivers must also learn how to *inhabit technology* in new ways, in the sense that the car has become a platform for multi-tasking. Car buyers find a certain level of excitement with each new gadget that is featured in the latest models. The marketing of 'smart-cars' stresses their enhanced capabilities for information and entertainment design – as intelligent mobile environments (Sheller, 2004). Cars today feature satellite radio with hundreds of commercial-free stations, flat screens for watching TV and video movies, and concert-hall club-quality audio systems.

1–2. Car Audio

Entertainment options have caused the car to move well beyond the mobile living room to a mobile den (Lutz & Fernandez, 2010; Thrift, 2004). Some on-board features of automobiles seem to make the car more productive, while others just seem to make it more pleasurable. In fact, we might even make the claim that the entertainment technologies as found in modern vehicles, along with better sound insulation and ergonomically designed acoustic interior, is not simply an *accessory* for driving but rather is more of the *purpose* for using today's car. Although in-car stereo began in the 1920s as a simple monophonic radio that barely received clear AM frequencies, after FM stereo broadcasts and recordings became the new industry standard, the industry was launched into a frenzy of inventiveness that invigorated the manufacture of powerful amplifiers, speakers, portable recording media, and satellite radio. These have all but enhanced car audio systems with sophisticated features that go far beyond any driver's dreams – an *audiomotive* science fiction of sorts. Simply said by one of the leading American experts on car-audio:

> The recent explosion in media options has made the DVD radios that were state-of-the-art a decade ago seem almost antiquated now. The advent of MP3 has freed music from a disc-based format so that now you're able to carry your entire music library on a small portable player such as an iPod. Alternatively, you can load hundreds of songs onto a single disc or even a USB thumb drive. Satellite radio has gained ground against traditional terrestrial radio, while high-definition (HD) radio promises to make AM and FM better and offer more content. Plus, in just a few short years, mobile video has turned 'Are we there yet?' to 'Are we here already?' (Newcomb, 2008, p. 8).

1–2.1. Historical Insights of Car Audio

After getting in the car and igniting the engine, for most drivers the next step is usually turning on the sound system. Almost instinctively, we reach for the stereo knob and crank up the volume. It is hard to imagine any vehicle without some kind of sound system. Yet, as Williams (2008) points out, there was a time when a radio was only an extra optional expense for consumers to consider in their purchase. Back then, if a radio was not installed in a vehicle, there was simply a blank plate covering a rectangle-shaped hole in the dash where it was to go.

1–2.1.1. The 1920s

In the 1920s driving was largely a silent event. Aside from the noise of the engine and the sound of thin tyres rolling over mostly unpaved roads, drivers had no other sounds to distract them (Lendino, 2012).[2] But, as people began using their car for leisure and motoring (i.e., motor-touring), whatever was to be found in the home eventually moved into the car. As this was also a time when the radio became more commonplace, it was inevitable that the radio eventually be used outside of the house, and the car was an obvious location. When it became highly fashionable for motorists to stop and picnic, not only were blankets arranged with fresh food, bakery, and drink, but one of the newest battery operated domestic radios was always found in one of the corners. Williams (2008) describes the first home radios as fairly large floor models encased in highly decorated wooden cabinets. Accordingly, early enthusiasts did what was possible to adapt these home-style receivers to operate in a vehicle. At first drivers placed them on the back seat, but in a very short period of time, radios encased in steel boxes were mounted under the car or on the running board (Goodwin, 2012). As radios became smaller and car interiors increased in size, the radio became fixed inside the automobile on the floor. Then, in 1919 the Westinghouse Company (USA) produced the first adapted portable battery radio for use in automobiles; unlike other receivers that employed headphones, this model allowed the drivers and their passengers to listen through

2 For the most part the historical review highlights developments that took place in North America. Details about Europe can be found in the English language in Goodwin (2010) and in the German language in Erb and Stichling (2012).

a high impedance cone loudspeaker fitted in the passenger door. In 1922, General Motors marketed a *Chevrolet* brand sedan car fitted with a Westinghouse radio receiver at a market price of $200, which came with an antenna that covered the entire roof, a set of batteries that barely fit under the front seat, and two huge speakers that were fixed behind the driver's seat. DeMain (2012) reflects on these configurations stating that they were about as convenient as taking a live orchestra along for a ride. By 1923 the Springfield Body Corporation (USA) began installing car radios as an optional feature, and the first American patent for a radio specifically designed for car use was granted in 1925 to the Heinaphone Company. Nonetheless, all of these custom-installed built-in radios were priced above the means of the average driver ($250 in 1926 which is roughly $2,800 at today's cost of living). In 1927 the Automobile Radio Corporation bought Heinaphone, and mass-produced a 2-dial tuning radio specifically for installation in automobiles under the name *Transitone* at a lower market price of $150 (Erb & Stichling, 2012). It is imperative to understand that until 1927 drivers could not listen to a car radio while the motor engine was running. That is, drivers had to stop, turn off the engine, and only then could they listen to the radio. This drawback was the result of increased mechanical ignition noise and insufficient audio power output.

The credit for the first commercially viable purpose-built automotive radio system is unanimously given to the American company Galvin Manufacturing, founded by Paul and Joseph Galvin in 1928 (Erb & Stichling, 2012). Galvin had to overcome several obstacles in the design and construction to develop automobile radios suitable for mass production and installation. Not the least was the fact that the available components to build car radios were bulky, and there was considerable electrical interference associated with the mechanics of car engines; in addition, bumpy roads plagued driving and thus reception coverage (Motorola, n.d.). With the assistance of engineering inventors Elmer Wavering and William P. Lear (of LearJet fame), these difficulties were overcome; production was initiated in 1929, and Galvin marketed his radio under the brand name 'Motorola' in 1930. *Motorola* was an obvious name choice created by Galvin himself (Lendino, 2012); 'motor' evoked cars and motion, while the suffix 'ola' was derived from the *Victrola* phonograph that was supposed to make people think of music. A vignette found on *This Day In History* web site[3] claims that in June 1930 Wavering and Lear fitted a Motorola radio and special speaker under the hood of Galvin's *Studebaker* for him to demonstrate their product at the Radio Manufacturers Association's annual meeting in Atlantic City. After an 850-mile drive from Chicago, Illinois, to Atlantic City, New Jersey, while parked outside the convention centre with the music volume turned up to the maximum, Galvin and his wife Lillian exhibited the radio to participants at RMA who came outside (at times coerced by Galvin himself) to see and hear the newly developed receiver. This uproar initiated Galvin's first sales orders that guaranteed the company would endure the immediate economic slump.

3 Also found among public access archives of The Motorola Corporation (Motorola, n.d.).

The introduction and consumption of car radios was apparent elsewhere around the globe. Although America was perhaps the first to make its mark in 1922, other regions were not far behind: Australia in 1924, Canada in 1932, Germany in 1932, the UK in 1933, Austria in 1934, The Netherlands in 1934, Switzerland in 1934, France in 1935, and Italy in 1935 (Erb & Stichling, 2012; Goodwin, 2012).

1–2.1.2. The 1930s

In two definitive historical surveys of the car radio (Cosper, n.d.; Norman, n.d.) there is a clear indication that despite the Great Depression, both cars and radios continued to sell throughout the 1930s, and that car radios were commonly installed in automobiles. Galvin's production line commenced in 1930 with the *Motorola 5T71* model placed on the market at a price between $110–$130; Williams (2014) claims that at the time this accessory was roughly 20% of the automobile (which sold for about $650). This model is considered to be the first commercially successful car radio. Galvin and Wavering themselves travelled throughout America selling radios and instructing dealers how to install them correctly. "With business growing, a fleet of Motorola sales and service trucks with factory trained sales engineers soon supported radio dealers with sales, service and installation" (Motorola, n.d.). In 1931 a car radio patent was issued to Galvin's partner Lear, but soon afterwards Galvin bought it. About this time several manufactures had already entered the car radio industry as viable competitors. Goodwin (2012) mentions: the Philadelphia Storage Battery Company (Philco), who bought-out the Automobile Radio Corporation (launching a new *Transitone* car radio priced just below $100); United American Bosch (subsidiary of Bosch, Germany, who in 1933 introduced the *Motor Car Receiver 80*); and Crosley Corporation (who produced the *Roamio*). Other makes and models listed by Erb and Stichling (2012) are: Advance Electric Co. (*Auto Radio*), Allied Radio Corp. (*Roamer*), Atwater Kent (1931, *MotorCar Radio*), Automatic Radio Mfg. Co. (1930 *Senior 224* and *Junior*), Carteret Radio Labs, Moto Radio, Hyatt Electric Corporation (*M5*), Roth-Downs Manufacturing (*Caradio*), and Fred W. Stein (*130A*). Then, in 1931 General Motors (GM) marketed the *Cadillac* as the first automobile brand to offer the radio with controls mounted on the steering column; three years later GM introduced dash-mounted controls in all of their high-end models. By 1932 there were already over 100 different car radio models available on American markets, and within one year roughly 100,000 cars had undergone installation. It is interesting to note that up to the early 1930s, the car radio was only somewhat powered by the car's own battery; power was required from an additional set of dry batteries placed elsewhere in the automobile. But, by 1935 car radios were designed to fully operate off the single existing car battery. Goodwin asserts that this was truly one of the most significant developments in car radio technology.

In 1935 General Motors contracted the Crosley Radio Corporation to manufacture car radios for their *Chevrolet* brand; a year later GM absorbed Crosley and launched their own Delco Radio Division to produce car radios (Goodwin, 2012). Over the next five years the following improvements to car radio design occurred: elimination of spark interference; automatic volume control; Delco's 'Wonderbar' that provided AM frequency scanning (via an electric motor to turn the tuning knob with the tuner stopping at the next available AM station); 1936 Galvin's push-button pre-set tuning system (to help drivers select stations without taking their eyes off the road); 1937 Ford's steel rod aerial; and 1938 Ford's telescopic aerial (DeMain, 2012; Tripp, 2005; Williams, 2008). Already from 1934, the cost of a car radio significantly decreased below the previous $100 standard; in 1934 the price was approximately $55 (Williams, 2014), and then from 1936, depending on the model, the price varied between $16–$89. Goodwin claims that as a result of this more viable market, over three million car radios had been installed in automobiles, and by the end of the 1930s 20% of Americans had installed car radios in their automobile.

1–2.1.3. The 1940s

GM's Delco Radio division produced its millionth car radio in 1940, and by 1946 over nine million people had installed a car radio in their automobile (DeMain, 2012). Throughout the 1940s, while car radios became more abundant and even an every day popular commodity, they could only receive AM radio frequencies. In 1947, the Galvin Manufacturing Corporation changed its name to Motorola.

1–2.1.4. The 1950s

The 1950s image of a couple on the open road in a convertible listening to early rock and roll music via the car radio is undoubtedly an essential part of American heritage (Lendino, 2012). AM radio frequencies dominated the airwaves in the early 1950s. However, in 1952 Blaupunkt introduced FM radio, adding bands to AM car radios (autos.com; Berkowitz, 2010). From the conflicting reports in the literature (Cosper, n.d.; Lendino, 2012; Norman, n.d.), it appears that FM listening remained relatively uncommon throughout the decade only to achieve mainstream popularity in the 1960s.

In 1953 Becker (Mexico) launched the first premium in-car radio; the model featured both AM/FM bands and a fully automatic station-search button with which the driver could sample each available radio station for a few seconds before choosing one (Lendino, 2012). Then between 1955–1956 the auto manufacturer Chrysler offered a music device in its high-end models called *Hiway Hi-Fi* developed by the CBS Electronics Division (Berkowitz, 2010; Goodwin, 2012; Williams, 2008; Williams, 2014). The competitive idea behind this gadget was to seduce potential buyers to think about taking *their own* music along for the ride rather than simply listen to the radio. Lendino describes this and other bizarre devices as in-dash turntables that played a 7-inch 45 rpm disk. Williams (2014) presents details of similar turntables manufactured and marketed by RCA Victor and Philips/Norelco (a model called *Auto Mignon*). Even Motorola produced a 45 rpm disc player for cars in 1956. According to Goodwin, a few auto companies began selling units that played special-pressed 7" microgroove 16⅔ rpm discs, and Williams (2014) asserts that Colombia Records produced most of them. These 7" car disks remained highly unpopular and were eventually taken off the market in 1961. Nevertheless, in the late 1950s the portable transistor radio was introduced, and that development led to several important changes, including: reduction of product size (small-scaled components), reduction of required energy (lower watt power), and reduction of market price (DeMain, 2012).

As a side note, in the 1950s Elmer Wavering developed the first automotive alternator replacing unreliable generators; this led to other car luxuries such as power windows, power seats, and eventually air conditioning (Tripp, 2005).

1–2.1.5. The 1960s

By 1961 FM stereo finally became the listening standard, and all manufacturers adopted an AM/FM-stereo double band tuning system (autos.com; Norman, n.d.). By 1961 GM's Delco Radio division produced 30 million car radios and then 50 million by 1968. The first all-transistor radio was introduced in 1963, which then spurred the aftermarket as an attractive alternative for purchasing more modern day car-audio technologies. According to Berkowitz (2010), among these models were the Becker 1963 *Monte Carlo* (the first fully solid-state car radio) and 1969 *Europa* (the first in-car stereo setup with a tuner amplifying two independent channels). By 1963, more than 50 million cars (i.e., over 60% of registered automobiles) were outfitted with a radio. It is absolutely unfathomable to consider that already at this point, over 33% of Americans who listened to music, did so while driving in their car (DeMain, 2012).

In 1963 Earl Muntz developed the *Stereo-Pac* 4-track cartridge to be used in automobiles; this product was rather short lived (Williams, 2014). Then in 1965 William Lear invented the 8-track cartridge tape player[4] (Tripp, 2005). The 8-track, made with Ampex tape and RCA Victor cartridges, was the main predecessor of the compact cassette music tape. The 8-track made its debut in vehicles through a co-joint effort between Motorola and Ford – marketed as *Stereo 8*. This system won over previously developed in-dash turntables as tape cartridges were more stable and gave the listener more control. Initially, *Stereo 8* players were optional in Ford cars, but then in 1966 it became a standard feature (Goodwin, 2012; History.com 'This-Day-In-History'). From 1967 GM, Chrysler, and American Motors provided 8-track players as an optional feature (Williams, 2014). It should be pointed out that from the mid-1960s most drivers had an AM/FM radio receiver installed in the dashboard, and many had an additional 8-track player fitted underneath it. Drivers could now listen to random music as heard through radio stations, and control the music they listened to by inserting their favoured album on 8-track cartridge (Lendino, 2012; Williams, 2008). Ironically, the 8-track cartridge tape player was doomed from the start, only to be gone by the early 1980s, replaced by the compact cassette tape (i.e., *Musicassette*) developed by Philips in 1962 (subsequently introduced to Europe in 1963 and America

4 Lear was also the later inventor of many technologies in the field of aviation, such as radio direction finders, the autopilot, automatic aircraft landing systems, and the LearJet.

in 1964). Initially the audio quality of musicassettes was poor, but by the early 1970s compact cassette tape not only caught up to the acoustic attributes of 8-track tape but overcompensated, causing a total eclipse of both 8-track tape cartridges and to domestic vinyl 33⅓ rpms LP phonographs (introduced in 1949).

1–2.1.6. The 1970s

In the early 1970s the aftermarket became the most popular solution for drivers seeking custom stereo outfits. Those who wished to upgrade their car's audio features beyond what was supplied as standard by auto manufacturers went to dealers representing brands such as Alpine, Blaupunkt, Kenwood, and Pioneer. At this time consumers sought high quality speakers and radio receivers that had their power output boosted by specially made amplifiers for car audio. The car now made an evolutionary leap from 'automobile' to 'mobile discotheque'. Cassette tapes finally made their way into the car by 1970, with Pioneer (Japan) introducing the first car compact cassette player complete with radio receiver in 1975 (Goodwin, 2012; Lendino, 2012; Williams, 2008). By 1977 cassette tapes became the most popular audio medium; they were smaller than 8-track tape cartridges, easier to store, and had an improved sound quality that not only matched, but also surpassed all other popular audio formats. Moreover, the cassette tape endorsed a new phenomenon referred to as 'the mix-tape' (Berkowitz, 2010). Nevertheless, Goodwin claims that for a very brief period of time from the mid-1970s, a quadraphonic 8-track cartridge introduced by RCA was popular among audiophiles as part of their car's radio-stereo system.

Three important developments to car audio occurred in the late 1970s. First, the popularity of FM radio triggered the manufacture of car receivers that could pick up FM signals better while driving. Then, in 1978 the first 12-volt amplifiers for car stereo systems arrived on the market. Finally, the first two developments created a demand for more powerful speaker systems and subwoofers. Lendino (2012) points out that between 1948–1978 car audio speakers were single full-range drivers fixed to the centre of the dashboard. But by the late 1970s, 2-way and 3-way speakers were developed; these split the incoming audio signal into separate frequency ranges, integrating separate high-range tweeters and midrange drivers in a plastic or metal bracket before reaching the low-frequency main woofer.

1–2.1.7. The 1980s

During the 1980s FM radio stations became so popular that AM music stations were abandoned by drivers. In an effort to regenerate themselves, sports channels and talk radio arose in the late 1980s, spreading to 400 stations in 1990s; by 2008 there were 2,056 stations (Lutz & Fernandez, 2010). With the expansion of FM bands on the radio dial, drivers had a more extensive number of stations that they could choose from. Hence, 'button-pushing' (a predecessor to remote-controlled *zapping* of TV channels) became a more common driver behavioural component of the in-car listening experience.

In 1982 Bose became the superior stereo system by partnering with GM's Delco Radio division; they produced an entire designer stereo system for Oldsmobile, Buick, and Cadillac brands (Berger, 2002; Berkowitz, 2010). This was also the era of the *Sony Walkman FM/Cassette-Tape Player*. Music had now become truly mobile in the sense that no matter what the mode of transportation (i.e., by foot or by wheels), people who felt the need now had the means to take music along. The truth is that the *Sony Walkman* created a new social behaviour that nourished a host of modifications in peripheral markets including the fashion industry. Listeners spent hours preprogramming compilations (i.e., mix-tapes) by re-recording the favourite music from their vinyl (33⅓ rpm LPs and 45 rpm singles) record libraries to cassette tapes. With great ease listeners interchanged tapes between their belt-mounted *Sony Walkman* to their dashboard-mounted receiver/tape player.

In 1982 Philips and Sony corporations introduced the Compact Disc (CD). This new technology initiated the subsequent shift away from previous tape-based in-cabin vehicular music reproduction (Lendino, 2012). CDs offered enhanced sound quality that endured thousands of re-plays (rather than getting worn-out over time as frequently occurred with tapes). Further, CDs allowed the driver to easily omit, repeat, and/or skip songs (rather than having to fast-forward or rewind the entire programme as was the case with tapes). Finally, the 'side-length' of CDs permitted there to be one continuous music programme (rather than having to flip to the other side of a tape or vinyl LP). Subsequently, in 1984 Pioneer introduced the first car CD player (Cosper, n.d.; Goodwin, 2012), and ironically (or perhaps not so) Motorola stopped manufacturing car radios that same year (Berkowitz, 2010). The in-car CD player, initially priced between $800 to $900 (Williams, 2014) remained the overriding format for in-car music reproduction for roughly 25 years. By 1985, high-quality sound reproduction stereo systems incorporated a dashboard-mounted 'head unit' to switch between radio, tape and/or CD audio sources. It wasn't long after, that auto manufacturers began installing brand-name CD players. Berkowitz points out that 1985 Becker's compact disc players were factory installed in all Mercedes-Benz cars. Then in the mid-1980s, multiple-disc CD changers appeared; these could store between five and ten CDs at one time, as well as switch between the CDs while driving. Yet, they were still far too expensive as an everyday peripheral for another decade. Throughout the 1980s and most of the 1990s, drivers purchased CDs as commercially manufactured album records as CD burners were not affordable for everyday consumers until the turn of the millennium. Hence, although many drivers installed both cassette tape and CD players, the majority simply waited till 1986 when a single combo unit appeared. While drivers appreciated the ultimate sound quality of CDs, they were still very passionate about their personalized pre-recorded mix-tapes seen as ideal for road trips.

1–2.1.8. The 1990s

Throughout the 1990s the cassette tape remained a viable format to drive with music, but the CD eventually overshadowed it. The increasing availability of home personal computers (i.e., PC) with attached compact disk-recordable (CD-R) burners provided everyday drivers the ability to make their own compilations. If we view the aftermath of the *Sony Walkman* on human behaviour as an ear-shattering earthquake, then surely the onset of *Sony Discman CD Player* (marketed in 1985) must be seen as a tsunami that in one momentous moment swept away all other music reproduction devices – including their own *Walkman*.

Although technology existed for small video screens to be installed in the car, and while Pioneer introduced the first car GPS navigation system in 1990 (Goodwin, 2012), these were not affordable for another decade. On the other hand, by the mid-1990s decreased prices of CD changers led to the common consumer practice of placing a multiple changer in the trunk or underneath the front passenger seat (Lendino, 2012). In 1996, DVD players first appeared in cars (Williams, 2008). Throughout the 1990s high quality speaker drivers manufactured by Bose, JBL, and Infinity dominated the market (Williams, 2014). These provided car audiophiles with clear quality, wider frequency response, and extended low bass frequencies (i.e., sub-woofers). Subsequently, the auto industry split over audio attributes and components: Ford, Lincoln, Toyota, BMW, and Lexus supplied speakers by Harmon-JBL; Subaru installed McIntosh; Jaguar and Lincoln-LS mounted Alpine; and Buick, Chevrolet, Hummer, Pontiac, and Volkswagen fitted Monsoon (Berger, 2002).

1–2.1.9. The 2000s

The AM/FM stereo car radio remained in use even beyond the turn of the twenty-first century. Once cassette tapes and compact discs were added to the receiver, dashboard mounted stereos became more commonly known as *central entertainment units*; these 'head' units simply functioned to control the sound output (see Figure 1.1). These state-of-the-art components commonly featured skip protection, built-in crossovers and equalizers, as well as Dolby Digital and DTS surround sound decoding. The top-of-the-line units contain DVD playback, satellite radio controls, digital HD radio, internal hard drives, Bluetooth capability, GPS navigation, and inputs for portable digital devices (autos.com).

In the 2000s there was a great burst in development targeting portable audio appliances. Most specifically, Apple introduced the *iPod* in 2001. This player allowed the storage of thousands of songs, and made it user-friendly to create custom playlists, all of which were easily searchable by song title, artist name, or album title. Although the *iPod* and other MP3 players immediately became sought after devices as accessories for in-car listening, they were not always simple to connect to car stereos (Lendino, 2012). Then, a new package offering hundreds of channels for drivers to listen to surfaced – the *Sirius XM*

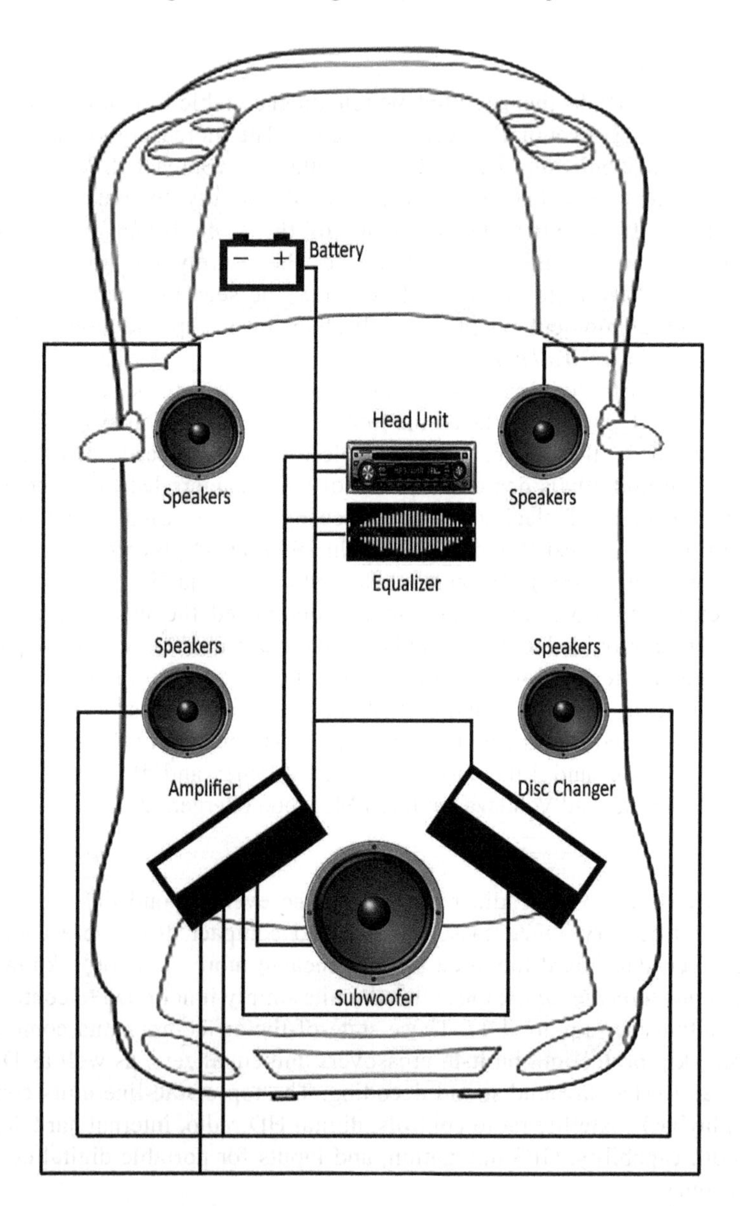

Figure 1.1 Car audio system

Notes: From the head unit, through the equalizer, to the amplifier; from the amplifier to the disc changer. from the amplifier to front and rear sets of speakers, and to posterior subwoofer. Adapted from Newcomb (2008).

satellite-based radio (Cosper, n.d.). By 2004, car stereos were capable of mono and stereo sound reproduction with multi-channel information that could route and re-route the appropriate signals to various speaker configurations located in the dashboard, doors, and back seat of the vehicle, with additional high-definition quality sound from 1,005 digital format satellite-based radio stations. Further, on-board features included: enhanced graphic animated displays with customizable colour schemes; standard controls for volume, tone (bass-mid-treble), speaker balance (left-to-right, front-to-back), channel and band tuning (AM/FM1/FM2, etc.); source selector (radio/CD/satellite/auxiliary-input); connectivity to external devices (⅛ mini PL for *iPod*/MP3 earphones jacks; USB socket and SD card slot); various playback formats and compression types (AAC, AIFF, .cda, MP3, MPEG-4, M4a, PCM, WAV, WMA); and enhanced security stealth modes with detachable faceplate (autos.com). It is interesting to note that the onset of digital MP3 players caused such ground-breaking interest and excitement that almost overnight, sales and usage of CDs declined and the compact cassette tape had all but vanished. By 2005 many car radios featured USB sockets that could read *iPod* and *iPhone* playlists. Thus, thumb scrolling by album name, artist, or song title became highly popular – albeit, today there is much evidence to show that this behaviour contributes to driver distraction more than any other music-related controller manipulation (described in Chapter 4). Further developments in the late 2000s include Bluetooth-enabled car stereos offering the capability to play music wirelessly from most digital sources such as *smartphones*.

Today's cars come with audio systems that are so comprehensive that some even come complete with in-car environmental sensors allowing it to finely tune frequency response curves that match specific car interior (Lendino, 2012). Moreover, some models offer built-in hard drives for storing entire music libraries (autos.com). Williams (2008) contends that our living room home media centre has simply moved into vehicles. Audio technology is no longer the domain of specialists who customize car audio systems – but rather everyday drivers.

1–3. Cars in Popular Culture: Motion Pictures Movies and Billboard Songs

The historic turning point between *dwelling on the road* and *inhabiting the car* came about with the onset of two developments: the radio was placed in the automobile cabin (Bull, 2001), and then motion pictures became the setting for cultural expression and engagement (Rees, 2002). Foremost, the beginning of automobile mass ownership in the 1920s occurred in parallel to the advancement of other domestic technologies, such as the radio, the gramophone, and the telephone. "Just as the home was becoming transformed into a space of aural pleasure and recreation for many, so the car was becoming the emblem of individualized freedom of movement" (Bull, 2004, p. 245). While the radio radically altered the nature of driving and the driver's experience, both motoring and cinema similarly offered a new awareness of the world whizzing-by in a unique perceptual sensation of speed. By the late 1920s, when radio airwaves and Hollywood movies were making their profound mark, American society was steadily on its way to being reshaped by the automobile (Eyerman & Lofgren, 1995).

1–3.1. Car Movies

Along the timeline when automobility was developing as a distinct culture, North America ultimately became the nation that provided the key cultural metaphors for vehicular experiences through sight, sound, and literature – all of which embrace the car as an icon. Urry (1999, 2006) upholds that American culture is altogether unimaginable without automobility, and Pascoe (2002) believes that "the best means to identify the culture that framed the American state of mind ought to be the road movie…" (p. 76). Certainly, the American dream was kept alive during the economic collapse of the Great Depression via the highway; one could still hope that a better life was just around the bend beyond the next curve of the road (Eyerman & Lofgren, 1995). These values were articulated in Hollywood films through the allegory of the journey, and this ultimately unique form of narrative developed into a genre that became known as the *Road Movie*. Many road movies revolve around the motorway where the car plays some part in the action within the context of escaping life, corruption, and social unrest, played out in some criticism of contemporary society. (For a wide-ranging listing of Road Movies see Appendix A.). The dream of hitting the road has a long cultural history with a specific iconography and aesthetic. According to Eyerman and Lofgren, it is a world of…

> gliding top down cars with sun scorched leather seats and music blasting,… sleazy motel rooms watched over by jaded clerks dressed in worn T-shirts, the TV crackling in the backgrounds. The gas station 'in the middle of nowhere', the bar hotel room and dingy bar, the corner café and the tired waitresses… (p. 54). A road culture of… gas pumps and motel signs,… [and] country stores,… where car travel itself became an adventure saga of magical quality (p. 59).

Among the road flicks listed by Urry are: *On the Road*, *Easy Rider*, *Alice Doesn't Live Here Anymore*, *Bonnie and Clyde*, *Vanishing Point*, *Badlands*, and *Thelma and Louise*.

Popular classic car films present car chases, anthropomorphised cars, and cars as central to teenage dating. In the movies, cars seem to be related to everyone and to all aspects of everyday matters, including: picnics, holidays, hitch hiking, joyriding, hot-rodding, and all sorts of foolhardiness (Miller, 2001). Accordingly, most anyone who has watched an action film anticipates the car chase – scenes of automobiles side-by-side, flung around at high speed. There will be a great deal of screeching tyres, the build-up with animated contorting faces on the edge of losing control, and then the predictable crash (of the antihero's car) with a great screen shot and digitally enhanced effects depicting exploding gasoline and airborne car parts inevitably dispersing in a meteorite-like fallout. Featherstone (2004) points out that the car chase and crash scenes have become a recognizable global vernacular that is central to car movie culture. Some examples are: *Bullitt*, *The French Connection*, *Gone In 60 Seconds*, and *The Fast and the Furious*. Yet, some of the automobile films are basically just one long car stunt, such as *The Blues Brothers*, *American Graffiti*, *Transformers*, and *Herbie Fully Loaded.*

Although cars are evident in most Hollywood movies, their use is varied and often for purposes other than transport. Some movies use automobiles for making love, as wheels for guns, as means of getaway, for car chases, for car crashes and cars on fire, as the place where villains are caught, as a contest of power, and to emphasize the risks of speed sporting (Mottram, 2002). It is interesting to note, as does Mottram, that since both automobiles and movies were distributed throughout the class system, fantasies about motorcars grew somewhat based on class needs. Eventually, the Road Movie of the 1970s became the major vehicle for a primary traditional American hero who was relocated from the Wild West, the backwoods prospecting sites, and the foreign battlefields of war. Accordingly, in the movies the car became the foremost mechanism within the conflict for social control and in establishing acceptable boundaries of popular morality. Besides the Road Movie, other genres of car-related movies are: 'Taxi Movies', 'Gang Movies', and 'Race Movies'. Examples of these as listed by Mottram and by Rees (2002) are: *The Getaway*, *Taxi Driver*, *Duel*, *Point Blank*, *Chinatown*, *Vanishing Point*, *Convoy*, *Two Lane Blacktop*, *Grand Prix*, and *Sugarland Express*. (For an extensive listing of Car Movies, see Appendix B.)

Whether we are driving them, or just watching them on the wide screen, the number of hours we spend immersed in car culture means that cars are everywhere and always on our mind (Lutz & Fernandez, 2010). Automobility presents a host of interactions between culture and technology, among them video and computer games. While games mirror motion picture animated movies, they also present numerous possibilities that allow for the participant to engage in a more active and personal process. Over the last 25 years, racing video games have become extremely popular at arcade centres and in the home. Car Video and PC games have a worldwide appeal and have given rise to a subculture that views computer gaming as a sport like any other, that is highly demanding for participants with skilled expertize and a competitive edge (Summers, 2012). Accordingly, gamers recognise the racing game genre as comprised of diverse subtypes that present unique identifiable attitudes including how music is incorporated into the milieu. Summers delineates five types of Car Video and PC games:

1. *Simulation.* Games in which the music background is absent during the races but Pop songs accompany menus, presented as if television coverage with title themes and introductory videos in the style of a sports programme. The music is used to sanction significance and excitement.
2. *Semi-Simulation.* Games in which the music background is heard throughout the races, presented as if a radio broadcast with the player controlling music selection. The music is used to place gameplay within a specific aesthetic and cultural context.

3. *Arcade*. Games in which the music is heard during the races experienced as a non-professional race with a light-hearted attitude, whereby the player controls the music selections. The music is original newly composed exemplars of contemporary popular dance styles.
4. *Street Racing*. Games in which the music background is limited to pre-existing well-known popular songs as if broadcast from an AM/FM radio programme, heard throughout the races. The music selections themselves capture the hard-core music styles and artists associated with Street Racing culture and sport (such as Hip-Hop, Heavy Metal, and Techno genres) – whereby car selection also modifies the entire music playlist.
5. *Fantastic-Futuristic*. Games in which the music is heard continuously throughout the game (races and menus) presented in a fictional sci-fi narrative. The track-specific music emulates 'sounds of the future' and the player is meant to have a 'futuristic experience'.

Summers claims that car video and PC games engender a *sonic portal* between the game world and the real world, as the music heard in real life appears to emanate from the game universe. Accordingly, the real potential of car video and PC games is in their potential link between the automotive experience and the psychological impact of music. (A list of driving-related car video and PC games is presented in Appendix C).

1–3.2. Car Songs

The image of a couple on the open road in a convertible in the 1950s, accompanied by the sound of the AM radio playing tunes which signified the early beginnings of Rock N' Roll, became permanently inscribed on the American psyche (Lendino, 2012). Accordingly, the car radio eventually became part of the American drive for everyone – girl, boy, woman, and man. The truth be told, the highway may have always been the site of the radio's most captive and attentive audience (Bull, 2001). Just as transistors redefined what was known about public spaces before the 1940s, the nature of automobile use was revolutionized by the arrival of the cassette deck in the 1970s, the FM stereo radio in the 1980s, and the compact disk in the 1990s. It was inevitable that twentieth-century music should describe cars – given their utmost domination over our daily existence. The car not only inspired the poetic lyrics of songs but remained a true inspiration even as the technologies of both music production and transportation underwent fundamental changes (Widmar, 2002). Already from the dawn of the motorcar, songwriters felt the need to convey the influence of the automobile on human behaviour and society. Accordingly, within a decade of the motorcar's appearance in American streets, Tin Pan Alley was already churning out car-related hits; between 1905–1907 roughly 120 songs were published. For example, a song titled *Love in An Automobile* (1899) indicated how helpful a car might be to would-be suitors. Already in the early 1930s American auto manufacturers made every effort to weld an association between the radio and individualized mediated listening in automobiles. This effort was highly successful leading to more than 30% of American cars fitted with radios over a five-year period between 1936–1941. By the mid-1950s many people became habitual listeners to music at work, and the young were taking radios with them everywhere. Moreover, the radio was often taken to locations far off the beaten track – to the beach where sunbathers marked their territorial boundaries with the sounds of their preferred music. Specifically, the most popular place for listening to music was in the automobile.

In popular culture, cars were commonly revered as symbols of the good life aspired to by all, and drivers turned up the volume for the songs of Cole Porter and Chuck Berry. But even before these two musicians, the lyrics of many records linked the motorcar to everyday life. Most specifically, the car itself was employed as a parable for the female body, and the act of driving became a metaphor for sexual behaviour. The definitive example is the 1936 lyric of Robert Johnson's landmark song *Terraplane Blues*; the singer promises to "check the oil, flash the lights, get under the hood, and open up a curious path". Gilroy (2001) highlights the fact that right from the start, automobiles were cherished objects for the serenade. Throughout the 1940s and 1950s, music entertainers not only viewed cars as a symbol for celebrity, but also respected the powerful expressive and lyrical possibilities that automobiles created. Gilroy lists several examples of car-revered artists, such as: Albert King who wrote about the neo-slave labour on the Cadillac assembly line; K.C. Douglas who sang about the love for his Mercury; Sam Cooke and Miles Davis who related to their journeys in a Ferrari; Aretha Franklin who sang about her Pink Cadillac; Jimi Hendrix who critiqued the crosstown traffic obstructing access to the city; Prince who sang about a Little Red Corvette; the New Power Generation who heralded their Buick Electra 225; and Bob Marley who protested working in a Chrysler plant. Field (2002) noted that car themes only appear in happy songs penned by Hank Williams, where the social possibilities of hot-rodding were celebrated.

Already in the 1950s, the car was seen as the setting for the most intense listening experience one could encounter. Unlike car-audio technicians, who consider the car to be a far-from-ideal listening environment compared to the home (Williams, 2010), from an acoustic designer's point of view, it is all the more possible to create a consistent pleasant listening experience in an automobile where environmental component and features such as cabin size, upholstery type, seat-padding density, and the number of inhabitants are all well-known. The music heard from a car radio was perceived as an all-pervasive all-consuming experience. In-car music listening seems to envelope the driver's own space, as well as to provide an infinite soundtrack for the external visual landscape seen through the windshield (Bull, 2001). But more than that, the car became *the* place in which parents and other beneficiaries of social traditions had no entitlement over the driver-listener (du Lac, 2008). This was not only an experience for ordinary young teen drivers in the 1950s, but as Field (2002) delineates, for many musicians alike. As an example, Charles Hardin Holley (aka Buddy Holly) claimed that he not only found the reception of AM car radio far superior than receivers in his parents' house, but in-cabin listening afforded him access to music that otherwise was kept far out of his reach – as was the practice of most parents of white children who banned them from listening to Rhythm&Blues music. These up-beat dance-oriented songs were prohibited, and further viewed as adverse 'race records'. Hence, for most young drivers in the late 1950s, driving with music clearly meant that the strict segregatory postures of the older generation were left behind. The freedom of mobility and listening to music in the car, paralleled autonomous self-governing thought, the development of cultural aesthetic taste and preference, and the initiation of liberal attitudes and moral political affiliation that eventually led to the freedom marches and anti-war sentiments in the 1960s. By 1953 teenagers were either spending their Saturday nights fixing hot-rods or dancing in clubs. Both of these encouraged a whole new generation of drivers to experience the new electric sounds coming out of the radios. The automobile, then, allowed youth to express themselves in a way that no other machine before had been able to do. Ultimately, these same spirits were those written down as lyrics for popular music. Without a doubt, the music genre that most celebrated the automobile was Rock N' Roll.

Widmar (2002) points out that Rock N' Roll is clearly an art form that can be called genuinely American in origin, and the automobile was a pivotal subject in these songs more than in any other music style. Rock N' Roll performers sang the praise of the car, and they themselves lived the car-cultured lifestyle to the fullest. Flashy automobiles transported Elvis Aaron Presley and Charles Edward Anderson (aka 'Chuck') Berry away from anything and everything that reminded them of their underprivileged American upbringing. For many musicians, cars said something about prosperity and affluence; their fixation (or perhaps passionate craving) for cars clearly shocked the nation as much as did their own lyrics, their vocal renditions, and their undulating stage movements (i.e., the pelvis thrusts of Presley or the duck walk of Berry). Not only did automobiles (and brand names) influence and find themselves in the lyrics of many popular songs but many groups were named after automobiles. For example, Widmar (2002) lists: *The Imperials*, *The Eldorados*, *The Continentals*, *The Cadillacs*, and *The Edsels*.

Over the years, cars have often been highly celebrated in songs. Artists in love with specific brands or models have immortalized them in the both titles as well as within the texts. Many listings of such songs appear regularly on the Internet (for example, see Urken, 2011). It is rather an oversight that only few musicological scholars have explored this collection in a serious manner. Actually, du Lac (2008) claims that car songs ran their course and did their thing; they were simply a trend that became out-dated and ceased. Accordingly, no one seems to be writing songs about cars anymore, but rather today's hits – by artists such as Audioslave, Tracy Chapman, R. Kelly, and Kanye West – only briefly mention automobiles as a metaphor for sex, status, or escapism. Yet, perhaps du Lac is only partially correct. That is, while artists and teens of the Rock N' Roll era had a romanticized fixation, and perhaps even obsession for cars, automobility was most certainly also about teen culture overcoming repressive attitudes and exercising independence in one's social milieu. Maybe, more than 60 years later, such basic requisites are not as relevant as they were in the 1950s. In addition, today there may be nothing too exotic about driving, and therefore the great romance is no longer valid. Du Lac infers that while the cars of the 1950s to 1960s were romanticized and celebrated as shinning objects of desire (with their metal-flake paint, sexy lines, and massive horsepower), the cars of the 2000s tend to be impersonal and inherently uninteresting. However, in opposition to the former opinion, Lezotte (2012) claims that while the classic car songs might have ceased to exist (as per the definition of du Lac), the automobile has nevertheless remained an *ideé fixe* in popular music. Albeit, today car songs are mostly sung by female artists who offer perceptions of how and why they drive. Lezotte proposes in the heyday of the Rock N' Roll car songs the function of the genre was not only to revel in the bliss ecstasy offered by the automobile itself, but also to pay homage to the legitimacy of the driver as being a male:

> The female singer-songwriter – who came late to both automobility and rock 'n' roll – has infused the automobile and popular music with multiple new meanings. Women from a variety of musical genres – rock 'n' roll, R&B, country, and pop – have called upon the automobile as a vehicle of freedom, escape, recollection, and rebellion, performed through the voice of women's experience. In the process, they have reconfigured the car song from a recounting of the white teenage male's rite of passage into a metaphor for the multiplicity of women's lives (Lezotte, 2013, p. 162).

Perhaps by joining the workforce during World War II, women had no other choice but to acquire a driver's licence; after all, men were shipped overseas, leaving their jobs and automobiles behind. Lezotte describes this situation as one in which women, as they got behind the wheel, experienced financial independence and freedom of movement as never before. But, then when the war ceased and GIs returned home, the men expected not only to fulfil the promise of a more prosperous life-style, but that their womenfolk would reassume a more dutiful and passive role as the companion in a domestic milieu. History indeed indicates that such expectations were only partially fulfilled, as women were not readily willing to abandon the enablement they had experienced. Hence, perhaps yesteryear's car songs were written especially for rebellious teenaged boys who dreamed of a getaway from their constricting home life, seeking some random adventure involving frolicking about getting girls and sexual escapades (that subsequently recoiled into further Rock N' Roll songs directly linked to the automobile). But, today's car songs no longer service the male-oriented teenaged rite of passage and have all but transformed into a graceful poetic (and perhaps starry-eyed) image that can be perceived by all drivers irrespective of gender. According to Lezotte, singer-songwriters no longer attribute importance to the automobile as an object for its own sake, but rather to the connotations denoted by the car. Lezotte concludes that the "definition of a car song is not only a tune that addresses the automobile, either literally or metaphorically, but also includes music played in and that becomes part of the car" (p. 171). To illustrate the breadth and width of car songs as a genre, and the fact that car-themed songs are ever-present in the popular culture of music beyond the 1960s – and well into the 2010s – the reader is referred to Appendix D and Appendix E.

Just as motor vehicles are referenced in popular music, so too music is a part of automobile culture and design. Today, everyday drivers anticipate taking their music along for the ride, and they have been doing so since the 1930s, while from the 1950s onwards car audio technology became an irrefutable feature in all vehicles. Moreover, just as motor vehicles have been at the centre of popular music cultures, music culture has in turn defined the attitudes of modern society about automobiles. Such mind-sets are rich sources of marketability, and the iconic graphics that reflect these attitudes have become part and parcel for brochures and products designed to acknowledge the linkage between cars-&-music. These have become especially fashionable among Auto Insurance Companies (see Figure 1.2). The impact of musical advertisement and audio branding of motorcars, as well as music celebrity-endorsers for motoring products, has had a knock-on effect influencing a host of other social concerns, and not the least are images of youth, gender, and courtship – all of which can be seen and heard on such web sites as *YouTube*. Du Lac (2008) states that the relationship between music and cars is as evident as ever – if only for the reason that *music still sounds great in a car*. Accordingly, as long as automobiles are accessible to consumers, drivers will always be cruising down some roadway with their favourite playlist of tracks blasting out of the cabin speakers in full volume, as they croon away in the best of karaoke style while simultaneously drumming the beats and fills on a steering wheel or fingering the hot-licks and riffs on an air guitar.

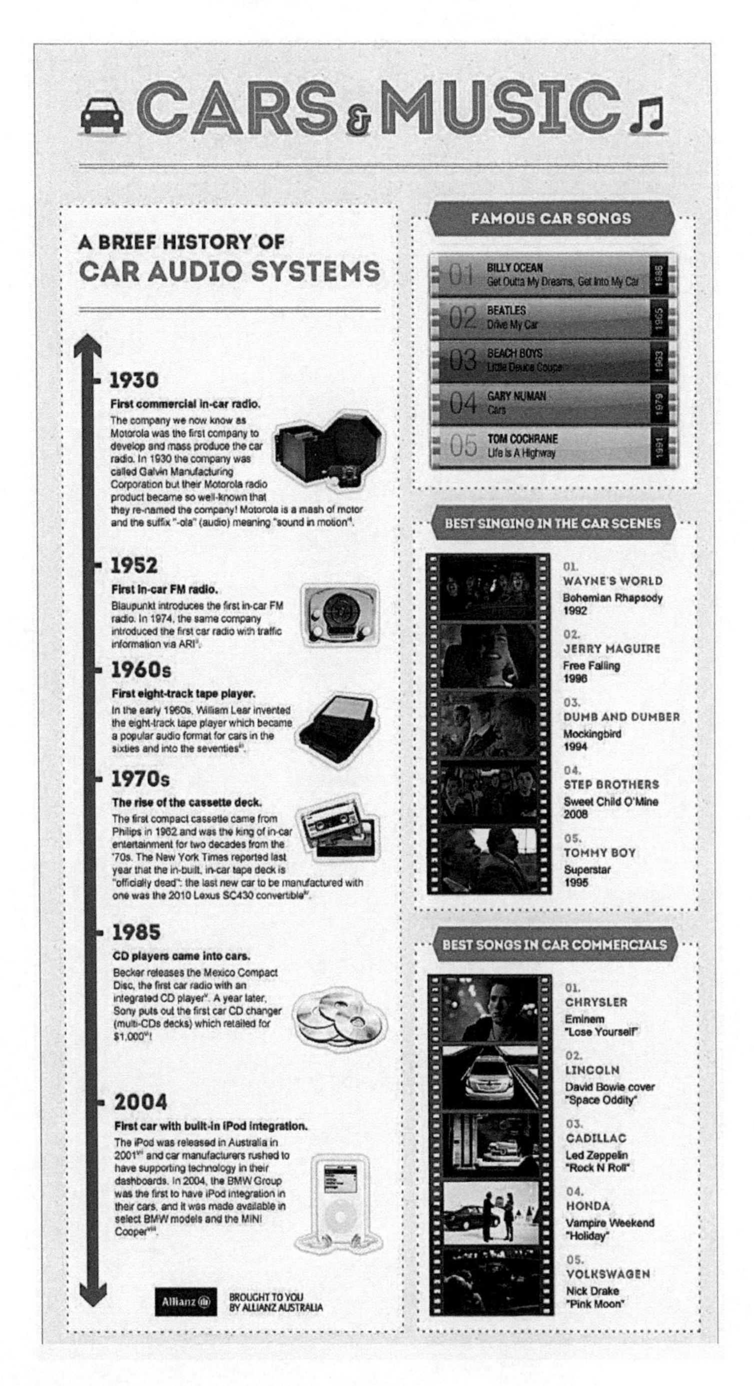

Figure 1.2 Car and music info-graphic

Source: Reprinted with permission from Allianz Australia Insurance, Ltd.

Chapter 2

Background Music in Everyday Life: The Listener and Effects of Music on Listeners

Chapter 2 is the second of two background chapters. The text provides a wide picture from the field of music psychology, and highlights several overriding issues concerning the use of music as a background to non-musical activities. Initially, five conceptual foundations about music listeners are outlined, and then the chapter focusses on background music as experienced in everyday life (with the exception of music listening in automobiles that appears separately in the Chapter 3). Nevertheless, as the text unfolds, we can envisage the impact that each fundamental concept might have on driving, whether related to the driver as a listener, the music selections as background for driving, or responses to the music itself while on the road. Each of these principles is considered in Chapters 4 and 5 when deliberating on the contraindications and ill-effects of in-car music listening. The chapter concludes by targeting research efforts that have employed the use of music in real-life everyday and commercial settings. It is expected that some insights from these studies will widen our understanding about the conceivable effects of music in other milieus – such as an automobile. The culmination of the chapter is the presentation of a model developed by David Hargreaves referred to as the *Reciprocal Feedback Model of Musical Responses* (Hargreaves, 2012; Hargreaves, MacDonald, & Miell, 2005; Hargreaves & North, 2010; Hargreaves, North, & Tarrant, 2006). This model is by far the most comprehensive conceptual scheme with which the determinants and aspects of human responses to background music can be described. The model accounts for a three-way interaction between the individual characteristics of the listener, the music being heard, and the environment in which the exposure of background music occurs.

2–1. Conceptual Foundations

Liking for music is a strong human trait, and for some people this admiration is as rewarding as food, sex, or drugs (Lamont & Greasley, 2009). Clearly, what happens to us when we listen to music has as much to do with who we are, as it does to the context in which we hear the music. The scientific literature points to several intervening factors that mediate how background music affects us. These are individual perceptual differences, the manner of response, if (or not) the music conveys a personal meaning, musical preferences and tastes, the level of engagement invested in listening to the music, and idiosyncratic personality characteristics.

2–1.1. The Music Listener

In his landmark book *The Musical Temperament*, Kemp (1997) claimed:

> It goes without saying that our preferences for particular pieces of music and composers may well reflect deeper aspects of our individual differences, particularly in terms of listening styles and perceptual processes (p. 37).

Each of us hears something unique in music, and it seems that the type of music we listen to can reveal quite a bit of informative material about who we are and how we think. Haack (1980) highlights several theories that underline individual differences among listeners. For example, Ortmann's 1927 expose delineates types of listeners based on cognitive styles that specifically relate to how people receive information about the world. Accordingly, there are three primary modalities, or styles of cognition, that can be employed when people listen to music: *sensorial listeners* are those who react primarily to the raw materials being heard; *perceptual listeners* are those who react to the fundamental relationships presented within the music being heard; and *imaginal listeners* are those who react to the associations activated by the music as it is heard. It is important to keep in mind that neither modality is better than the other, but rather each reflects the unique way in which different listeners assemble knowledge from nature. Another model exploring differences among music listeners is Meyer's 1956 taxonomy in which two basic perceptual positions have been proposed. Listeners are characterized as either employing 'absolutist' or 'referentialist' perceptual tenacities (for more details see: Hodges & Sebold, 2011; Meyer, 1956a, 1994; Radocy & Boyle, 1979). Accordingly, while listening to music some of us experience intrinsically generated knowledge. In this case we would be considered an *absolutist* or *isolationist* in the sense that we perceive music value and meaning as resulting exclusively from the absolute sounds – and nothing more. Some absolutists may be 'formalistic' in that the meaning of music is conveyed solely from the formal aspects of the structural relationships in a piece, while other absolutists may be 'expressionistic' in that they recognize that the structural relationships in the pieces are capable of stirring emotions that subsequently express meanings and connotations. However, while listening to music some of us experience extrinsically generated knowledge. In this later case, we would be considered a *referentialist* or *contextualist* in the sense that we view the meaning of music as resulting from more than the sounds themselves – and perceive the extra-musical ideas, emotions, images, and stories that surface from the music as of primary significance. Most referentialists are *expressionists* as they consider all music to express human experiences through extra-musical connotations. To summarize thus far, perhaps the overriding differences between music listeners are their apparent cognitive styles, which involve three concentrations: the 'mode of interacting' with the environment (a process of data entry, often referred to as *sensation*), the 'platform used to inspect' the information received (a process of organizing the data, often referred to as *perception*), and the 'format of interpreting' the organized data (a process of establishing meaningful connections between bits of information, often referred to as *cognition*).

However, all of the above descriptions are further predisposed by one of two autonomous cognitive-based perceptual attitudes. Namely, as listeners we endorse a specific approach that demonstrates *field-dependence* or *field-independence*. Kemp (1996) brings out that those of us who are *field-dependent* seem to be more easily overwhelmed by the general organization and structure in the music. In this case, the various elements in the piece (such as instrument voices, textures, and section parts) become fused together *as if* there is only one complete whole unit that has been experienced. Alternatively, those of us who are *field-independent* seem almost effortlessly to be able to breach the surface cues in the music with a more analytic stance. In this later case, many more specific details about the various music components are encoded, and subsequently we are capable of following each unique music part as a contributing element in creating the whole piece. It should be noted that as far as enjoyment is concerned, neither of these perceptual listening styles have any advantage over the other – albeit listeners who endorse field-independence encode music with significantly richer detail. Music psychologists look upon these approaches similarly as they would other behavioural characteristics, such as personality traits and temperaments.

Radocy and Boyle (1979) delineate other theoretical models that explore individual differences among music listeners. For example, Adorno's 1976 attempt to understand music as a preoccupation among everyday people led to a sequential hierarchy of musical involvement, in which seven levels of music engagement were described – from the 'most intensive' to the 'least engaged':

1. *Expert listeners* are those with formal training who can understand the syntax and grammar within the language of music.
2. *Good listeners* are those without formal training but who can still understand the language of music without an extensive knowledge of its syntax and grammar.
3. *Culture consumers* are those who listen to music often, and have amassed much information about music, but feel that the syntax and grammar of music are immaterial for their engagement.
4. *Emotional listeners* are those who use music to regulate their affect, whereby their intention of listening to music is to match, stimulate (intensify), or sedate (pacify) their mood, all the while disregarding the more intellectual aspects of music.
5. *Resentment listeners* are those who use music solely as a form of protest against socio-political conditions.
6. *Entertainment listeners* are those who are addicted to listening to music as a background for other activities.
7. *Indifferent* or *unmusical listeners* are those who seem to be disengaged from music listening altogether.

At this venture, it is important to point out that not only are we different with regards to our cognitive *mode* of listening (Ortman, 1927), the *resolution* at which we perceive music (Kemp, 1996, Meyer, 1956a), and the *intensity* to which we engage in music listening (Adorno, 1976), but also as far as to the *function(s)* to which music listening serves us in our daily lives. In one of the first studies on this subject, Behne (1997) demonstrated the existence of diverse manners involving music listening based on function. Behne simply asked large samples of German music listeners (who were not professional music performers) to respond to the statement: 'When I listen to music, I… [fill in the blank]'. Their responses fit into one of nine overriding clusters, whereby each revealed some type of independent use of music listening that he or she referred to as *utility*. The typical responses were: When I listen to music…

1. *Associative function*: 'I invent a story, as if I were watching a movie'
2. *Compensating function*: 'it changes my mood'
3. *Concentrated function*: 'I like to close my eyes'
4. *Diffuse function*: 'I like to do other things besides just listening'
5. *Distancing function*: 'I try to understand the words of the vocal part'
6. *Emotional function*: 'I pay attention to what types of feelings are expressed through the music'
7. *Sentimental function*: 'I remember things from the past'
8. *Stimulative function*: 'it makes me feel excited, even aggressive'
9. *Vegetative function*: 'I sometimes feel my heart beat faster, my skin prickling, butterflies in my stomach'

Behne's research pointed to new directions in exploring the foundations of everyday listening to music. It seems that not only are we quite unique in *how* we listen to music, but also *why* we listen to music. Yet, the final aspect we need to understand about our responses to music has to do with the *way* we react to the music.

2–1.2. Music Listening Response Theories

When reflecting upon human engagement with music, it is paramount to target 'response theories' that serve as the foundation for understanding the effects of music on listeners (Radocy & Boyle, 1979). Accordingly, there are three major forms of reactions that can be observed and tested on music listeners. These are: 'instinctive', 'physiological', and 'motor' responses. The *Instinctive Response Theory* furthers a claim that listeners perceive music based upon an instinctive human predisposition for aural impressions. That is, we respond to music through a pre-determined genetic endowment that represents a set of inborn human behaviours, and these compliment other inherited traits from our familial bloodline. For example, it seems that we can keep time (i.e., clap hands, tap a beat, dance with a partner) whether or not we have formally learned an instrument. Such capabilities are intuitive. Then again, if our lineage includes a musical pedigree, then perhaps our responses to music would be inherently subtler. The *Physiological Response Theory* furthers a claim that listeners perceive music as based upon recurring physiological cyclic phenomena and the processes that are recruited to keep physiology in a steady state. That is, our responses are actually established by on-going biological activities (such as the rate of human heartbeat, nervous discharge, synaptic firing, etc.); these react to changes in the external environment and then attempt to return to a relative stable internal environment in spite of exposure to latent characteristics in the external environment. For example, we can feel the pace of a song becoming faster or slower (as we perceive the deviation of an internal speed or a time-linked proprioceptive impression), and this alteration cues a sympathetic physiological reaction stimulating or sedating our heartbeat (and other bodily functions) to which a parasympathetic resolution (return to static state) ensues. Finally, the *Motor Response Theory* furthers a claim that we perceive music based upon muscle action. Our responses to music are established by the brain and central nervous system that co-jointly act to engage voluntary muscle activity. For example, through a process called 'rhythmic contagion', we walk in synchrony, clap or march in unison, and even dance with a partner to a pulse beat heard in the song. Of course, it is most plausible that as listeners we respond to music in a more complex fashion. Perhaps we engage in the perceptual organization of musical stimuli, and respond through all three response-modes simultaneously: innate instinct + physiological echo + motor discharge. We might also assume there are other elements to consider such as *emotion* and *learning*.

Much of what listeners experience with music is directly related to affective responses. *Affective Responses* involve a host of effects, including psychophysiological reactions, mood responses, associations, intrasubjective experiences, and verbal messages (Abeles, 1980; Haack, 1980; Radocy & Boyle, 1979; Nater et al., 2006; Webster & Weir, 2005). A description of each follows:

1. *Psychophysiological reactions* involve any combination of ten biological responses. For more details, see: Bartlett (1996), Hodges (2010), Hodges and Sebald (2011), or Sloboda (1991). Physiological reactions include:

a. Biochemical processes – adrenocorticotropic hormone ACTH, beta-endorphins, blood glucose, cortisol, dopamine, epinephrine, growth hormone, interlukin-1 -6 -10, melatonin, mu-opiate receptor expression, natural killer cells, neutrophils and lymphocytes, norepinephrine, secretory immunoglobulin A, serotonin, stress hormone, testosterone.

b. Brain function – activation of limbic system and cortex, brainwaves, cerebral blood flow, event-related potentials.

c. Cardiovascular activity – blood oxygen saturation, blood pressure, blood volume, circulation, heart rate, pulse rate, respiration.

d. Facial gestures – frown/eyebrows, smile/mouth-cheeks, under eye.

e. Frisson behaviour – blushing, chills, crying, goose-bumps or pilomotor/piloerection response, flushing, lump in throat, prickly feeling back of neck, shivers, sweating, tears, thrills, tingling on spine, trembling.

f. Gastric motility – contractions, metabolism.

g. Muscular contraction – knee-jerk patellar reflex, numbness, tension, trembling.

h. Pupillary activity – constriction, dilation, startle eye blink.

   i. Sensations of body casing – electrodermal responses such as body temperature, finger temperature, skin resistance, skin conductance, skin temperature.
   j. Skeletal movement – foot/hand tapping, head nodding, swaying.
2. *Mood responses* involve elicitations and sentiments of emotion that may be perceived or induced. It should be pointed out that our mood responses (as listed below) have been shown to be somewhat mediated by the biological sex of the listener (Nater et al., 2006; Webster & Weir, 2005). For more details, see: Hodges (2010), North and Hargreaves (1997b), Panksepp and Bernatzky (2002), Sloboda, O'Neill, and Ivaldi (2001), Terwogt and van Grinsven (1991), Trainor and Schmidt (2003), or Webster and Weir (2005). Mood responses include:
   a. Affect – positive or negative affect reflecting high versus low valence.
   b. Feelings – alert, angry, anxious, aroused, bored, comforted, connected, detached, distressed, drowsy, energetic, fearful, generous, happy, irritable, insecure, interested, in the present, involved, lonely, nostalgic, relaxed, sad, secure, tense, or tired.
   c. Dispositions – a general sense of soothing or agitated.
3. *Associations* that surface from listening to music clearly involve re-experiencing an event or occasion in which the same or similar music has been previously heard. These may also involve 'connotations' that reflect associations between the music experience with some other non-music experience (Meyer, 1956b, 1994).
4. *Intra-subjective experiences* occur when the sounds themselves evoke some kind of imagery or imaginary scenes.
5. *Verbal messages* involve language and are more apparent when songs are heard that present an underlying refrain or unremitting 'hook', offering a subtext to an underlying social, political, or religious issue.

To summarize, music listening response theories provide evidence that is indicative for the *way* we react to music. In many cases, the way we react says something about the meaning we derive from the music content itself. To understand how this occurs, and why some music is more meaningful than others, it is important to clarify how music affects listeners in non-musical settings.

2–1.3. Finding Meaning in Music

When we listen to music, some pieces seem to be more meaningful to us than others. Music pieces that are meaningful for one listener may simply be unappreciated by another, and may even be considered to be nothing more than noise. Much of the difference between people has to do with the extent to which they understand the music that is being heard, and then subsequently, the extent to which the music events can be anticipated as it unfolds. Music is ultimately comprised of components and aural attributes, such as pitch (melody), harmony, duration (rhythm), loudness, and timbre. Together, these offer us an architectural structure whose *complexity* runs the gamut from over-simplified and transparent to highly intricate and dense. It may be that the degree to which we can understand and anticipate the occurrences in a music composition is directly related to the extent to which we can find meaning for the piece. That is, the magnitude of information is crucial, and relates to the *information theory*. It seems that *capacity* has much to do with the amount of uncertainty that is within the music; it is a measure of coverage by which the message in the music is clearly received. Hence, too much information can lead to greater listener uncertainty of what may be expected in the next moments or measures in the music, while the lower the degree of coverage the greater the build up of tension. Obviously, then, an optimal level of information needs to be covered to achieve the ideal level of anticipation. Abeles (1980) points out that arousing expectations, building suspense, and feeling the resolution is the process in which musical meaning is brought about in most Western music. This process is essentially a structural ploy that has been carefully planned out by the composer and strategically carried out by the performer, with the sole purpose of temporally inhibiting (i.e., delaying) gratification towards intensifying pressure. Then, exactly at the pre-planned musical moment, the music steers us to a climactic discharge of pent-up energy (Meyer, 1956a, 1994). This engagement affords us the impression that the music was 'felt' in a more personal powerful manner, and therefore it has meaning. However, if such manipulations do not lead to a plausible resolution of the pent-up energy, go beyond our ability of comprehension, or surpass our tolerance levels for frustration, then we will feel overwhelmed. In this later case, we might loose any semblance of musical understanding or meaning from the music altogether. Thus, the optimal amount of information required for each individual listener to find meaning in music has to do with two functions: (1) the relationship between our aptitude to perceptually organize music into mental schemas, and (2) the complexity (or redundancy) of musical structures as they interrelate to our previous experience with the given music style of the piece being heard (referred to as the *Expectancy Theory*). We might then summarize, that a listener can only acquire meaning for music when their capacity for understanding is optimal but with a slight demand for increasing mental challenges within a highly familiar music culture.

All theories and concepts outlined thus far, and indeed within the entire context of this book, solely relate to responses from music as heard among listeners of Western cultures. It is customary to view listeners who are born and bred in geographical regions that fit this background as having a *Western ear*. Such a stance is also true of the great majority of theoretical essays, clinical descriptions, and applied empirical works on the effects of music. This tightly bonded (albeit biased) focus accounts for our long-term familiarity, and hence implicit knowledge, of various musical styles associated with Western culture. We do have a lifetime of personal experience with the specific conventions and nuances that have been employed by music composers and performers in Western art culture for over the past 600 years of music consumption. For the most part, there are many music styles that are popular with listeners in Western societies. An overriding catalogue of roughly 30 music genres has been repeatedly explored within the context of a social psychology of music (for example, see: Chamorro-Premuzic & Furnham, 2007; Colley, 2008; Litle & Zuckerman, 1986; North & Hargreaves, 2006, 2007a; North, Hargreaves, & Hargreaves, 2004; Rentfrow & Gosling, 2003). The catalogue of music genres represents styles that are widely known across North America and Western Europe rather than localized to a specific geographical region (see Table 2.1).[1] In general, we might assume that all music genres are somewhat equal in their effects on human psychology and physiology. Albeit, that for one listener a particular style will be perceived as unique and meaningful, while for another listener a totally different music genre will provoke these feelings. Yet, several studies have pointed out that a few specific genres can be seen as Problem Music (such as Hard Rock, Hip-Hop, Rap, and Punk music styles) as these seem to cause more harm than other genres (North & Hargreaves, 2006). Further, exposure to songs with violent lyrics has been seen to cue aggressive thoughts, feelings, and perceptions, as well as increase deviant conduct, and minimise pro-social behaviour (Anderson et al., 2003; Barnet & Burris, 2001; Godwin, 1985). Moreover, Krahe and Bieneck (2012) found that aversive Techno music without lyrics (that is, void of cognitive sematic content) often leads to negative moods that prime aggressive behaviour. Finally, two specific genres (Heavy Metal and Country music) have been linked to an increased risk for suicide ideation, as well as to actual rates of suicide (Lacourse et al., 2001; Rustad et al., 2003; Stack & Gunndlach, 1992).

2–1.4. Musical Preferences and Musical Tastes

There are many affective responses that can be experienced when listening to music (McCraty et al., 1998; Radocy & Boyle, 1979). Among these, are: *simple feelings* (pleasant *vs.* unpleasant, positive *vs.* negative, enthusiasm *vs.* repulsion), *activity feelings* (hunger appetite, romantic sexual desire), *attitudinal expressions* (aesthetics, morals, religion, social), *moods* (anxious, cheerful, grief), and *emotions* (anger, embarrassment, fear, happy, sad). In addition, other behaviours interconnected to listeners' emotions are: musical appreciation, musical attitudes, musical interests, musical values, musical preferences, and musical taste. The last two are of great importance to the current discussion. As defined by Abeles (1980; Abeles & Chung, 1996) *Musical preference* is an articulated choice of one style or musical piece over others. Musical preferences can be modified by countless variables, including: age, aptitude, education, ethnicity, familiarity, geography, intelligence, musical training, nationality, personality, political orientation, race, social group, social class, socioeconomics, and stage of life (for a review,

1 In North and Hargreaves's three-part study (2007a, 2007b, 2007c) covering more than 2,000 participants, a number of genres were ruled out as less than common among everyday listeners, including: 20th century, Ambient, Baroque, Choral, Drum&Bass, Early Music, Electronic, English Folk, Funk/Acid Jazz, Heavy Metal, Irish Folk, New-Age/Relaxation, Psychedelic Rock, Punk, Reggae, and World Music styles.

Table 2.1 Main musical styles and genres

	Music Style/Genre	Representative Artists
1	Alternative	Incubus, Korn, Tool
2	Blues	B.B. King, Robert Johnson, Johnny Winter
3	Classical	J.S. Bach, L.V. Beethoven, W.A. Mozart
4	Country & Western	Hank Williams, Jr., Johnny Cash, Dolly Parton
5	Dance	David Guetta, PitBull, Don Omar
6	Death Metal	DeiCide, Cannibal Corpse, Obituary, Burzaum
7	Disco	Donna Summer, Sister Sledge, Chic
8	Easy Listening, Light-Instrumentals	Kenny G, Montovani Orchestra, George Zamfir, Richard Clayderman
9	Electronic	Air, Massive Attack
10	Folk	Joan Baez, Bob Dylan
11	Funk	James Brown, George Clinton-Parliament, Funkadelic
12	Fusion	Dave Weckl, Herbie Hancock, Weather Report
13	Golden Oldies	Frank Sinatra, Nat King Cole
14	Gospel	Luther Van Dros, Mahila Jackson, Andrew Gouche
15	Gothic	Cradle of Filth, Type O Negative
16	Grunge	Nirvana, Pearl Jam, Alice In Chains
17	Hard Rock, Heavy Metal	AC/DC, Iron Maiden, Metallica, Pantera
18	Hip-Hop	Eminem, Jay-Zee
19	Jazz	Stan Getz, Duke Ellington, Charlie Parker, Dave Brubeck
20	Opera	G. Puccini, G. Verdi
21	Pop, Chart, Billboard, Top-40	Jenifer Lopez, Backsteet Boys, Madonna, Michael Jackson, Justin Timberlake, Beyoncé, Britney Spears, Christina Aguilera
22	Progressive Rock	Yes, Genesis
23	Punk	Clash, Sex Pistols, Fugazi, Rancid, Offspring
24	Rap	Tupac Shakur, Notorious B.I.G., N.W.A.
25	Rhythm & Blues	Howlin' Wolf, Muddy Waters
26	Rock	Guns N' Roses, Aerosmith, Rolling Stones, Led Zeppelin, Deep Purple
27	Rock N' Roll	Elvis Presley, Chuck Berry
28	Soft Rock	James Taylor, Carole King
29	Soul	Stevie Wonder, Aretha Franklin
30	Sound Tracks, Broadway, Theatre/Movies, TV	My Fair Lady, The King And I, Cats, Chicago, Little Mermaid, The Lion King, Smash, Glee
31	Speed Metal	Slayer, Metallica, Megadeth
32	Swing Big-Bands	Benny Goodman, Tommy Dorsey
33	Techno	Skooter, Plastic Man, Aphex Twin
34	Trance	Infected Mushroom, Skazi

see Russel, 1997). As an example, much research has shown that upper-class well-educated listeners tend to prefer high-brow music styles such as Classical, Opera, and Jazz Big-Band genres, whereas working-class less-educated listeners tend to prefer music styles that are considered to be low-brow such as Country, Gospel, and Rap. Other studies have shown that residents of urban neighbourhoods veer towards Jazz, Classical, and Contemporary-Rock music styles, whereas those living in suburban and rural surroundings tend to listen to Classic Rock, Country, Folk, and Oldies. Other studies have shown that sometimes ethnic groups such as African Americans display stronger inclinations for Jazz and Soul, whereas American Caucasians may exhibit greater predispositions for Rock, Country, and Classical music (Rentfrow & McDonald, 2010; Rentfrow, McDonald, & Oldmeadow, 2009). Nevertheless, as musical preferences are somewhat transitory, reflecting short-term commitments, they are not always consistent for long periods over the span of a lifetime. Namely, musical preference is no more than us liking one piece of music compared to another – at any given point of time. Nonetheless, it is apparent that such short-term experiences of preference for particular pieces or styles (that are subsequently re-experienced repeatedly in the same manner), do inform our longer-term judgments of taste. *Musical taste*, then, is a more stable overall attitude about what kind of music we prefer, representing an enduring pattern over long periods of time (Hargreaves, North, & Tarrant, 2006; Lamont & Greasley, 2009). It is of interest that musical taste is often viewed in a similar fashion as personality traits or temperaments. Whereas musical preferences are discernable by short-term decisions concerning which piece to listen to, musical taste is identifiable by exploring the real-life patterns of music engagement over longer periods of time. Namely, which music records (CDs and downloads) do we prefer to purchase, and which music concerts (bands and performers) do we prefer to attend. Lamont and Greasley point out that musical preference and musical taste have been explored with infants, children, teenagers, and adults. Accordingly, the literature is full of studies that have employed a diverse assortment of empirical methodologies, such as: looking-listening paradigms, operant music listening recorders, measurements of duration (listening time) to specific channels variegated by music styles, behavioural and verbal responses, rating scales, semantic differentials, self-report measures, individual preference nominations, single-episode structured interviews, and repeated-exposure multiple-episode in-depth interviews.

It is fascinating that people intuitively understand what aspects of their being display what kind of information to others. For example, we know that some of our personal features (i.e., physical appearance, non-verbal conduct, facial gestures), and even behaviour such as our own housekeeping skills (i.e., arrangement and tidiness of spaces including work, office, bedroom, and kitchen), are all ample materials that others can see and learn something about who we are. Ironically, there is widespread belief that the musical styles and pieces we listen to are highly obvious and potent forms of communication with which information about our identity is conveyed to others. The truth is that research studies have indeed demonstrated that personalogical material solicited from musical preferences is both quantitatively and qualitatively much more exhaustive than any other information that might surface directly from other sources – including photographs or brief videos.[2] Survey studies often report that people feel their musical preferences supply even more information about themselves than the clothes they wear, the movies they watch, the food they eat, the websites they visit, the hobbies they engage in, or how they spend their leisure time. Perhaps, as listeners we feel this way because of our own past experience when we have attempted to get acquainted with someone. Namely, in our effort to construct a representation of what someone else is like, we often seek details that we ourselves perceive as 'clues' on a host of topics – *as if* some bits of information that form a 'diagnostic impression'. It may be of interest that people do consider music preferences as a substantial source of evidence providing a clear accurate unobstructed message about a host of our feelings, thoughts, traits, preferences, and personalities. Hence, at least intuitively, musical preferences can be seen as reflecting 'who we are', 'who we want to be', and 'how we want others to perceive us' (Rentfrow & Gosling, 2006, 2007; Rentfrow, McDonald, & Oldmeadow, 2009). But are such intuitions quantifiable? Does knowing our musical preferences really entail knowing aspects of our social group, attitudes, beliefs, and values? In a comprehensive review of the research literature Rentfrow and colleagues point to many studies that demonstrate men as perceiving women who prefer Classical music as being more attractive and sophisticated, or women as perceiving men who prefer Heavy Metal music as being more aggressive and rebellious. Other studies reviewed found that listeners who prefer Chart Pop styles were perceived as being more physically attractive, conventional, energetic and enthusiastic, while those preferring Classical music styles were perceived as more conservative, intellectual, religious, and traditional. From all of these investigations it would seem that impressions formed through awareness of someone's musical preferences rely on two overriding sources. First, specific *features* that are found in the music itself (such as a slow tempo) often lead to interpretations about a person's behaviour (such as being a relaxed person) or to their personality traits (such as having an introverted temperament). Second, there are supposedly well-known stereotyped traits that we associate with certain music *genres*; such attitudes seem to have resulted from years of programmed conditioning, and hence can be seen as a form of Hollywood indoctrination through film and TV exposure. For example, there are plenty of things we know (said cynically), and therefore feel, about fandoms of Heavy Metal, New-Age Meditation, or Religious music styles. Whereas both of the above may seem to be over-generalizations without much evidence, the truth is that several studies implemented by Rentfrow and colleagues demonstrated that such perceptions are actually valid, reliable, and cross-cultural. Accordingly, there appears to be widely

2 It should be pointed out that notices and alerts to the public about materials placed on social media such as *Facebook* have always included cautionary warnings regarding photos and video clips. Rarely, if ever, do we consider if others might target listings of music exemplars and/or musical preferences seeking personal information about us.

acceptable conceptions that fans of Classical music are friendly, hardworking, introverted, intelligent, and artistic people who drink wine; while it is also widely acceptable to conceive of Rap music fans as extraverted people, who value social recognition, and drink beer and/or smoke marijuana.

Given the above, it may be no surprise to learn that music choices do express information about social class, ethnicity, and nationality, as well as link to personality typologies. Moreover, studies have indicated that musical taste has been explicitly associated to characteristic temperaments and observable behaviour patterns (Abeles & Chung, 1996; Carpentier, Knobloch, & Zillmann, 2003; Delsing et al., 2008; Kemp, 1996, 1997; Lewis & Schmidt, 1991; Litle & Zuckerman, 1986; McCown et al., 1997; Rentfrow & Gosling, 2003, 2006, 2007; Rentfrow & McDonald, 2010; Rentfrow, McDonald, & Oldmeadow, 2009; Radocy & Boyle, 1979; Schwartz & Fouts, 2003; Zuckerman, 1994). Perhaps the earliest attempt to study such associations was implemented by Cattell and Anderson (1953) who had a notion that personality could be inferred from music preferences. Cattell subsequently identified twelve *music preference factors* that were interpreted as revealing unconscious aspects of personality. From the mid-1950s onwards, most researchers do claim that we are inclined to choose specific kinds of music because we have characteristic needs, or that we choose to listen to music because there are constitutional issues that may be either reflected or satisfied by the music we identify with. In a truly outstanding study, North and Hargreaves (2007a, 2007b, 2007c) took on the task of quantifying how music may be a means of identifying social groups in the real world. Their study points to a wide variety of connections that surface between particular musical preferences and various lifestyle choices. Astonishing, the study found that music preferences and tastes point to differences among listeners variegated by combinations of thirteen factors, including: *interpersonal relationships* (marital status, number of partners, children); *sexuality* (promiscuity, orientation); *living arrangements* (on own/with partner, own/rent a house/flat); *beliefs on moral, religious, and political issues* (liberal/conservative affiliation, guns, taxation, nuclear weapons, prayer); *criminal history* (anti-authoritarian behaviour, delinquency); *use of substances* (drugs); *use of art and media* (newspapers, books, TV, radio, Internet, smartphones, mobile devices); *leisure time activities* (pubs/bars, clubs, cinema, nature trips); *patterns of travel* (vacations at local/cosmopolitan locations, number of yearly trips); *personal finance* (cash/credit payments, level of income); *education* (level of certified education); *employment* (salaried/independent work status); and *health behaviours* (visits to doctors, personal hygiene, tobacco/alcohol, diet).

Finally, men and women also seem to differ in both their musical preferences and musical tastes (Nater et al., 2006). That is, music preferences have been linked to biological sex (male/female) as well as to psychological gender (masculine/feminine). For example, Colley (2008) indicates that young men seem to admire heavier contemporary music styles (such as Blues, Heavy Metal, Psychedelic Rock, Rock, and Southern-80s), while young women seem to like lighter music styles and chart tunes (such as Classical, Country-Pop, Late 70s Disco, Folk, Gospel, Jazz, Mainstream-Pop, Reggae, R&B, and Soul). Similarly, Schwartz and Fouts (2003) found that heavier music styles that are very loud and fast paced music based on electric guitar and drum pounding discordant sounds (such as Classic Rock, Hard Rock, Heavy Metal, and Rap) best reflect the concerns of young men, which are: aggression, anger, antisocial interpersonal behaviour, criminal intent, decreased respect for women, dominance, hypersexuality, machoism-hypermasculinity, rebellion, risk-taking, and sensation-seeking. On the other hand, lighter music styles that are soft and medium-paced sounds based on acoustic guitar and keyboard rhythmic melodies (such as Ballads, Dance, and Teen Pop) best reflect what is on the minds of young women, which are: autonomy, cooperative behaviours, emotions, identity, a reflective demeanour, relationships, increased responsibility towards others, and sociability. Considering the above, we might then recognise the fact that those of us who share similar preferences and tastes for listening to music are more apt to be compatible, intimate, and experience far less tension leading to conflict. Hence, perhaps longevity, intensity, and satisfaction of relationships may be somewhat predisposed by music behaviours (Rentfrow & McDonald, 2010).

2–1.5. Music Engagement and Personality

The structure of human personality has been extensively studied for over 50 years, and the relevant literature demonstrates the existence of five overriding dimensions that have been referred to as 'The Big Five.' These are: *extraversion* (sociability and energy), *agreeableness* (friendliness and warmth), *conscientiousness* (organized and dependable), *neuroticism* (anxious and emotionally unstable), and *openness* (intellectual and creative). One question that has intrigued both researchers and music listeners alike entertains the possibility that certain personality dimensions indicate the way we use background music (Chamorro-Premuzic & Furnham, 2007, Delsing et al., 2008; Rentfrow & Gosling, 2003, 2006, 2007; Rentfrow & McDonald, 2010). Is the *why* and *how* we choose to listen to certain music genres (or specific pieces) mediated by our personality and temperament? Hargreaves and North (2010) raise several prospective avenues for inquiry: Do listeners of specific typologies use music more as a vehicle to 'reflect' or 'compensate' for their personality styles? Do listeners with idiosyncratic characteristics use music as a means to regulate their levels of arousal? To highlight these and other issues, the field of music psychology has targeted four overriding areas for research investigation, including: prototypicality and fandom, extroversion and introversion, disposition and temperament, and handedness.

2–1.5.1. Prototypicality and fandom

Our preference for a particular music genre is based somewhat on the degree to which certain music stimuli activate mental representations. The more we are exposed to the typical structural features that make up what we refer to as a specific music style, and the more the music style is capable of increasing our arousal activation, the greater is our reaction towards the same music genre in future exposures. That is what is meant when we refer to *musical preference*. Subsequently, the reciprocal liaison between the underlying operational qualities within the structure of a specific music genre and a set of typical traits within a person's disposition is designated as *prototypicality*. For example, several studies (Litle & Zuckerman, 1986; McCown et al., 1997; Rentfrow & Gosling, 2003) indicate that those of us with higher levels of sensation-seeking prefer intense and stimulating music styles (such as Heavy Metal, Rock, Punk), while those with higher levels of extroversion prefer energetic party music styles, antisocial individuals prefer arousing music styles that emphasize rebellion (such as Alternative Rock, Heavy Metal, Rap, Rock), and those who are more open to new experiences prefer more sophisticated music styles (such as Classical, Jazz). Other researchers (Kemp, 1996; Nater, Krebs, & Ehlert, 2005) found that impulsive, tough-minded, and sensation-seeking individuals prefer more aggressive genres (such as Electronic, Pop, Rock), while athletic individuals prefer intense and stimulating styles.

It is interesting to note that prototypical aspects of *fandom* for styles such as Classical, Electronica, Jazz, Pop, Rock, and Rap indeed come to the surface (Rentfrow, McDonald, & Oldmeadow, 2009). Accordingly, fans of Classical and Jazz music have been found to be both high in agreeableness and emotional stability, while Jazz fans have also been found to be higher in extraversion and openness but lower in conscientiousness. In comparison to Rap fans, listeners preferring Classical music are higher in political conservatism, intelligence (intellect and wisdom), religiosity, and artistic ability. Moreover, prototypical fans of Electronica and Rock are both higher in extraversion and openness, moderate in agreeableness, and lower in conscientiousness and emotional stability, while lower in conservatism, religiosity, intelligence, and attractiveness. Nevertheless, listeners preferring Rock music are lower in athletic facility and higher in artistic ability than fans of Electronica. Finally, the prototypical Pop and Rap music fans demonstrate higher in extraversion, moderate-to-high emotional stability, and low openness – with Pop fans higher in agreeableness than Rap fans. On the other hand, listeners preferring Rap music are higher than all other listeners regarding social recognition and self-respect while they are lower than all other listeners regarding beauty and forgiveness. As stated earlier, group membership and social class have also been linked to musical preferences. Rentfrow et al. report that prototypical Classical music fans are significantly more often found among upper-middle class and higher-class groups, while prototypical Rap fans are significantly more often found among working-class and lower-middle class groups, and prototypical Electronica and Pop fans are significantly more often found among lower class, working class, and lower-middle class groups.

2–1.5.2. Extroversion and introversion

Several studies have shown that listeners who are *extroverted* as well as *open to new experiences* tend to prefer upbeat energetic and intricate music styles that incorporate vocals and words, than other popular conventional music genres. These listeners seem to use background music in a more rational and intellectual fashion, indicating higher levels of cognitive processing, and therefore exhibit an overall preference for more complex music genres (such as Classical, Jazz, R&B, and Soul). On the other hand, *introverted* listeners tend to prefer music that can better serve their emotional needs. Here, then, we target *music complexity*. For example, research points to differences of interaction between music complexity and cognitive tasks (such as reading comprehension, observation, immediate recall memory, and delayed recall memory) as variegated by the personality dimensions of extroversion or introversion. Accordingly, as the level of complexity featured in the music increases, on-task behaviour as measured by actual performance scores increases for extroverts – all the while decreasing for introverts (Cassidy & MacDonald, 2007; Furnham & Bradley, 1997; Furnham & Allass, 1999). The findings presented above linking personality structures with inclinations and preferences for explicit music components and specific music genres appear to generalize across geographical regions and age groups.

2–1.5.3. Dispositions and temperaments

In their landmark study, Rentfrow and Gosling (2003) conducted a wide analysis of music preferences accounting for fourteen music styles. Consequently, four specific temperaments surfaced – each corresponding to a characteristic profile of musical genre – that have been seen to be highly reliable across time, populations, geographical region, as well as having withstood replication by Delsing et al. (2008). The music dispositional and temperamental profiles described by Rentfrow and Gosling are:

1. *Reflective and complex*. Listeners are open to new experiences, emotionally stable, tolerant of others, creative, imaginative, value aesthetic experiences, with self-perceived intelligence, political liberalism, and verbal ability, as well as adhere to more clever, pleasant, dreamy, and romantic positive affects. Those of us who exemplify this profile seem to prefer Blues, Classical, Folk, and Jazz musical styles.
2. *Intense and rebellious*. Listeners are also open to new experiences and emotionally stable, but are also more curious about different things, take risks, with self-perceived intelligence, physical activity, athleticism, and verbal ability, as well as exhibit a more enthusiastic, energetic, and loud affect. Those of us who exemplify this profile prefer Alternative Rock, Heavy Metal, and Rock musical styles.
3. *Upbeat and conventional*. Listeners are extroverted, socially outgoing, helping to others, agreeable, cheerful, contentiousness, reliable, with self-perceived attractiveness, athleticism, relatively conventional ideals, political conservatism, as well as with an overall enthusiastic mood. Those of us who exemplify this profile prefer Country, Pop, and Religious musical styles, as well as have a positive appreciation for Soundtracks.

4. *Energetic and rhythmic.* Listeners are extroverted, agreeable, politically liberal, full of energy, forgiving, with conservative ideals, self-perceived attractiveness, athleticism, as well as a more talkative demeanour. Those of us who exemplify this profile prefer Dance, Electronica, Funk, Hip-Hop, Rap, and Soul musical styles.

2–1.5.4. Handedness

Finally, it may be of interest to note that one recent study found that mixed-handed listeners are likely to listen more intently when experiencing new unfamiliar musical styles, and subsequently develop wider musical preferences and tastes compared to strong single right-handers (Christman, 2013). Accordingly, strong right-handedness is associated with decreased cross-hemispheric cerebral interaction and decreased cerebral cognitive flexibility. Namely, those of us with mixed-handedness are significantly better at updating beliefs when presented with new information, demonstrate increased ability for switching attention between two or more points of focus (including perspectives and alternative interpretations), and hence are more open to new experiences and divergent creative thoughts. All of these abilities are components of emotional tolerance and cognitive flexibility. This study by Christman is perhaps the first to empirically indicate a totally unique feature among people as regards their musical preferences and tastes. Accordingly, mixed-handers seem to be significantly more open to an increased number of musical styles including those that are often considered to be less popular genres (at least from a perspective of commercial sales) such as: Ambient, Avant-Garde, Blues, Bluegrass, Electronica, Folk, Funk, House, Jazz, Reggae, Soul, and World. On the other hand (pun intended), strong right-handers who seem to demonstrate music preferences of a more narrow range typically prefer the more firmly established set of popular music styles, such as: 80s Pop, Alternative Rock, Classic Rock, Country, Heavy Metal, Hip-Hop, Modern Pop, Modern Rock, R&B, and Rap.

2–2. Background Music in Everyday Life

Although there are still instances when we listen to music for its own sake,[3] one characteristic feature of twenty-first century lifestyle is that we most often listen to music as a background accompaniment for a range of everyday activities. Background music is heard both inside the home (with housework, eating, and reading), as well as outside of the home (when walking about, or in transport). As a matter of fact, music may be so omnipresent and pervasive that we are not always

3 In a study by Sloboda, O'Neill, and Ivaldi (2001) deliberately listening to music for its own sake was found in only 2% of all listening episodes, while in a study by North, Hargreaves, and Hargreaves (2004) 12% was reported. Therefore, even at the highest estimations, listening to music for its own sake is unfortunately a rather infrequent and untypical event in today's Western culture.

aware that music is there in the background – especially when we eat, drive, read, shop, or work (Kampfe, Sedilmeier, & Renkewitz, 2011). Bull (2000, 2003, 2005) demonstrated that we wake up with radio, go walking with songs, drive to music tracks, and relax or go to sleep accompanied by reproduced musical sound. Music follows us to work and back, and it is everywhere in between. Certainly some people differentiate between *background* music and *foreground* music based on aspects of the performance:

> Foreground music includes original artists and lyrics, whereas background music uses studio musicians playing instrumental… [and] tends to be more restricted in its range of tempos, frequencies, and volume (Yalch & Spangenberg, 1990, p. 57).

Indeed, for the most part we consider background to be a subordinate and inferior genre, and often refer to it unfavourably with depreciating names, including:

Ambient Music	Mood Music
Beautiful but Kitsch Music	Piped-in Music
Business Music	Playbacks
Canned Music	Popular Music
Easy Listening Music	Sonic Soundscapes
Elevator Music	Table Music
Light-FM	Timeless Pop
Light Classics	Wallpaper Music

Yet, common to all of these labels is the fact that they denote a musical endeavour in which the listener is engaged in some task or activity other than attending to the music. Moreover, such seemingly derogatory portrayals assert that listeners who employ music in this capacity actually intend to *hear* the music while not really *listening* to it. Therefore, the phenomenon of music being used in this way has been termed *aural cover* – to connote something that is in the background (like wallpaper) yet blending so well within the environment that it attracts little to no attention. Although most musicians (composers, conductors, performers, and music teachers), as well as amateur music-loving connoisseurs, downplay the value of background music (on both cultural and aesthetic levels), one welcomed commentary has attempted to balance out such dissonant objections – but unfortunately is no more than tilting at windmills:[4]

> It seems odd that so much social stigma should be associated with the use of music… [For example] the use of turquoise emulsion paint in supermarkets does not devaluate Picasso's blue period, so why should piped music supposedly devalue 'real' music? Similarly, it is rare to hear of people complaining about the décor in a supermarket, yet it is perfectly acceptable to hear complaints

4 An English idiom that means 'attacking imaginary enemies', as was the case in an episode from Cervantes' novel *Don Quixote*. Since the early seventeenth century when the novel was written, the phrase has come to be used as vernacular expression for 'entering into a fight for a lost cause'.

> about the music; and it is difficult to understand why this particular aspect of the aesthetics [among] environments should be subject to such strong feelings. The relationship between music and [contexts] may be controversial, but it is important because it [reflects] the everyday circumstances in which people hear music (North & Hargreaves, 2010, pp. 924–5).

2–2.1. Music in Everyday Life

Over one hundred years ago the use of music as an everyday background had already been recognized as serving a highly valuable function. The early twentieth-century French pianist and music composer Erik Alfred Leslie Satie (1866–1925) coined the term 'Furniture Music' as a marque of what he maintained to be one of the most important functions of music for mankind. Satie claimed there is a need for unique music to be created specifically as part of our surrounding environment; he felt that music could mask and/or neutralize noises that are indiscreetly forced into our consciousness (Lanza, 1994). Since the turn of the millennium, many investigative studies have systematically addressed the importance of music in everyday life (Chamorro-Premuzic & Furnham, 2007; Rentfrow & Gosling, 2003; Hargreaves & North, 1999; Kemp, 1996; North & Hargreaves, 1997a, 2000; North, Hargreaves, & Hargreaves, 2004; Tarrant, North, & Hargreaves, 2000). We might have assumed that the most prevalent everyday exposure to background music is in fact experienced when we are placed 'on-hold' while waiting to talk to someone over the telephone. Or perhaps, the most often experienced background music are segments of TV sit-coms or weekly screened soap dramas – or even a sequence of televised commercial advertisements. But, actually, the reality is quite different. Music has become fully embedded into the course of our 24-hour daily existence. For example, a wealth of music psychology research studies (Greasley & Lamont, 2011; Sloboda, 1999, 2005; Sloboda, O'Neill, & Ivaldi, 2000, 2001; Sheridan, 1998 [2000]) have found that self-chosen music is used to supplement any number of day-to-day home-based non-musical activities, including: waking up, dressing, deskwork, exercising, housework, reading, games, waiting (doing nothing), arriving home, cooking, eating (having a meal), washing, bathing, sitting in bed before going to sleep, and sleeping. To investigate the everyday involvement with music, Sloboda cued the responses of 500 UK representative correspondents in the 1998 Sussex Mass Observation survey with questions such as: 'Do you use music in different ways?' and, 'Are [the music pieces] linked to particular times, places, activities, or moods?' The findings demonstrated that activities, which were accompanied by music, were predominantly domestic or solitary and most frequently included housework or driving. In a series of follow-up studies investigating specific functions of music in everyday life, Sloboda, O'Neill, and Ivaldi studied diary-type journals of entries written subsequent to hearing a random pager signal seven times per day for one week. These studies confirmed that while some situations involving everyday activities offer little room for music involvement, most other types of activities are open to background music, and music exposure is more likely to occur when the person is alone in a situation associated with the opportunity for personal choice over the music. While Sloboda found that deployment of background music is a predominately domestic and solitary affair, he also pointed out that music pieces are also often heard to complement events that engaged others, such as socializing with friends and romance. Furthermore, music is often brought along as an accompaniment outside the house on walks, in public transport, at shopping markets and malls, for activities such as running and cycling, and when driving in a car. Similarly, Rentfrow and Gosling found that music was listened to when alone, as well as when 'hanging out' with friends, getting ready to go out, and when exercising.

There are indeed an increasing percentage of people who deliberately choose to listen to music. Especially as the distribution, marketing, and availability of recorded music increases, and with mounting unrestricted downloadable music obtainable over Internet, an increased number of listeners view themselves both as *participants in music* and as *investors in music*. In fact, the mobility of music means that we choose to hear music more often than before, and we listen to music in a wider variety of situations and contexts than ever before. Sloboda, Lamont, and Greasley (2009) point out that while music engagement may be passive (in the sense that such participation does not involve the investment of developing musical skills towards the actual performance of music), there has never been a more opportune time for active involvement in music than during the 2010s. Today we are highly adept in understanding and have a great proficiency in moving music between foreground and background in our everyday lives. That is, for the most part, as listeners we have developed specific capabilities in choosing appropriate music for different contexts. Sloboda et al. highlight the fact that as music background often accompanies a non-musical activity, music selections are mostly chosen by the listener on the basis that they can reliably fulfil the intention of achieving certain outcomes (such as to enhance the activity). Moreover, we seem to handpick specific pieces differentially. Accordingly, Sloboda et al. feel that there are four main recurring music functions that serve listeners' choices:

1. *Distraction* – to engage unallocated attention and reducing boredom
2. *Energizing* – to maintain arousal and task attention
3. *Entrainment* – to synchronize task movements with the rhythmic pulse of the music
4. *Meaning enhancement* – to draw out from the music background elements that add significance to the task and activity.

In fact, North and Hargreaves (2000) tested such a concept among participants of relaxation training versus spinning (stationary bicycling) exercises. By using two versions of the same music exemplar (that had been manipulated for tempo and intensity), each track emulated either low-arousal music (a slow-paced 80bpm piece reproduced at a decreased volume 60dBs) or high-arousal music (a fast-past 140bpm piece reproduced at an increased volume 80dBs). Their results demonstrate that during relaxation training participants favoured low-arousal music, while high-arousal music was preferred as a music background during the bike workout.

North and Hargreaves conclude that people prefer and choose to listen to the most appropriate music that helps them attain the goal of successfully completing a task by maintaining the particular levels of arousal suitable for a specific context. Such a differentiation denotes that some music selections are perceived more typically appropriate for specific activities while others are not. Through everyday personal experiences, as well as from shared experiences from those who are close to us, most ordinary people without formal music training learn about selecting music that is typical for the context, and align their musical preferences accordingly.

In comparison to other lifetime daily activities, Rentfrow and Gosling (2003) found evidence that we *listen to music* for more total number of hours per week than the amount of time we spend engaged in a hobby, watching TV/movies, or reading books/magazines. North, Hargreaves, and Hargreaves (2004) found that exposure to music is more frequent on the weekdays between the hours 11:00 to 21:00, while Sloboda, O'Neill, and Ivaldi (2000, 2001) documented that there is at least a 44% chance of music being present during any two-hour period between 8 a.m. and 10 p.m. on any day. Accordingly, just about half of all listening episodes take place at home as accompaniment to household chores, intellectual tasks, eating, and active listening to music, while the other 50% of episodes occur outside the home – with the single most often reported place of listening to music being inside an automobile while driving. Sloboda, Lamont, and Greasley (2009) summarized five main categories of music listening in non-music related contexts:

1. *Travel*. Most people travel everyday, and music is most often chosen to accompany travel (including walking, driving in the car, public transport) than in any other setting. Although music can enhance driving performance through its capacity to assist with appropriate levels of arousal, concentration, and as a counteraction for drowsiness (see Chapter 3), music can also distract drivers by hampering attention to key visual signals and therefore decrease driving performance (see Chapters 4 and 5). Listening to music while in transit (by bus, train, plane) helps pass the time and causes enjoyment. In general, most music exposure in transit is fixed field (i.e., via headphones), which allows us to mask unwanted sound (such as noise or conversation of other travellers), as well as to function as a mechanism against anxiety (from being in extremely close proximity to strangers).
2. *Physical work*. Music accompanies domestic chores such as everyday routines and manual labour like washing, cooking, cleaning and grooming; they are simple repetitive low-demand tasks. Music is often used in these activities as accompaniment to distract, energize, entrain, and to entertain the listener.
3. *Brain work*. Activities involving solitary tasks of private study, reading, writing, and other forms of thinking, as well as word processing, web-surfing, and e-mailing – all employ music accompaniment. Music is not intended to be listened to per se, but rather to enhance concentration, as well as to contribute to our well-being by reducing stress, blocking out unwanted noise, and increasing perceived quality of working environment.

4. *Body work.* Music has power to alter bodily processes such as physiological states and behavioural movements. The listener's choice and preference of the music is most often biased by the appropriateness of arousal potential of the music for the specific activity. In general, people prefer listening to high arousal music during exercise, but prefer low arousal music during relaxation, yoga, and pain management.
5. *Emotional work.* Listeners employ music for mood management, reminiscence, and the presentation of identity.

Certainly these uses of music in our everyday life have much to do with the availability and convenience of taking preferred music with us everywhere we may wish to go.

2–2.1.1. Mobile music

The accessibility of music could not have come about without the technological advancements in music reproduction over the past 25 years such as music cassette tapes, compact discs (CDs), digital video discs (DVDs), DAT tapes, mini-discs, MP3s (*iPods*), palm-pilots, pocket PCs, laptops, tablets (*iPads*), flash drives, and smartphones (*iPhones*). Hi-tech has certainly generated positive opportunities for the advancement of our species. Over a decade ago, Kallinen (2002) was keen to observe that pocket devices were quickly changing how we preoccupy our time outside the home and office – such as in cafeterias, public places, and on business trips. In this connection, Bull (2005) points out that the range of mobile sound media that support users in effectively preserving their sense of intimacy while experiencing everyday mass-transport transfer through a depersonalized and publicly controlled urban environment is far greater than anyone ever expected. Bull's notion is that music technology quite successfully has performed the ultimate benefit for travellers by simultaneously assimilating them into the world, while at the same time creating spaces of freedom. Bull alleges that through the mediation of technology, in the form of both futuristic music (the styles themselves) and hi-tech music reproduction devices (such as the latest Apple *iPod*), we have realigned our perception of mobility and how we spend our time moving about. Such efforts regulate more meaningful everyday experiences of the temporal spaces that exist in our daily routines – including traffic.

Nevertheless, concerns have also been raised about our current day *over-indulgence in music listening* and the idiosyncratic behaviours that such activities seem to promote. Foremost, Mobile Music (Gopinath & Stanyek, 2014) has made an overwhelming presence in the twenty-first century, with the principal feature of 2010s social life being coloured by the presence of *smartphone* technology and on-board applications. Together these not only provide the primary means for personal communication (speech conversation, network e-mailing, text messaging, and Web browsing) but also serve the user with hundreds of state-of-the-art automated capacities. For example, one major role of the *iPhone* is its function as a pocket-sized database depository of recorded music, which can be organized through user-defined searchable playlists; the device has music reproduction features that offer extremely high resolutions achievable by connecting inner-ear speakerphones through an auxiliary output socket. By simply having a smartphone in our pocket, we can listen to music everyday, all the day, anytime, anywhere, and everywhere we go. No matter what the circumstances may be (withstanding depletion of the battery), the quality is as good, and even better than listening to music seated in our own living room or in an acoustically designed concert hall. Mobile music "has become the soundtrack to everyday life, and thus a central part of personal development and identity for many people" (Hargreaves & North, 1999, p.73). It is no wonder, then, that we listen to background music when we cycle, drive, study, travel, or work. Nevertheless, today it has also unfortunately become quite common to see people wandering about in an apparent stupor that demonstrates that they are somewhat oblivious to their surroundings – with wires dangling from their ears (Heye & Lamont, 2010). Certainly, the advent of the *iPod* in 2001 significantly increased the use of a technology that separated listeners from society (Burns & Sawyer, 2012). Perhaps this type of behaviour is not entirely new. For example, living-room isolation was fairly common in the 1940s and 1950s when television became an integral fundamental aspect of domestic habituation. Then, when Sony released the portable *Walkman* stereo tape cassette player in the 1980s, users expressed the same reclusive conduct but in a more ambulatory fashion. Yet, as Bull points out, smartphones and other mobile music players are completely different. First, the personal stereo (both cassette tape and compact disc player) caused users to face great challenges in having to select what limited music they could take with them, while the *iPod* and *iPhone* expand such possibilities with seemingly unlimited numbers of tracks and playlists. Second, the personal music stereo (both the *Walkman* and *Discman*) was used as an 'in-between device' serving the listener to engage in music listening exclusively from door-to-door, whereas the *iPod* can be employed with a docking cradle to home hi-fi devices, with a USB cable to PC computers (at offices or in classrooms), and also with an audio cable to automobile head-units. Finally, digital miniaturized music technology has all but triggered an upsurge in maladaptive conduct among a younger generation of music listeners – and there is a spillover among younger drivers. On the one hand, mobile music listening has become a form of self-mediated escape from the stresses of everyday life; listening to music filters the infinite stream of aural stimuli that enters our conscious awareness through a more controlled auditory environment. On the other hand, mobile music listening has generated a behavioural stance that supports social isolation; listening to music pushes away the people and events in front of us, subsequently creating an alternative space that only remotely resembles our actual surroundings.

Considering the above, it would seem that through self-involvement with listening to personalized playlists we often generate a virtual 'auditory bubble' (Bull, 2005) with which we maintain some sense of security by not interacting with others or the environment. That is, the use of mobile music technology allows us to prompt a personalized narrative of the world mediated by a more controlled soundscape that serves to facilitate the temporal and spatial aspects of our experience. Although Heye and Lamont (2010) consider this 'bubble' to be somewhat permeable and may effectively heighten our awareness of the surroundings, clearly such a sonic envelope does change our perception of the outside world via an apparent realignment of landscape and soundscape. Certainly, when considering driving through traffic, such perceptions could lead to misconstrued acuities and accidents.

2–2.2. Functional Music

Music listening outside the home is not only heard in the streets themselves, but most often in commercial settings such as nightclubs, pubs, restaurants, shops, shopping malls and supermarkets (North & Hargreaves, 2005; North, Hargreaves, & Hargreaves, 2004; North, Hargreaves, & McKendrick, 1999a, 1999b; Sloboda, 2005). In addition, we find background music in public places including houses of religious worship, exercise gyms, cultural centres, buses, trains, and waiting rooms. In the workplace, music seems to strengthen on-task performance, increase inspiration, encourage positive mood, intensify concentration, manage personal space, neutralize distraction, offer relief from stress, and perhaps serve as a vehicle to escape from the work and the office environment (Haake, 2011; Oldham et al., 1995). In this connection, survey studies have confirmed that roughly 40% of workers listen to music through headphones, and that on average office workers listen to music for roughly three hours per day (which represents 36% of the work week).

Historically, music has always served a number of basic functions in society (Hargreaves & North, 1999; Hodges & Sebald, 2011; Radocy & Boyle, 1979). In 1964 the music anthropologist Alan Merriam recognized ten major music functions for all of human society. These are:

Aesthetic Enjoyment	Continuity and Stability of Culture
Communication	Enforcing Conformity to Social Norms
Emotional Expression	Integration of Society
Entertainment	Symbolic Representation
Physical Response	Validation of Social Institutions and Religious Rituals

A brief look at research on the functions of music indicates that human engagement with music has been fairly stable for over 600 years, and even in the second decade of the twenty-first century we still use music for the same purposes that our ancient ancestors did. For example, Gregory (1997) outlined 14 traditional function roles of music since Biblical times, including: *ceremonies and festivals* (civil and religious); *communication* (messaging); *court* (royal, national, political affiliation); *enjoyment* (emotional, aesthetic); *games and dancing* (social milieu); *healing and trance* (music therapy, music medicine); *identity* (ethnic and/or group membership); *lullabies* (parental bonding, child development); *romance* (intimacy); *salesmanship* (advertisement); *story telling* (holy scriptures, ballads); *symbols* (representation); *warfare* (battle, interrogation, confrontation, security); and *work* (work songs, background music).

It should be noted that the uses of background music involve the listener on two distinct levels (Radocy & Boyle, 1979). *Receptive behaviours* are covert in the sense that they are internal processes reflecting perception and cognition. For example, while listening to music we may initially engage in a mental search to establish if what is heard is familiar or unfamiliar music. Then, we might become involved with intricate memory processing targeting the recognition of a melody (either by name of the tune or the name of artist/band performing the piece). Subsequently, we may engage in auditory discrimination tasks that involve judging between the various instruments playing the piece or even perhaps in a musicological analysis discerning the various structural sections (such as verse/chorus). Alternatively, *production behaviours* are overt responses to background music that appear at lower levels of consciousness. Most often production behaviours are time-linked behaviours involving increased action (such as singing, tapping, clapping, walking, moving, eating, athletics, home chores) or decreased action (becoming quiet, relaxing, sitting, resting, sleeping). The truth is that many production behaviours are essentially *re-production* actions based on patterns within music that mirror the beat, rhythm, tempo, or melody line. Radocy and Boyle state that the character of the response is co-dependent on a tri-part interaction involving the *type of listener* involved, the *context* in which the listening takes place, and the desired *production behaviour* or the nature of the task that the listeners are expected to perform. Accordingly, listener-responses may be *direct* or *indirect*. Namely, a *direct reaction* may be viewed as a conditioned response when there is a direct link between the reproduced behaviour (for example, a mood or affect) and the music (via clear associations or memories). Alternatively, an *indirect reaction* includes the influence of information processing (i.e., cognitive activity) on subsequent evaluations that provoke mood congruent thoughts effecting later behavioural performances (Alpert, Alpert, & Maltz, 2005). Certainly, the purpose of this book is to explicitly take issue with both receptive and production behaviours among drivers while they listen to music and drive a car in traffic.

It is clear that many different kinds of music generate or support various types of behaviours. When the music background is most appropriate or *congruent*, the listeners might be able to heighten behaviour related to the specific task (such as reading, studying, working, eating, driving, loving, praying, or shopping). That is, when the music *fits* the context the resultant behaviour is increased and subsequently improved performance can be seen. However, should the music selections in the background be inappropriate or *incongruent* to the context, then the music can prompt *ill-effects* and decrease the listener's capacity – degrading the required response and performance. In this light, Kellaris and Kent (1992) suggest:

> [In some environments] certain common practices involving music may actually be counter productive. For example, playing peppy music in major keys to customers in check-out lines may actually augment their perceptions of waiting times... Conversely, playing amodel new age music – an increasingly common practice in yuppie restaurants – may diminish diners' perceptions of time passage (relative to what they would experience if exposed to more conventional music), and thus exacerbate feelings of being rushed (p. 374).

How background music is seen as fitting the context, for purposes other than aesthetic enjoyment of the music itself, has much to do with overriding conceptual underpinnings about specific characteristic features in the music (for more details, see: Yeoh & North, 2010a, 2010b, 2012). For example, the emotional valence of background music is often seen as cueing dispositions of affect (such as, 'positive' *versus* 'negative', or 'happy' *versus* 'sad'), which may or may not be appropriate. In a classic music study, Hevner (1936, 1937) differentiated associations and behaviours between two music subtypes based on music elements; the first type was comprised of fast tempo, loud dynamics, lively and varied rhythm, major tonality, and a high register (perceived as happy, graceful and playful music), while the second type was comprised of slow tempo, quiet dynamics, unvaried rhythm, minor tonality, and a low register (perceived as sad, dreamy and sentimental music). Such theoretical conceptions eventually spearheaded more than 50 years of research exploring musical effects based on a continuum between energetic 'stimulating' music and gentle 'sedative' music. *Stimulative music* increases psychophysiological responses while *Sedative music* decreases them (for a more detailed review of the research literature, see: Hodges, 2010; Hodges & Haack, 1996; Hodges & Sebald, 2011). Findings highlighting these two diametrically opposed music types demonstrate variances in drawings, body posture, and physiological responses. More specifically, *Sedative mus*ic has been seen to decrease blood pressure (Chafin et al., 2004), reduce frustration with calm tenderness (Caspy et al.,1988), increase the relaxation response (Scheufele, 2000), and escalate verbalization while dwindling decisiveness in small work group settings (Stratton, & Zalanowski, 1984).

2–2.2.1. Stimulative music

Stimulative music emphasizes rhythm rather than melody or harmony. This music has an overriding beat that is ever present, diverse, and always in the forefront. The music type arouses the listener as it is composed of strong energizing components enhancing bodily energy and emotional effects. Stimulative music is characterized by a loud fast-paced temperament comprised of highly accentuated staccato detached rhythmic percussive-like sounds. Many pieces that generate stimulative processes present music that is dissonant with a wide pitch range, disjunctive melodic lines, and abrupt unpredictable changes (Gaston, 1968; Haack, 1980; Hodges & Sebald, 2011). A traditional example of *Stimulative music* is the military march; such music has served mankind throughout history to spur-on and motivate troops into combat.

2–2.2.2. Sedative music

Sedative music emphasises melody and harmony rather than rhythm. This music has an underlying beat that is weak, subdued in the background, and unvaried. The music type soothes the listener as it is composed of soft serene relaxing components, evoking physical sedation, and intellectual contemplative behaviours. Sedative music is characterized by a quiet slow-paced music temperament, comprised of legato-sustained flowing conjunct narrow-range melodic lines. Many pieces that generate sedative processes present music that incorporates very slight gradual predictable changes (Gaston, 1968; Haack, 1980; Hodges & Sebald, 2011). A traditional example of *Sedative music* is the lullaby; such music has served womankind since the dawn of evolution in an effort to pacify infants and cuddle them to sleep.

2–2.3. Industrial music

In the 1930s, *Muzak Corporation* (USA) developed a music product that was broadcasted and received in offices, lobbies, medical clinics (such as surgical theatres and waiting rooms), production-line industries, public arenas (such as bus terminals, subways, theme parks), and commercial settings (such as elevators, restaurants, retail shops, supermarkets, airlines). The purpose of *Muzak*'s programming was to provide a psychological lift by moderating, and hence alleviating, tension, boredom, melancholy, and fatigue. According to Lanza (1994) *Muzak* positioned itself at the forefront in producing music that directly addressed modern day life. Specific music features (such as tempo and timbre), as well as music complexity (i.e., the overall architecture, rhythmic activity, melodic and harmonic structure, voicings, and number of instruments), were all employed in such a way to reflect specific time zones of daily life. In this way, *Muzak* sought to adjust musical characteristics to the human fatigue cycle. *Muzak* programmed its 15-minute block broadcasts, and subsequent cassette tapes that were distributed among tens-of-thousands of subscribers worldwide, so that the progression of music selections adhered to an ascending curve whereby the instrumental pieces gradually increased in volume, tempo, and complexity.

Many radio stations have incorporated *Muzak*'s success in their programming. For example, Budiansky (2002) describes the National Public Radio Station (WETA-FM in Washington, DC) as adhering to a format that promoted Classical music predominately for background listening, and created playlists that targeted 'day parting' (i.e., providing particular sounds for listeners who tune in at different times of the day). Accordingly, the music programme director searched for pieces with bright and smooth orchestral textures that just by the nature of the instrumentation required less demands of listeners' attention in order to allow them to concentrate on other things they may be doing at the same time while music was heard in the background. In the morning when people were getting ready for work, they would want a good start to their day with uplifting music; for the rush hour energetic lovely selections would be preferred; by 9 a.m. the selections were geared for workday hours with medium tempos that were not too aggressive or too relaxed; in the afternoon a mix of energetic drive time (for the commute home) with some relaxing tranquil sounds for those who just need to sit back after the day's end for a rush-hour relief were presented. Obviously, the issue is not simply one of tempo, but also the desired mood of the music that was at the forefront of music selection.

Based on available research data reported by *Muzak* (Hodges & Haack, 1996), their programming caused the following effects: (1) supermarket shoppers walked slower and bought more; (2) restaurant diners ate faster causing an increase in customer turnover; (3) telephone company employees were involved in less non-essential conversations and activities; (4) corporate employees came to work on time and had fewer sick days (i.e., decrease in absenteeism); (5) accounts payable sections of business offices were more focussed with decreased errors; (6) publishing company employees increased productivity; (7) electric utility company employees increased key punch activity with decreased errors; (8) travel agency employees decreased airline agent turnover; (9) main city centre branch bank increased earnings; (10) publishing house book editing division increased accuracy; and (11) work-related problem solving abilities increased.

Beyond *Muzak* a host of independent researchers have also targeted the use of background music in many industrial and commercial contexts (for a review, see: Garlin & Owen, 2006; Kampfe, Sedlmeier, & Renkewitz, 2011; Hargreaves, MacDonald, & Miell, 2005; North & Hargreaves, 1997c, 2010; Turley & Milliman, 2000). Experimental studies have been carried out in bars, banks, computer assembly plants, exercise and relaxation clubs, restaurants, shops, and on-hold telephones. Among other avenues, these have explored customer activity, purchasing and affiliation behaviours, and waiting time. In the main, investigators perceive that music can be somewhat 'controlled' as sounds seem to range from 'loud' to 'soft' and 'fast' to 'slow', involving either 'vocal' versus 'instrumental' or 'simple' versus 'complex' textures, and reflect either Classical traditional or Contemporary urban music styles (Broekemier, Marquardt, & Gentry, 2008). For the most part researchers of occupational and commercial psychology feel that music is relatively easy to manipulate, and has a predictable appeal based on age and lifestyle (as described earlier in sections 2–1.4 and 2–1.5). One especially popular setting for investigation has been among hotelier services. In an innovative exploration, Magnini and Parker (2009) outline the six basic functions of music within this environment:

1. Background music on hotel web sites influences visitors' arousal, interest, and bookings.
2. Background music on telephones influences a guests' satisfaction during a wait-on-line.
3. Background music throughout the premises influences the overall brand image.
4. Background music throughout the premises influences guests' perception of decor (in rooms and lobby).
5. Background music throughout the premises influences employees' productivity and overall level of service.
6. Background music throughout the premises increases guests' spending.

Areni (2003) surveyed over 100 managerial executives of hotels, restaurants, and bars; based on their career-long observations about background music, 13 implicit notions came to the surface:

1. Background music sets the atmosphere.
2. Background music must match the customers' demographic profile.
3. Background music must match circadian rhythms associated with specific times of the day.
4. Background music might also distract customers.
5. Background music can either invite customers into the establishment or drive them away.
6. Classical music in the background is perceived as 'up-market'.
7. Background music makes customers stay a little longer.
8. Repetition of music items is not effective and can be destructive.
9. Silence is deadly.
10. Background music can either encourage or discourage inappropriate behaviour.
11. Background music can manage (and even influence) customers' perception of time.
12. Most customers eat by the beat, that is, eating behaviour is tempo-related.
13. Background music can block out annoying ambient noise.

Certainly, music influences behaviour due to its effect on information processing, emotional and physiological changes, as well as through a process of classical conditioning (Caldwell & Hibbert, 2002) whereby people (consumers and employees) develop a more or less 'trained' response through repeated exposure. In order to explore the effects of music many researchers employ pieces variegated by their psychophysiological character (specifically stimulative *versus* sedative qualia). Nevertheless, other independent variables and listening conditions, based on various levels of distinctions, have also been employed when investigating the effects of music on non-musical behaviour – especially those studies targeting driving behaviour and vehicular performance. Among these are:

Familiar Music	*versus*	Unfamiliar Music
Well-liked Music	*versus*	Un-liked Music
Positive Music	*versus*	Negative Music
Pleasant Music	*versus*	Aversive Music
Happy Music	*versus*	Sad Music
High-Complexity Music	*versus*	Low-Complexity Music
High-Arousal Music	*versus*	Low-Arousal Music
Preferred Music Music	versus	Non-Preferred Music
Listener-Chosen Music	*versus*	Experimenter-Chosen Music
Listener-Mediated Music	*versus*	Forced Listening Music

In general, studies investigating background music have targeted and measured five principal dependent variables. These are: *affect* (arousal, emotion, mood, nostalgia, pleasure); *financial return* (gross margin, quantity purchased, rate of spending, sales); *attitudes and perception* (brand loyalty, experience satisfaction, liking, intention to recommend, perception of service quality, perception of visual stimuli, price sensitivity, product evaluation); *temporal effects* (consumption time, decision time, duration of music listening, estimated versus actual duration, service time, unplanned time); and *behaviour* (affiliation, frequency of patronage, impulse behaviour, in-store traffic flow, items examined, items handled, number of customers entering premises, number of customers leaving before served, customer-recommended service, speed of behaviour, store choice). In two meta-analytic literature reviews, both Garlin and Owen (2006) and Kampfe, Sedlmeier, and Renkewitz (2011) point out that although many studies have indeed yielded statistically significant effects, for the most part these are moderate in *effect size.* Moreover, Kampfe et al. negated Behne's (1997) previously proposed concept that the availability of mass music media everywhere in our environment has caused people to become so used to hearing music in the background that it's impact on everyday behaviour, if any, has progressively decreased over time – dwindling down to a present day *null effect*. Unequivocally, the meta-analyses points to at least six clear truisms that can be generalized for all studies:

1. Presence of background music, especially pieces that are familiar and well-liked, have a positive effect on patronage and pleasure.
2. Familiar music of a slow-pace and lower volume results in consumers staying longer at a venue.
3. Well-liked slow-paced quieter music causes consumers to perceive shorter time durations.
4. Music tempo has the greatest effect on arousal than any other musical feature.
5. Presence of background music has a positive impact on sports by increasing performance.
6. The impact of background music on reading comprehension and memory tasks (including the contents of advertisements) is negligible at best and more often negative in direction.

The effects of background music on listeners in everyday contexts are detailed below. The listing consists of the major studies, arranged by date of publication, carried out in real-world natural environments. These findings have been vetted to exclude any investigation that employed foreground music (i.e., involving focussed music listening), clinical investigations (music therapy, music medicine), investigations employing one dedicated physiological marker as the exclusive dependent measure, and investigations with null effects. Also excluded from the listing are studies that presented videos as a sole observational procedure, simulation studies employing life-like mock-up environments, studies that rely entirely on self-report narrative descriptions of perceived intentions (rather than genuine behaviours), and studies that recruited children as participants. Finally, the listing also excludes all explorations involving automobile themes, driving simulations, and on-road driving as these are presented in the following chapters.

2–2.3.1. Time in context

(1) Customers spend more time in a store with soft background music (Smith & Curnow, 1966); (2) Supermarket customers spend more time in food aisles (i.e., in-store traffic flow) when slow paced music is heard (Milliman, 1982); (3) Once food is served, restaurant diners take more time to complete their meal when accompanied

by slow-tempo music (Milliman, 1986); (4) Shoppers spend more time in the supermarket when they like instrumental music heard in the background regardless of the tempo or volume (Herrington & Capella, 1996); (5) Restaurant diners spend more time seated at a table eating when accompanied by slower Jazz standards than faster Jazz standards (Caldwell & Hibbert, 2002); (6) Restaurant diners spend more time seated at a table eating when the background music is perceived as close to their musical preference and musical taste (Caldwell & Hibbert, 2002); (7) Classical music changes the atmosphere of a popular restaurant causing some patrons to leave earlier in the evening resulting in fewer customers remaining at 11:00 p.m. than on nights when Easy Listening, Pop, and Jazz musics are heard (Wilson, 2003); (8) Drinking songs such as 'The Drunken Sailor' encourage patrons to spend more time in a bar compared to Top-40 Hits or Disney Tunes (Jacob, 2006); (9) Pedestrians spend more time at open air market stalls when joyful Classical music is heard (Gueguen et al., 2007); (10) The combination of loud Pop/Dance music with a vanilla scent aroma stimulates increased levels of satisfaction among young fashion shoppers resulting in spending more time in a fashion store (Morrison et al., 2011).

2–2.3.2. Time perception

(1) Younger men perceive that they spend more time shopping in a menswear department when exposed to background instrumental music rather than foreground Top-40 music, while older men claim exactly the opposite (Yalch & Spangenberg, 1990); (2) University students over-estimate the time duration of Paula Abdul-style Dance music pieces in a major tonality compared to the same music in a minor tonality (Kellaris & Kent, 1992); (3) Gymnasium members are least accurate in estimating the time spent in workout regimes when slow tempo Pop music is heard in the background (North, Hargreaves, & Heath, 1998); (4) University students waiting in line for class registration estimate longer queue times when fast-paced versions of instrumental Jazz-Funk covers are heard rather than when slower-paced versions of the same pieces are heard (Oakes, 2003); (5) Students waiting in line estimate shorter durations of time have passed when either familiar contemporary Dance music or Billboard-Chart Hits are heard rather than unfamiliar traditional Country and Western music or uncharted 1950s Oldies (Bailey & Areni, 2006); (6) Casino gamblers estimate longer times spent captivated (i.e., temporarily overcome) by slot machine gambling when slow-paced loud music slightly masks the natural ambient club noise (Noseworthy & Finlay, 2009); (7) University students perceive they have spent less time returning home at the end of the day (either walking or taking the bus) while listening to an MP3 player or *iPod* (Heye & Lamont, 2010).

2–2.3.3. Waiting time

(1) The average waiting time to be seated in a restaurant takes longer when slow-tempo music is heard in the background (Milliman, 1986); (2) Callers of a state-wide protective services Abuse Hotline remain on-hold longer when accompanied by modern Fusion-Jazz or Country music than when listening to relaxing meditational music such as Pachelbel's Cannon with added ocean surf sounds,

which caused the greatest level of lost-calls (Ramos, 1993); (3) Callers from the general public remain on-hold longer when instrumental cover versions of Beatles songs are heard (North, Hargreaves, & McKendrick, 1999a); (4) University students wait longer for a scheduled activity when New-Age ambient music is heard than when no-music is heard (North & Hargreaves, 1999a); (5) Everyday callers perceive shorter on-hold waiting times when music is heard regardless of music genre (i.e., Country, Classic Rock, or Jazz) than when no music is heard (Gueguen & Jacob, 2002; Kortum, Bias, Knott, & Bushey, 2008).

2–2.3.4. Waiting experience

(1) University students waiting for a scheduled activity undergo a more positive experience with increased positive mood when Classical music is heard (Cameron et al., 2003); (2) Everyday customers are more satisfied with telephone-related service provision when they like the music selection heard while placed in an on-hold wait queue (Kortum, Bias, Knott, & Bushey, 2008).

2–2.3.5. Sales

(1) The daily gross dollar sales of purchases in a supermarket increases with slow-tempo music (Milliman, 1982); (2) Restaurant customers purchase more drinks from the bar with slow-tempo music (Milliman, 1986); (3) The overall estimated gross margin per dining customer is higher with slow-paced music (Milliman, 1986); (4) Restaurant wine-cellar sales of more expensive wines increases with Classical music than Top-40 styles (Areni & Kim, 1993); (5) Department store wine bottle sales from Germany (but not France) increases when German folk music is heard, while wine bottle sales from France (but not Germany) increases when French folk music is heard (North, Hargreaves, & McKendrick, 1999b); (6) University cafeteria sales are higher on days when Classical or Pop-Chart music is heard than when Easy Listening or no music is heard (North & Hargreaves, 1998); (7) Gift Shop impulse buying increases when ambient scents are successfully matched with Classical music (i.e., lavender fragrance + slow tempo relaxing pieces; grapefruit fragrance + fast tempo energizing pieces) than when fragrances and music exemplars are mismatched (Mattila & Wirtz, 2001); (8) Restaurant diners spend more money per meal (including food and drinks) when the background music is perceived as close to their musical preference and musical taste (Caldwell & Hibbert, 2002); (9) Drinking songs such as 'The Drunken Sailor' increase the amount of money spent in a bar than Top-40 Hits or Disney Tunes (Jacob, 2006); (10) Restaurant diners spend more per meal (including food and drinks) when accompanied by slower Jazz standards than faster Jazz standards (Caldwell & Hibbert, 2002); (11) Restaurant diners spend more money per meal when Classical music is heard than when either Pop music or no music is heard (North, Shilcock, & Hargreaves, 2003); (12) The combination of loud Pop/Dance music with a vanilla scent aroma stimulates increased levels of satisfaction among young fashion shoppers resulting in spending more money in a fashion store (Morrison et al., 2011).

2–2.3.6. Atmosphere of context

(1) University student cafeteria diners are more aware of the music, and subsequently consider it as the most essential part of the restaurant, when 'disliked' instrumental New-Age music is heard than when 'liked' music is heard (North & Hargreaves, 1996a); (2) University student cafeteria diners 'feel-good', and subsequently evaluate the restaurant atmosphere more positively and verbally indicate their desire to return in the future when 'liked' instrumental 'New-Age' music is heard (North & Hargreaves, 1996b); (3) University student diners perceive the atmosphere of the same cafeteria as a 'fun and upbeat' atmosphere with Pop music, as 'sophisticated and upmarket' with Classical music, and as 'downmarket and cheap' with Easy Listening music (North & Hargreaves, 1998); (4) Patrons in the main lobby hall of a city centre bank perceive the hall as a 'dynamic, upbeat, and inspirational' atmosphere with Classical music, as 'dis-encouraging' with Easy Listening instrumental arrangements of well-known pieces, and as 'weary, subdued, and dis-encouraging' without background music (North & Hargreaves, 2000); (5) Business professional patrons of a city centre bar perceive the bar atmosphere as 'peaceful' with quiet background music, but as 'invigorating' with loud foreground music, regardless if the exemplars heard were instrumental Classical music or original Pop-Chart songs with vocals (North & Hargreaves, 2000); (6) Buyers perceive the atmosphere of a gift shop as 'pleasurable' when ambient scents and background Classical music are matched (i.e., lavender fragrance + slow relaxing music, or grapefruit fragrance + fast energizing music) than when mismatched (Mattila & Wirtz, 2001); (7) Customers in a mall outlet of a national chain specializing in trendy mid-priced clothes for young men and women perceive their experience of service and sales personnel more positive when pleasurable music is heard (Dube & Morin, 2001); (8) Restaurant diners indicate they enjoy the venue, have intentions to return again, and will recommend the restaurant to others when the background music is perceived as close to their musical preference and musical taste (Caldwell & Hibbert, 2002); (9) Young fashion shoppers experience increased pleasure, arousal, and satisfaction when they smell a vanilla scent aroma with loud Pop/Dance music (Morrison et al., 2011).

2–2.3.7. Musical fit

(1) Classical music is a more fitting genre to tempt consumers to purchase lavish costly wines than Top-40 music (Areni & Kim, 1993); (2) Instrumental cover versions of Beatles songs more appropriately encourage callers to remain on the phone longer when placed on-hold than the original Beatles songs themselves (North, Hargreaves, & McKendrick, 1999a); (3) Traditional German folk music is a more fitting prime for sales of German wines than traditional French folk music classical music, and visa versa (North, Hargreaves, & McKendrick, 1999b); (4) High-arousal (fast-paced increased volume) music more appropriately fits a

biking workout than low-arousal (slow-paced decreased volume) music while the opposite is true for relaxation exercises (North & Hargreaves, 2000); (5) Music that fits the emotion of a radio advertisement (for cosmetic surgery) increases marketing students' recall of content (clinic brand name) and benefit (detailed purpose of treatment) (Oakes & North, 2006); (6) Chinese university students more often choose Malaysian food (*popiahs*) versus Indian food (*samosas*) after Malaysian folk music is heard, but choose Indian food versus Malaysian food after Indian raga music is heard (Yeoh & North, 2010a).

2–2.3.8. Gastronomic activity

(1) Cafeteria patrons eat faster (i.e., number of bites per minute) when fast music is heard (Roballey et al., 1985); (2) A can of soda is consumed quicker when fast music is heard (McElrea & Standing, 1992).

2–2.3.9. Cognitive/motor tasks

(1) Fast-paced symphonic music that modulates from slow to fast tempo, decreases and even delays perception of exercise-induced fatigue, yielding better workload and efficiency among stationary cyclists in progressive prolonged exercise regimes (Szabo, Small, & Leigh, 1999); (2) Slow ballads (both instrumental and sing-along) disrupt word processor typing fluency by decreasing the words typed per minute/hour (Ransdell & Gilroy, 2001); (3) Fast-paced Classical music increases reading time and efficiency (i.e., content recall) of news items when read on a pocket PC (or PDA) in a noisy cafeteria compared to slow-paced Classical music (Kallinen, 2002, 2004); (4) Computer software information system programers decrease time-on-task and increase quality of work when they self-choose the music they listen to from a library of multiple genres (Lesiuk, 2005); (5) University marketing students recall more content (product name and commercial offer) of radio advertisements for a home improvement store when slow-tempo instrumental music is heard (Oakes & North, 2006); (6) High-arousing Pop songs reduce ability to store information (i.e., memory, attention, and concentration) for later recall (Cassidy & MacDonald, 2007); (7) High-tempo high-intensity (i.e., fast, loud) instrumental Classical music disrupts reading comprehension as demonstrated by less correct answers on a multiple choice questions (Thompson, Schellenberg, & Letnic, 2012); (8) Listening to advertisements with music background causes reduced memory for product description and information (Ziv, Hoftman, & Geyer, 2012).

2–2.3.10. Service gratitude

(1) Restaurant patrons leave increased tips (even where a 12% service tax is automatically added to the bill) when songs with pro-social lyrics (empathy) are heard compared to neutral instrumental pieces (Jacob, Gueguen, & Boulbry, 2010).

2–2.3.11. Product image

(1) Listening to advertisements with an optimistic valence music accompaniment leads to a more positive product image (Ziv, Hoftman, & Geyer, 2012).

2–2.3.12. Community service/helping behaviour

(1) University students more willingly volunteer to hand out leaflets for a charity event when up-tempo 'uplifting' British Top-20 singles are heard than when aversive 'annoying' Avant-garde Computer Music is heard (North, Tarrant, & Hargreaves, 2004).

2–2.4. The Reciprocal Feedback Model of Responses to Music

To conclude the chapter, a model developed by David Hargreaves is brought forward. The model is known as *The Reciprocal Feedback Model of Responses to Music*. This model is by far the best conceptual scheme with which we can describe the different determinants and aspects of human responses to background music. It accounts for the three-way interaction between the *music* being heard (the actual pieces), the background *environment* in which the listening takes place (the real-world situation), and the individual characteristics of the *listener* (distinctions based on sex, education, and temperament). Namely, by accounting for the related musical, contextual, and personal factors, Hargreaves feels that the shared give-and-take of casual influences of all principal components can surface, and consequently lead to fresh insights (see Figure 2.1). Hargreaves claims that "music is heard and used in 'non-musical' contexts, and listeners' responses to style, genre, and quality are inevitably affected" (Hargreaves, MacDonald, & Miell, 2005, p.10). As can be seen in the Figure 2.1, music pieces of various styles, which present an overall architecture perceived on a continuum of complexity and familiarity (that are recognized as comprised of typical features) may or may not be an appropriate fit for a particular context in which music is experienced. Today, listeners experience music not only in the festive artistic atmosphere of a concert hall or in the entertaining extravagant comfort of their living room but rather in everyday situations that most often involve other activities – such as driving a vehicle. Then, the listener who is a distinctive being (based on individual components, such as age, biological sex, culture and nationality, education and musical training, gender stereotyped behaviour, musical preference and taste, perceptual-cognitive styles, and personality temperament) relates to the music heard in a given situation *as if* it is a definitive music exemplar reflecting their overall accumulative music listening experience going back to earliest childhood. These individualities and *envelopment* exert powerful weights on a listener's response to music (that can link to cognitive, psychophysiological, and/or affective behaviours). Finally, behavioural effects are also predisposed by two other factors: the presence or absence of others and the engagement in some other ongoing activity. Clearly, "any one of the three main determinants of musical response can simultaneously influence the other two, and these influences can work either way" (Hargreaves, 2012, p. 543). Hence, the feedback loop can affect reactions that may be seen as either positive or negative. "The reciprocal feed back relationship between situations and contexts and the listener refers to the interaction between the effects of music on a listener in a specific situation, and the ways in which individuals in contemporary society use music as a resource…" (Hargreaves, North, & Tarrant, 2006, p. 135–6). It should be pointed out that *The Reciprocal Feedback Model of Musical Responses* has also been adapted for musical communication and music performance (Hargreaves et al., 2005), aesthetics and liking for music (Hargreaves & North, 2010), as well as for musical imagination and composition/improvisation of music (Hargreaves, 2012).

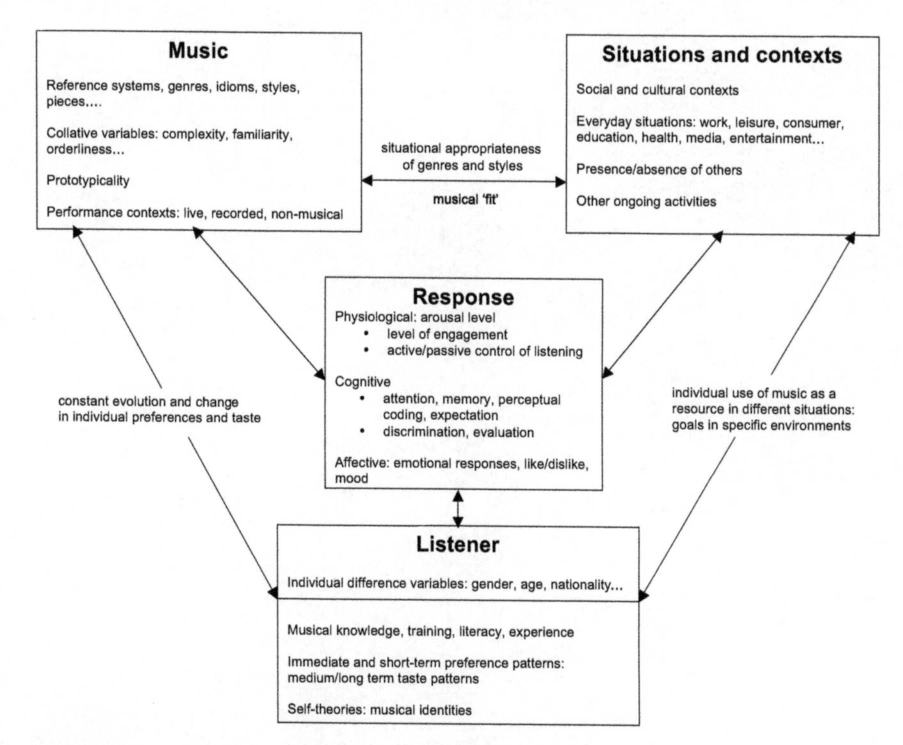

Figure 2.1 Reciprocal feedback model of musical response
Source: Hargreaves (2012). Used with permission of SAGE.

Chapter 3
In-Car Audio Culture: The Benefits of Driving with Music

Chapter 2 reveals that the use of music in everyday life has taken the centre stage of music science research. Many studies described in the previous chapter exemplify how real people employ music in particular spaces, and these indicate that the effects of music are not easily disassociated from the specific contexts in which they are used. One model, developed by Hargreaves (2012, see Figure 2.1) describes the reciprocal interactive process whereby musical responses are both triggered from the music itself, the environmental context in which the music is heard, as well as accounts for covariates (such as personality traits, musical preference and taste, musical fit, presence of others, and other ongoing activity). Certainly, we not only do things *to* music, but most of the time, we do things *with* music. DeNora's (2000, 2003) ethnographic studies confirmed that people ride, eat, fall asleep, dance, romance, daydream, exercise, celebrate, protest, purchase, worship, meditate, and procreate – with music playing in the background. Nonetheless, it is somewhat absurd that the popular location where individuals seem to be when they listen to music is not in the comfort of their living room, nor is it shared with social agents such as intimate partners, extended family, or friends. Rather, the circumstance most frequently reported while listening to music involves unaccompanied vehicular driving. This phenomenon was initially reported by Sloboda (1999, 2005; Sloboda, O'Neill, & Ivaldi, 2000, 2001), and then confirmed by Rentfrow and Gosling (2003). These studies have found that people do indeed listen to music when alone, when hanging out with friends, when getting ready to go out, and when exercising. Yet, in comparison to other lifetime daily activities such as hobbies, watching TV/movies, or reading books/magazines, the studies found that we listen to music more often when we are seated inside an automobile driving alone. It should not, then, be surprising that drivers outfit their car as an audio-environment. Chapter 1 detailed the developments of the car-audio: a central entertainment unit has all but replaced the once-upon-a-time standard AM/FM car-radio receiver that had become the legendary radio/tape-cassette player. Newcomb (2008) summarises this development elegantly:

> In the not-too-distant past, you had just a few choices in music formats: AM, FM, CD, or cassette. But MP3s, the iPod, and satellite radio have changed the way people listen to music in the car. And now it's not uncommon for in-dash radios to sport USB drives or even SD card slots that allow you to play dozens of digital music files on these handy devices. Plus, it's possible to burn as many MP3 and WMA tracks on a single CD as a clunky old CD changer once held on multiple

> discs. Now, devices such as phones and portable media players (PMPs) that use Bluetooth Advanced Audio Distribution Profile (A2DP) technology to wirelessly wing music to a compatible car radio are starting to become available (p. 9).

Today's drivers, especially the young, customize their automobiles by installing a host of audio-components including compact disk players, changers, amplifiers, equalizers, and speakers of various frequency ranges. More specifically, the sound systems of cars in the 2010s feature capabilities such as digital HD sound reproduction with volumes up to 110dBls (Shawcross, 2008), multi-channel information that can route and re-route the appropriate signals to assorted speaker driver configurations, enhanced graphic animated displays, standard volume and timbre controls, channel and band tuning, selection between on-board sources and external devices or memory cards, a multitude of audio formats and compression types, DVD playback, internal hard drives, Bluetooth capability, GPS navigation, and enhanced security stealth modes with detachable faceplate.

In this chapter thoughts by drivers towards their use of music while driving on the road will be presented. Suffice it to say that for the most part, drivers feel that listening to music enhances their driving skills as well as the vehicular performance.

3–1. Driving with Music

Everyday drivers anticipate taking their music along for the ride, and they have been doing so since the 1920s when mass ownership of the automobile paralleled the growth of domestic technologies such as the radio, gramophone, and telephone. In the 1930s Paul Galvin introduced the car radio to Americans just when the *home* was being transformed into a space of aural pleasure and recreation. The *car* then became the emblem of individualized freedom of movement, and already by 1935 manufactures began associating the radio with personalized mediated listening in automobiles (Bull, 2004).

3–1.1. The Autosphere

Cars have indeed become powerful listening environments, as well as highly developed mobile sound machines. The automobile is one of the few places where you can listen to whatever you like, as loud as you like, without having to be concerned about disturbing others. Bull (2001) concluded that the cabin of the car allows drivers to acclimate in a less constrained manner, and indulge in their 'aural whims' with no inhibitions. That is, drivers can listen to non-conformist unpopular (and even what may be considered avant-garde) individualized music selections. In addition, they can sing at the top of their vocal capacity, which they cannot do in their home as houses have other occupants (and neighbours) to prevent such behaviour. Since the 1950s mediated sound has become the definitive experience of car habitation. Music listening is undeniably a most overwhelming component of driving a car.

Music provides the driver an experience of presence within a *sonic envelope*. The car is the ultimate auditory bubble, which functions as a personalized listening environment, seemingly more or less resistant to the outside. Oblad (2000) was the first to report that when a driver liked the music, the sound level could never be high enough; the drivers felt 'near' or 'inside' the music, and perceived the experience as 'impenetrable'. We might actually reflect on the cabin as an *autosphere* in which the mediated role of musical sound provides drivers with a feeling of warm-heartedness similar to what they feel about their home – in contrast to feeling a sense of stone-cold un-involvedness that is often perceived in urban spaces. Thus, drivers seem to use music in an effort to make vast public spaces conform to a more personalized perception of a cosy private space (Bull, 2004; Featherstone, 2004).

After entering the car, and switching on the engine, the first thing most drivers do is to turn on the sound system. Williams (2008) claims that almost instinctively we reach for the stereo knob and crank up the volume – whether it is an audio book, talk radio, music CDs, or *iPod*/MP3 player. Drivers have described the cabin environment as being energized as soon as the music system is turned on. In many cars, simply switching on the engine automatically powers the radio/media centre. Bull (2001) offers an explanation for this behaviour, claiming that most drivers feel discomfort if spending any time in a car without music – a position in which the driver is left alone with only the sound of the engine. His explanation, based on narrative reports by drivers, is that nowadays rarely are people in a position of being alone without the accompaniment of their favourite music tracks, as these are stored in personal music stereo devices or mobile phones. Bull's estimation has recently been confirmed by an *Arbitron/Edison* research survey (Rose & Webster, 2012), which found that 61% of Americans own a portable digital media device (i.e., smartphone, tablet, or MP3 player), and that 84% of these reported that such devices are most always nearby – within an arms length. Hence, 'unaccompanied solitude' is perhaps an experience that is quite difficult to fathom after Year 2000, and subsequently a situation that may be emotionally challenging. Bull (2004) contends that the use of music while on the move (whether walking in the streets, on public transport, or in the automobile) represents wide-ranging social transformations of everyday life that have occurred over the last 50 years. Schwartz and Fouts (2003) point to demographic statistics for Year 2000 that indicate that typical American teenagers spend over 10,000 hours listening to music between 12 to 17 years of age, and this accumulated duration is quite equal to the total hours of classroom teaching mandated between 1st to 12th grade. Hence, music listening, at least in the USA, seems to have become a paramount component of development in areas relating to personal and social life throughout the teen years and even later in early adulthood. These stages of human developmental overlap with attaining a driving licence and acquiring (at least virtual ownership of) a first automobile.

The automobile has become the decisive listening environment – one that is difficult to replicate in any other domestic or public space – even when accounting for use of personal stereos with state-of-the-art headphones. The sounds of *in-car music* create an 'auditized space' that masks the hums and revs of the mechanical engine, as well as the echoic resonance of the outside urban environment. In such a music-accompanied autosphere, drivers can mediate their experience of being in the car alone; they can use the time as one for concentrated listening to music, for personal solitude, and for reflection. A car is potentially the most comprehensive acoustic chamber we can experience on an everyday basis – one that is topped only by recording studios, high-performance sound laboratories, and anechoic chambers. Therefore, sound immersion technologies found in the automobile cabin not only facilitate driver behaviour towards the management and control of the mobile environment per se, but also enhance their efforts to regulate moods and thoughts.

The autosphere provides drivers a temporal space and relaxed atmosphere to reflect and deliberate thoughts, to listen to the news radio when there is not enough time to read a newspaper, or to make contact with others over the phone (Akesson & Nilsson, 2002). In-cabin music accompaniment is often viewed as offering drivers a temporary respite from other everyday demands and realities. It may seem extraordinary, but most drivers claim that they quite intuitively have the ability to cut themselves off from the world around them, albeit at the same time for safety reasons, remain aware of every inch between them and the car in front of them (Bull, 2001; Miller, 2001). Driving with music allows people to find themselves while at the same time to lose themselves, within the music rhythms and melodies they hear while driving a car. Moreover, many drivers feel that they can coordinate the music they intend to drive with to match their own mood as well as to heighten the nature of the journey.

3–1.1.1. Attractions and expectations

The relationship between the music, the driver, and the automobile, was initially explored in 1992 by Oblad; she continued in later years to investigate 'the automobile as a concert hall' (1996, 1997, 2000a, 2000c; Oblad & Forward, 2000). Oblad presumed that more than just an attraction, individuals have specific expectations when they play music in the car. It is not necessarily the music that drivers want to listen to, but rather they simply want to spend time in the car with accompanying music. Oblad extensively surveyed drivers on their motivations for driving with music. She inquired if drivers simply want to be 'left in peace', if the car emulates an extension of their home, and if drivers use music for self-regulation. A comparison of how music is used in the home versus the car was central to exploring what the car may offer as an environment for musical experiences that homes do not. Oblad found that the feeling 'to be left in peace' indeed covered a wide range of interpretations – from unconscious escapism to a conscious effort to change the environment. She found that the young(er) drivers tend to view *to be alone* quite similar, but yet at the same time very different, from *to be by yourself*. The former was viewed as to 'feel' alone or involuntarily be without company, while the latter was interpreted as to voluntarily 'spend time' alone for your own sake. The car interior was perceived as a living sphere similar to a private room wherein one creates sequences of existential value. Oblad noted:

> A young woman expressed that she preferred to take her car to and from work in spite of traffic congestion, rather than taking the subway: '*You sit there calmly, listening to music or smoking a cigarette. It's more relaxing than travelling by public transport...you can think and you are left in peace for a while. As a matter of fact, I think this is quite nice*' (Oblad, 1997, p. 639).

> An elderly man, Kurt, talks about his car as a 'music studio' where he can play the pieces at such high volumes that it '*almost influences my heart rhythm. When I do like that, I then feel that I am inside the music, it is so suggestive, I am carried away... it is a feeling of self-esteem, an upper you could say. It is positive and I follow the music*' (Oblad, 2000a, pp. 128–9).

While 85% of Oblad's Swedish sample reported to have at least one particular room in their home that they could be in alone and do as they pleased, 50% claimed that when they want to be 'left in peace' the car is the optimal place to be. She concluded that this finding is a clear indication that the car truly provides some autonomous value and benefit beyond what homes can offer.

Another central question that Oblad investigated concerns intense musical experiences in everyday life. In this connection she found that over 33% of the respondents identified the car as the environment in which their strongest musical experiences had occurred. This finding may imply that neither the music nor the car are individually strong enough to account for such experiences, but rather it is the interface of driving with music accompaniment that contributes greatly to such intense experiences. Oblad postulated that the existence of an interactive co-dependent relationship between driving and music is developmental in nature and is conceived early in one's driving history during their mid- to late teen years. Among the factors she lists as constituting the basis for such strong music experiences, are: 'feeling near or inside the music', 'perceiving the experience as impenetrable', 'viewing the car as an exceptional acoustic environment', and 'contributing to the music making' (that is tapping along or singing to the music). Regarding the last point, many participants in Oblad's (1997) sample claimed that they enjoy opportunities to drive alone because of their 'desire to sing'. Almost all of the drivers identified themselves as *car singers* who preferred to sing familiar songs from their favourite tapes while alone in the car without anyone listening; it should be noted that they also admitted they think they are poor vocalists.

Finally, Oblad examined the function of music in the automotive setting. Accordingly, drivers play background music in the car, to: 'create an atmosphere', 'to stay awake', 'to have company', 'to feel good', 'to get physical and mental energy', 'to reduce unwanted noise', 'to achieve individual body rhythm', and 'to reach an alternative reality'. The findings also point out that the strongest reason to play music in the car was simply to 'access memories' of particular situations, and these usually surface with attached feelings and moods of the remembered initial occurrence. This later point was emphasized by Sheller (2004) who describes how listening to hours of 'retro' music while driving long distances enables drivers a format to recover events and feelings from the past, and then to rescript them as an *emotional mélange* in the present moment. Even US national

TV broadcasts such as *Fox News* have reported on music's ability to open various associations and emotions related to one's motorcar biography (Manni, 2005). Accordingly, hearing 'Start Me Up' (Rolling Stones), 'Lust For Life' (Iggy Pop), or 'Rock and Roll' (Led Zeppelin) brings back memories of hanging out in the high school parking lot and even driving your first car. In this connection, Lezotte (2013) feels that songs as heard on the car radio have most certainly become integrated into our collective memory of the automobile and those who were in it. It should be pointed out, that the drivers who participated in Oblad's late 1990s studies most frequently listened to home-produced cassette tape collections (i.e., mix-tapes). These were played repeatedly as a loop, especially on long trips. The music they played most often in the car was familiar popular hit songs with vocals, and varieties of rhythmic Rock music. Oblad claimed that many drivers were aware of their own reactions to specific melodies, and they chose the pieces differentially based on situational events: "Drivers know the prerequisites of the car; they usually know what music to choose and how this music will affect them in the car" (Oblad & Forward, 2000, p.132). The drivers themselves described the music they listened to as influencing both rhythms of driving and concentration, as well as their perceptions of relaxation and stimulation. Further, they reported that their favourite tunes seemed to dominate over all other circumstances of driving, and these tracks were played repeatedly over and over again from *loud* volume to even *louder* intensities. The data points out that driver-preferred music seemed to cause the drivers to surpass all other traffic rules and driving behaviour patterns – even the ones they usually follow when other music was heard in the background.

Oblad (2000a, 2000c) developed a model that follows in-car listening as a dual-loop process (see Figure 3.1). Accordingly, listening to music while driving is initiated by both external and internal motivations that mediate the music choices. Musical preferences for in-car listening are somewhat based on availability of CDs in the cabin. At this point, drivers can reject playing music, but in most cases they do choose to play music while driving. After the first sounds are heard, mental negotiation ensues regarding intentions to leave the chosen music on, to choose different tracks, or to even turn off the music. Subsequently, the driver may question how the background music might be furthering adaptive (or maladaptive) psychophysiological changes that further facilitate (or hamper) driving performance.

3–1.2. In-Car Listening

The 'coat-of-arms' of an Autocentric Culture, which is the product of a society passionately preoccupied with automobility (Bull, 2004; Sheller, 2004) may just be *in-car listening*. Although seemingly trivial, the extent to which music has become a fundamental component of the driving experience among 72–100% of drivers is evident, while on the other hand, the car may be simply be a requisite for existentially strong experiences with music.

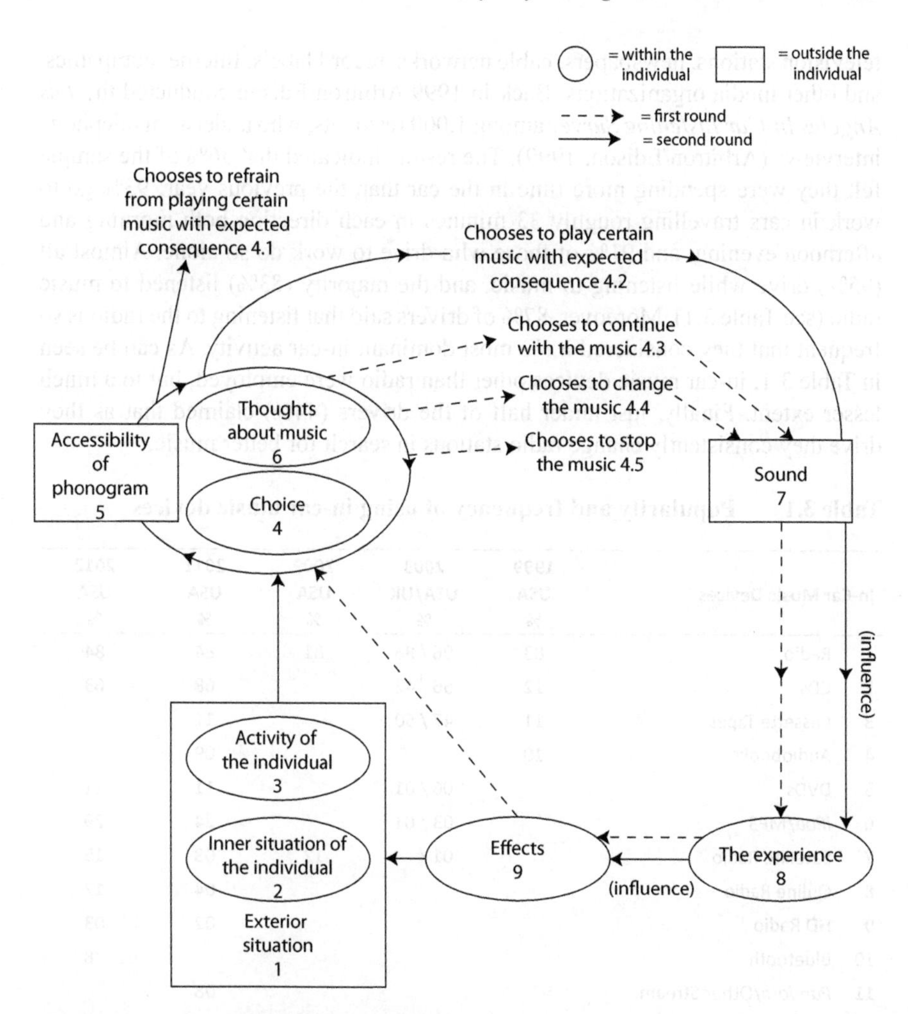

Figure 3.1 In-car listening model
Source: Reprinted with permission of the author.

3–1.1.2. Popularity and frequency

Music has truly become a principal constituent of driving for millions of people. Today the opportunity for self-selected music listening while driving has increased through the presence of radio, cassette tape, compact disc, and MP3 players – all prevalent components of in-vehicle entertainment systems (Dibben & Williamson, 2007). The most comprehensive longitudinal study on the popularity and frequency of usage of car radio has been undertaken by a research team comprised of two American organizations: Arbitron Inc. and Edison Research. *Arbitron* is a media and marketing information service firm that primarily serves radio, television, cable, advertising agencies, advertisers, retailers, and online media; *Edison* conducts survey research and provides strategic information to radio stations, television stations, newspapers, cable networks, record labels, Internet companies, and other media organizations. Back in 1999 Arbitron/Edison conducted the *Los Angeles In-Car Listening Survey* among 1,000 residents, who underwent telephone interviews (Arbitron/Edison, 1999). The results indicated that 36% of the sample felt they were spending more time in the car than the previous year; 93% go to work in cars travelling roughly 33 minutes in each direction both morning and afternoon/evening, and 91% of those who drive to work do so alone. Almost all (95%) drive while listening to music, and the majority (83%) listened to music radio (see Table 3.1). Moreover, 87% of drivers said that listening to the radio is so frequent that they considered it *the* most dominant in-car activity. As can be seen in Table 3.1, in-car music devices other than radio were employed, but to a much lesser extent. Finally, just under half of the drivers (44%) claimed that as they drive they consistently change radio stations in search for better music.

Table 3.1 Popularity and frequency of using in-car music devices

	In-Car Music Devices	1999 USA %	2003 USA/UK %	2009 USA %	2011 USA %	2012 USA %
1	Radio	83	96 / 86	81	84	84
2	CDs	12	56 / 42		68	63
3	Cassette Tapes	11	47 / 50		11	
4	Audiobooks	10			09	
5	DVDs		06 / 01		11	11
6	*iPod*/MP3		03 / 01		24	29
7	Satellite Radio		01 /	12	08	15
8	Online Radio				04	12
9	HD Radio				02	03
10	Bluetooth					28
11	*Pandora*/Other Stream				08	
12	Ford Sync (Music App)				01	

Source: Adapted from Arbitron/Edison Studies, 1999–2012.

In 2003 the research group conducted two studies: the *Arbitron National In-Car Study* (Arbitron/Edison, 2003; Bouvard, Rosin, Snyder, & Noel, 2003) and *Shifting Gears: The UK In-Car Study* (Arbitron/Edison, 2003). In the American study there were 1,505 telephone interviews, and in the British study there were 1,105 telephone interviews; both were conducted on random samples. The studies found that Americans spent an average 11 hours in the car on weekdays and four hours per weekend (total 15 hour per week), while the UK average was nine hours per weekdays and four hours per weekend (total 12 hours per week); in the USA 15 hours per week represented roughly 14% of total waking hours. More Americans (93%) than British (70%) consistently drove to work each day, while public transportation use was 10% higher for British residents. One explanation might be that more British drivers (73%) than Americans (70%) feel that congestion and commuter times have greatly increased; in 2003 Britains commuted 35 minutes in each direction over a distance of 14 miles, while Americans were driving to work for 28 minutes over a distance of 16 miles. These findings were similar to American *AAA* sponsored studies (for example, see: Stutts et al., 2005) that documented 81% of Americans drive to and from work on a regular basis, with an average one-way commute distance of 17.8 miles, estimated at an average of 245 miles per week – totalling 12,700 miles per year. The Arbitron/Edison study found that 33% of Americans always drove with music, and this percentage was significantly higher than for British drivers (18%), French drivers (24%), German drivers (15%), and drivers from The Netherlands (13%). Among those who drive while listening to music, the majority (USA: 96%; UK: 86%) use music radio, and roughly half (USA: 57%; UK: 45%) claim that as they drive they consistently change radio stations in search for better music. The findings indicate that song choice (i.e., selecting the radio channels) is more often in the hands of adults: USA (Adult 64%, Child 21%, Co-Joint 5%); UK (Adult 42%, Child 34%, Co-Joint 12%). Clearly, these findings point out that the power for in-cabin audio is not overwhelmingly controlled by the driver, nor are such driving behaviours alike in all cultures. As can be seen in Table 3.1, a host of other in-car music devices were also employed in 2003; among Americans CDs were in use more than cassette tapes, while the opposite was true of drivers in the UK. In this connection, especially young(er) American drivers between the ages of 16 to 24 had the strongest preference for CDs. The 2003 survey highlights that 61% of USA drivers were very interested in enhanced car radio displays – especially those that denoted the title and the performing artist of the song. It should be pointed out that the same study was replicated six years later in 2009 (Williams, 2009) among more than 1,600 American participants. That study found a 6-year increase of 31% in the amount of time spent in the car. Namely, people aged 18+ spent more than three hours per weekday and two hours per weekend day (for a total 19 hours per week), travelling an average distance of 224 miles per week. Finally, 81% of the respondents felt that AM/FM radio was still maintaining its place as a significant component of in-car media despite changes in audio technology (see Table 3.1); 12% of drivers reported to subscribe to newly available in-car satellite Internet radio (such as *Sirius XM*).

In 2011 the research group implemented *The Road Ahead: Media and Entertainment in the Car* study, interviewing more than 1,500 people above the age of 18 via telephone (Arbitron/Edison, 2011). This study found that people were spending an average four hours per weekday and twelve hours per weekend in the car as drivers (or passengers) – for a total 32 hours per week. Further, the 2011 study found that 84% of drivers listened to AM/FM radio; a larger percentage of female drivers were more passionate about listening to the radio than male drivers

(59% *vs*. 41%). Yet, as can be seen in Table 3.1, other devices were becoming much more prominent – especially CDs and peripheral music players. As a matter of fact, although AM/FM radio was still dominant (64%), and CDs were employed by roughly a quarter of drivers (21%), the survey in 2011 indicated that drivers were engaging digital audio much more; *iPod*s and cellphones (8%), *SiriusXM* (5%), HD Radio (1%), and built-in hard drives (1%). Targeting the young(er) drivers in the sample, the researchers found that more than half of drivers aged 18 to 24 years old actively engaged in coupling external *iPods*/MP3s to the central entertainment-media unit as a format to reproduce their pre-programmed music playlists whenever they drove, and that 19% employed *Pandora* streaming via their mobile phone as a music background to accompany driving. Nevertheless, attitudes of drivers towards platforms of in-car devices were shifting quicker than expected. For example, in 2011 the percentage of drivers who were already expressing their passion for post-millennium in-car music technologies and devices, were: satellite radio (54%), *iPod*/MP3 player (45%), audio books (38%), HD radio (37%), *Pandora* streaming via mobile device (34%), DVD player (32%), AM/FM stream via mobile device (30%), built-in hard drive (30%), AM/FM Radio (28%), CD player (23%), non-*Pandora* streaming via mobile device (16%), cassette player (16%). Finally, 45% of drivers claimed that they would be interested in new infotainment features, such as: voice controlled stereo systems (41%), pause/rewind/replay radio (40%), separate front/back tuning and volume (39%), and built-in Internet radio device (36%).

In 2012 Arbitron/Edison conducted *The Infinite Dial 2012: Navigating Digital Platforms* (Rose & Webster, 2012; Webster, 2012). Accordingly, this survey targeted over 2,000 Americans. The study found that while a total 88% of Americans reported to own a cell phone, smartphone ownership was roughly 44% (which was a 300% increase over two years since 2011). Further, there was a 25 to 50% year-over-year jump in on-line radio use. That is, drivers who connected their cell phone to the car stereo system as a method to listen to on-line streaming radio increased from 6% (in 2010) to 11% (in 2011) to 17% (in 2012). The 2012 study points out that AM/FM radio use still maintained dominance over all other in-car music devices, there was an eclipse of the cassette tape (and audiobooks), and a significantly negative correlation between CD usage and other digital music devices – especially after the popularity of Bluetooth linkage (see Table 3.1). Hence, the Arbitron/Edison studies indicate that the popularity of certain in-car audio technologies after the millennium had boosted the usage of digital platforms for everyday drivers, and the frequency of usage points out that there is no other possibility but to see *driving with music* as the most prominent driver-preferred in-vehicular cabin environment. Lastly, most recently Arbitron/Edison (2013) presented figures illustrating that in year 2013, 139 million Americans (53%) owned a smartphone, while 54% of these owners (or 75 million people) use the device to download and listen to music, and 44% (or 61.2 million people) use the smartphone to listen to online radio – in their home, work, and *automobile*. It would seem, then, that notwithstanding the explosion of other hi-tech car-audio options (Johnson, 2012), even today most drivers continue to *tune-in* to the radio while driving, and therefore one might conclude AM/FM radio *rules the road* of in-car entertainment. An interesting piece of data from Arbitron/Edison is that 72% of young(er) drivers aged 16 to 24 years perceived the radio as the third most significant source to learn about new musics; ranked above the radio as the most important means of exposure were 'Friends & Family' and *YouTube*. Having said that, the radio out-ranked '*MTV*', *Facebook*, and *iTunes*.

The picture that surfaces from all the above is that while traditional means of listening to over-the-air local AM/FM radio stations via standard broadcast signal on a regular radio may still entertain many drivers, a new trend of listening through the Internet (via smartphones or other mobile devices) has become increasingly more mainstream. As a matter of fact, recently the radio has further specialized into three varieties: 'personalized online radio' where the driver can create custom-made programmes (for example focussing on an artist), 'on-demand music' where a driver can pay a subscription fee (for example to access a specific library or music collection on demand), and 'streaming live radio' where a driver can listen to a wide variety of live radio stations that are not localized to one geographical region. As a result, Edison Research was commissioned to examine these trends on behalf of the *Streaming Audio Task Force* – a consortium of industrial partners including *Pandora*, *Spotify*, and *TuneIn*. *The New Mainstream* report (Edison, 2013) details data from a July 2013 national online survey of 3,016 Americans. Accordingly, Americans are listening to radio more frequently when they are in a car/truck (88%), than when at home (77%), or at work (38%), while walking about (37%), exercising in a gym (33%), or on public transportation (23%) (see Table 3.2). As can be seen in the table, while in a car almost all (94%) listen to music from a traditional AM/FM car radio, nearly a fifth (19%) also listen to Internet radio – and of these 40% prefer Personalized Online, 35% Streaming Live, and 24% subscribe to On-Demand Music channels.

Over the years there have been other American studies, such as those by Quicken Insurance (2000), Rentfrow and Gosling (2003), and Stutts et al. (2001a, 2001b, 2003, 2005). In a survey commissioned by Quicken Insurance, 91% of the sample reported to drive with their preferred music cassette tapes and CDs playing in the background, and 71% claimed they sing along with music (52% when they are alone, 40% in the presence of passengers). Rentfrow and Gosling analysed the everyday music activities among 3,500 Americans; they found that the activity most frequently listed as involving background music was 'driving a car' – an activity far more frequently involving music than studying, exercising, being with friends, or going to sleep. Finally, Stutts et al. conducted the *Naturalistic Driving Study*, investigating the on-the-road behaviours of 70 American drivers. From the analyses of in-cabin videotapes, the researchers confirmed an overall 72% incidence of listening to background music, with 'listening to radio' representing roughly 86% of all music listening events; only four drivers did not listen to some form of car-audio at all.

Table 3.2 Popularity and frequency of using in-car music radio

	Home %	Work %	Walking %	Exercise Gym %	Public Transport %	Driving Car/Truck %
Listen to Any Radio	77	38	37	33	23	88
1. Traditional Radio	74	63	54	55	61	94
2. Internet Radio	55	60	65	67	61	19
2a. Personalized Online	71	60	63	68	57	48
2b. Streaming Live	41	40	33	32	29	35
2c. On-Demand Music	29	26	21	25	29	24

Source: Adapted from Edison (2013).

There have also been several studies within the British Commonwealth. For example, Dibben and Williamson (2007) conducted a UK national survey that posed questions about in-vehicle music behaviour to a representative sample of 1,780 British adults aged between 18 and 50+ years old; over 1,000 of these drivers had accumulated a four years 'no-claims discount' on their insurance. The results of this study indicate that 65% of the drivers reported to listen to music while driving (music radio 69%; CDs/tapes 61%), while talk radio, conversation and silence were reported among 29% of the drivers. The musical style most often listened to by the drivers was Chart Pop music. When comparing drivers between age groups, drivers 50+ years of age reported more often to drive in silence than young(er) drivers between 18–50 years of age. We might wonder if this is the result of driver adaption to aging (which may involve reducing risks in face of age-related declines in skills), or if this difference is simply an artefact of being a member of an older generation who might appreciate a quiet drive more than members of the younger generation. Further, Dibben and Williamson found that more than half of the drivers reported to sing along with the music while driving; female drivers sang more often than male drivers, and drivers aged 50+ reported to sing significantly less frequently. Finally, roughly 20% of drivers aged 18–50 claimed that the car was the only location where they were able to listen to the music they wanted.

In a previously published UK study, the research team North, Hargreaves, and Hargreaves (2004) sought initial normative data on what people listened to in naturalistic everyday circumstances. The researchers looked at the following issues: When people listen to music, where are they and why are they listening to music? In that study, 346 British residents who owned a mobile phone were sent text messages at random times once per day for 14 days; on receiving the message, each participant completed a questionnaire. The study found that the greatest number of episodes involving music occurred while participants were on their own, that Pop Music was heard most frequently, and the activity mostly reported as an event with accompanied listening was 'driving an automobile' (12%). In a third study by Young and Lenne (2010), 287 Australian drivers from Victoria (ages 18–83 years old) were surveyed; the study found that 94% of drivers listened to either music radio or to music CDs. Finally, a most recent UK survey carried out in 2012 by *72 Point* on behalf of Allianz 'Your Cover' Insurance (Allianz Insurance, 2014; *Claims Journal*, 2012; NBLawGroup, 2012; PRNewswire, 2012; Sinclare, 2012), polled 1,000 motorists between the ages 18–60 (*Md* = 40 years old, 50% female) about their driving habits. The study found that 70% of British drivers listen to music in their cars for roughly seven hours per week – a frequency reflecting an overall 72% of all automobile trips, or over 13 billion (i.e., 13,621,072,920) hours of listening to music while on English roads per year. Other outcomes of the Allianz study were: 61% of drivers feel that listening to music while driving creates the illusion that they will reach their destination quicker; 47% claim that in-car music helps them relax when behind the wheel; 42% admit that they thoroughly enjoy a sing-along when travelling; 25% claim that music helps them concentrate more when driving; and 13% state that playing music enhances their driving by blocking out road noise (see Table 5.5).

We seem to have countless beliefs about the *power of music* in all aspects of human enterprise.[1] Especially when it comes to driving an automobile, most of us assume that listening to music will enhance the overall atmosphere of the driving experience, tame our temper, and improve driving capabilities as well as control over the vehicle. Further, we often feel that we can adequately match the mood of the journey through music selections (Sheller, 2004). For example, if we wish to decrease driving stress, music can sooth and relax us (Dibben & Williamson, 2007). As a result of such attitudes, CD compilations for driving have become highly popular accessories for road trips.

3–1.2. Music Tracks for Driving

In light of a rather high incidence of audiophile drivers, motor-magazines long ago began implementing surveys and 'reader's polls' to promote favourite driving-tunes. "Music makes a good drive even better, a long road trip more fun, a daily commute more bearable. Think of your best times behind the wheel, and chances are there's a soundtrack that goes along with it" (Newcomb, 2008, p. 7). In a recent 2012 *AutoTrader* Survey (AutoTraderBlog, 2012), 1,200 readers between 16 and 50 years of age were questioned about driving songs. We would have assumed that since most people have their own individual music preferences, then no such consensus about favourite driving songs would even surface. Yet, the following list of tunes for driving did arise although they might actually be much more characteristic of British drivers than not.

---

1 While such knowledge may seem to be solely intuitive, music therapy research has demonstrated clear clinical effects of music on anxiety, physical fatigue, and well-being (for a review, see Standley, 1986, 1992, 1995).

***Autotrader*'s 2012 Top-10 Driving Songs**

1 'Sultans Of Swing' (Dire Straits)
2 'We Will Rock You' (Queen)
3 'Another One Bites The Dust' (Queen)
4 'Good Vibrations' (Beach Boys)
5 'Go Your Own Way' (Fleetwood Mac)
6 'All Right Now' (Free)
7 'Born In The USA' (Bruce Springsteen)
8 'Rolling In The Deep' (Adele)
9 'I Heard It Through The Grapevine' (Marvin Gaye)
10 'I Can't Get No Satisfaction' (Rolling Stones)

Remarkably, the 2012 study presents a list of driving favourites, in which four hit songs from the 1970s are among the top six preferred songs. Further, the study found that 927 drivers (76%) turn up the volume when their favourite driving song comes on the radio, 732 drivers (60%) sing along if they are alone in the car, 437 drivers (38%) use the steering wheel as an impromptu drum, 85 drivers (7%) 'jig-about' to the song, 24 drivers (2%) play the *air guitar*, and 12 drivers (1%) sing into an imaginary microphone. Further, Motavalli (2012) who explored published driving playlists, mentions two UK surveys. The first is the 'Top 10 Best Driving Songs' published by *About.com* (Lamb, 2012):

1 'Radar Love' (Golden Earring)
2 'Stuck In The Middle With You' (Stealers Wheel)
3 'Let's Go' (The Cars)
4 'Layla' (Derek And The Dominos)
5 'Authority Song' (John Mellencamp)
6 'Mexican Radio' (Wall Of Voodoo)
7 'Get Outta My Dreams, Get Into My Car' (Billy Ocean)
8 'Drive My Car' (Beatles)
9 'Little Deuce Coupe' (Beach Boys)
10 'Take It Easy' (The Eagles)

In addition, Motavalli describes the results of a survey implemented by *Halfords* (a UK retailer of car parts and national provider of MOT yearly car licensing) published in the *New Musical Express* music magazine (*NME*, 2011). Accordingly, the survey did not look at the most popular tracks, but arranged music songs based on driver safety:

Top-5 'Calming Tracks'	**Top-5 'Blood Pressure Raising Tracks'**
1 'The Four Seasons' (Vivaldi)	1 'Sabotage' (Beastie Boys)
2 'Breakdown' (Jack Johnson)	2 'Firestarter' (The Prodigy)
3 'Someone Like You' (Adele)	3 'To Be Loved' (Papa Roach)
4 'Yellow' (Coldplay)	4 'Stronger' (Kanye West)
5 'Landslide' (Fleetwood Mac)	5 'Prelude in C Sharp Minor' (Rachmaninoff)

Auto enthusiasts have always taken pride in having expertize that they can share with naive amateurish drivers. The same can be said about 'audiophile' drivers who are on the one hand connoisseurs of car-audio technologies, while on the other hand serve as self-proclaimed musicologists who have gleaned years of experience about the most fitting music tracks for cruising. A quick web search will offer access to an abundant number of listings by those hopelessly devoted to changing autospheric driving experiences for millions of drivers. Even reputable car manufacturers have turned to publishing suggested playlists for road trips or scenic drives. For example, Volvo (2008; Swade, 2008) promoted a list of tunes as their *2008 Volvo Summer Recommendations*:

1 'Born To Run' (Bruce Springsteen)
2 'S.O.S' (ABBA)
3 'Ramble On' (Led Zeppelin)
4 'Me And Bobby Mcgee' (Janis Joplin)
5 'Dancing Queen' (ABBA)
6 'Jackson' (Johnny And June Carter Cash)
7 'Take A Chance On Me' (ABBA)
8 'Paradise By The Dashboard Light' (Meatloaf)
9 'Stuck In The Middle With You' (Stealer's Wheel)
10 'Mamma Mia' (ABBA)
11 'Born To Be Wild' (Steppenwolf)
12 'Highway To Hell' (AC/DC)
13 'Low Rider' (War)
14 'Running On Empty' (Jackson Browne)
15 'Southbound' (Allman Brothers)
16 'Waterloo' (ABBA)
17 'Highway 61 Blues' (Bob Dylan)
18 'Little Deuce Coupe' (The Beach Boys)
19 'Knowing Me, Knowing You' (ABBA)
20 'Drive My Car' (The Beatles)
21 'On The Road Again' (Willie Nelson)

Besides the obvious selections of Swedish pride (i.e., Abba), the list also suggests *Car Songs* (found in Appendix D and E). Perhaps such a manoeuvre by Volvo seems to be no more than a marketing ploy to attract consumers in an effort to boost auto sales. But in actuality, the provision of Volvo's playlists was part of a long-term war between the American auto industry giant General Motors (GM) and European exports – a war waged through music! In 2001, GM conducted a national campaign including billboards of a 1963 Red Corvette Sting Ray with the caption: '*They don't write songs about Volvos*'.[2] GM compiled a list of 200 songs that mention *Chevrolets*; the songs mostly refer to the golden age of the 1950s when *Chevys* personified a newly found freewheeling spirit of the Rock N' Roll era. This was followed in 2003 with a television commercial that debuted as a spot at the *45th Grammy Awards* ceremony showing old footage of the Beach Boys performing 'My 409', followed by Don McLean singing 'American Pie', and then Prince performing 'Little Red Corvette' – after which the camera pans

2 A framed replica can be purchased at: www.Chevymall.com (item #CCL8141).

outside the Kodak Theater (later renamed the Dolby Theatre) to show GM's latest Chevy model. In retaliation, Volvo Cars of North America released a list of the all time greatest songs about Volvos titled 'More Great Car Songs Are about Volvos, Expert Reports' (Johansson, 2003; KristinS, 2009). Truth be told, few if any well-known songs appear on this listing, with an exception of a lesser known song by Sheryl Crow titled 'Volvo Cowgirl 99' – the B-side of one of her early-90s hits. Six years later KristinS well-wished all Volvo owners to hold their head up as there *is* music out there about Volvos; she specifically called for the driving public to become acquainted with a song called 'Volvo Driving Soccer Mom' by *Everclear* – an American Alternative-Rock Post-Grunge band.

A series of rosters published by Britain's *RAC Foundation* are noteworthy. In the years 2002 and 2004, the *RAC* circulated both recommendations, as well as advisory cautions, for music tracks to accompany driving. The *RAC* publically announced warnings against Wagnerian music[3] as the most dangerous soundtrack for driving an automobile (BBC, 2004; RAC, 2004; USA Today, 2004). Accordingly, when drivers listen to music with a pounding beat rather than more relaxed tunes they are at greater risk, and ironically, it may be of no consequence what kind of music style the driver prefers. The *RAC* claimed that it does not matter if drivers listen to romantic opera, classical symphonies, or the latest rave music – the bottom line is that up-tempo music has been shown by Brodsky (2002) to cause drivers to have double the amount of accidents as those drivers who listen to slower music. While the British automobile club advocated that in general people should more carefully choose the pieces they listen to when driving, they also officially sanctioned two specific listings.

Top-5 'Most Recommended' Driving Tunes	**Top-5 'Most Dangerous' Tunes for Driving**
1 'Come Away With Me' (Norah Jones)	1 'Ride Of The Valkyries' (Wagner)
2 'Mad World' (Gary Jules)	2 'Dies Ire From Requiem' (Verdi)
3 'Another Day' (Lemar)	3 'Firestarter' (Prodigy)
4 'Too Lost In You' (The Sugababes)	4 'Red Alert' (Basement Jaxx)
5 'Breathe Easy' (Blue)	5 'Insomnia' (Faithless)

A second UK organization to recommend driving tunes was *ACF Car Finance* who commissioned a *Billboard* Chart Study (ACF, 2009; Betts, 2009). Accordingly, musicians were recruited to re-rank weekly 'Top-10 Hits' in order of their temporal velocity (based on the research findings by Brodsky in 2002). *ACF*'s most recommended *Safe Driving Tracks* (i.e., safest driving tempos) were the top five songs from the weekly hits, while the bottom five songs were cited as the most dangerous weekly *Billboard* songs to drive with. The tempo-ordered listing, with Billboard position, was:

3 Wilhelm Richard Wagner (1813–1883), German composer, conductor, and theatre director. Wagner is known primarily for his four-opera cycle Der Ring des Nibelungen (The Ring of the Nibelung), as well as for other music dramas such as Die Meistersinger von Nürnberg (The Mastersingers of Nuremberg).

'Top-5' *Safe* Driving Tracks	**Billboard Position**
1 'Love Story' (Taylor Swift)	5
2 'Dead And Gone' (T.I./Justin Timberlake)	6
3 'Whatcha Think About That' (The Pussycat Dolls)	10
4 'Tshirt' (Shontelle)	9
5 'My Life Would Suck Without You' (Kelly Clarkson)	4
'Bottom-5' *Dangerous* Driving Tracks	
6 'The Fear' (Lilly Allen)	7
7 'Just Cant Get Enough' (The Saturdays)	2
8 'Right Round' (Flo Rida)	1
9 'Poker Face' (Lady Gaga)	3
10 'Just Dance' (Lady Gaga)	8

We can't but help notice that the three top Billboard listings (based on sales) fall in the Bottom-5; these exemplars are obviously of a higher tempo (i.e., fast-paced music) and therefore ranked as more dangerous to accompany driving.

Finally, the data from a most recent and highly publicized study undertaken by UK insurance price comparison web site *Confused.com,* was analysed by the London Metropolitan University (Brice, 2013; Dolak, 2013a, 2013b; Jolley, 2013; Katic, 2013; Philipson, 2013; Presta, 2013; Rao, 2013; Sanchez, 2013, TheMayFirm, 2013). The investigation centred on eight drivers who were continuously measured online during two trips totalling 500 miles. The first 250-mile trip served as a baseline for normative driving data, while the second run assessed how the drivers reacted to different playlists variegated by music genres. The drivers were monitored via an innovative GPS technology telematics smartphone application called 'MotorMate'; this program assesses changes in driving performance and behaviour such as speed, acceleration, and braking. According to researchers at LMU, music that is noisy, upbeat, and increases heartbeat, causes excitement and arousal leading to drivers concentrating more on the music than the road, while on the other hand, songs with an optimum tempo of 60–80 beats per minute are safer. The study found that songs from music styles such as Hip-Hop, Rap, Dance, and Heavy Metal led to driving that was more aggressive, with faster accelerations and last-minute braking. The study also found that Classical music caused young(er) drivers to behave more erratically, and Soft-Rock was found to be safest for driving. Confused.com complied a playlist of 50 songs they recommend for driving (Confused.com, 2013a, 2013b), as well as list of songs that drivers should be wary of.

Top-10 'Safest Songs'

1	'Come Away With Me' (Nora Jones)	6	'Cry Me A River' (Justin Timberlake)
2	'Billionaire Feat.Bruno Mars' (Travie Mccoy)	7	'I Don't Want To Miss A Thing' (Aerosmith)
3	'I'm Yours' (Jason Mraz)	8	'Karma Police' (Radiohead)
4	'The Scientist' (Coldplay)	9	'Never Had A Dream Come True' (S Club 7)
5	'Tiny Dancer' (Elton John)	10	'Skinny Love' (Bon Iver)

Top-10 'Most Dangerous Songs'

1. 'Hey Moma' (The Black Eyed Peas)
2. 'Dead On Arrival' (Fall Out Boy)
3. 'Paper Planes' (Mia)
4. 'Walkie Talkie Man' (Stereogram)
5. 'Paradise City' (Guns N' Roses)
6. 'How You Remind Me' (Nickelback)
7. 'Hit The Road, Jack' (Ray Charles)
8. 'Get Rhythm' (Johnny Cash)
9. 'Heartless' (Kanye West)
10. 'Young, Wild And Free' (Snoop Dogg & Wiz Kahlifa Featuring Bruno Mars)

High consumer interest in having the 'right' kind of music creates a great demand for listening options. For example, the 2003 *Arbitron National In-Car Study* (Bouvard, Rosin, Snyder, & Noel, 2003) delineated that the Top-5 American preferences for *drive-time* radio music stations, were: Religious, Alternative-Rock, News/Talk, Rock, Contemporary Hits, and Oldies. However, only one decade later Arbitron's *Radio Today 2013: How America Listens to Radio* (Rodrigues et al., 2013) listed the Top-5 USA drive-time radio music genres as: Country, Talk Info-Personality, Pop Contemporary Hits, Adult Contemporary/Soft Adult Contemporary, and Classic Hits. Truly, commercial CD markets have not been left behind. Long ago a host of music publishers jumped on to the market in an effort to cater for a diverse population of drivers by creating a wide range of specialty music CD compilations for road trips. The largest commercial stockpile of music CD collections for driving is directly available through *Amazon On-Line Shopping* (amazon.com). Compilations such as *Best of Driving Rock, Classic FM Music for Driving, Greatest Ever Driving Songs, Top Gear Anthems: Seriously Cool Driving Music*, and *Trucker Jukebox* are but a few examples of collections that have remained best sellers for over a decade. (For a more detailed representational listing of CD song collections specifically packaged for accompanied driving, see Appendix F.) Remarkably, the British *Direct Line Car Insurance* (Direct Line, 2010) has even endorsed several of these CD compilations in a national posted recommendation to their client base:

> Most drivers listen to some kind of music while they drive. But does it make a difference to your driving whether you listen to Mozart, Megadeth or Marvin Gaye?... Why not invest in a good radio (not one that needs constant tuning), auto-changing CDs and get your MP3 playlist loaded up to last for the duration of your journey? These are all much safer than the need to change the radio station or CDs manually... Obviously, whichever music you listen to is a matter for personal taste. If you hate jazz, then it will irritate you to have it on; if you love rock 'n' roll then that's probably the best option for you – just try and choose songs that are easier on the ear and turn the volume down so that you can hear yourself think (and hear other road users). It's certainly not going to appeal to everyone (indeed some may drive you mad!), but various canny publishers have come up with some special 'driving' CD compilations including *Classic FM Music for Driving, Greatest Ever Driving Songs, Best of Driving Rock, Top Gear: Seriously Cool Driving Music* and *The Ultimate Driving Experience*. They might just prompt you to make your own optimum driving music list that you can stick on your favourite tunes when you hit the road. Whatever soundtrack you choose, at Direct Line we want you to drive safely.

Nevertheless, with the exception of Hawkins (2012) who conducted a musicological study of the *Top Gear* anthology, no empirical research has ever gone into any of the above-mentioned commercially available compilations. Yet, the public is undeniably predisposed to purchase music by ratings, celebrity endorsements, and chic advertisement. Although there is a considerable market for driving compilations, and that driving-related boxed collections seem to sell well as reflected in the number of albums found in-stock at online warehouses, Dibben and Williamson (2007) feel that today's drivers tend not to invest in commercial anthologies but rather choose to listen to their personalized MP3 playlists. The basis of their conclusion is that 87% of drivers participating in their UK survey (i.e., 1,549 respondents) claimed that they listen to the same music exemplars while driving as they do in their home. Accordingly drivers nowadays do not typically outfit the auditory environment of the automobile with distinctive car songs, but rather listen to the same music that they like in general. Perhaps Dibben and Williamson have touched on a new behavioural stance that needs further exploration. Or, perhaps they have been somewhat misguided as CDs of songs for driving continue to be a popular commercial item – supposedly collating the best of the best songs onto one CD for driver convenience. Hawkins conceives of these compilations as 'sonic objects', which not only express speed and machoism, but also are symbolic agents representing practices of automotive-related consumption and reflect the ideology of 'listening to the sound of driving'. This last point is remarkably thought provoking. If as stated earlier, that music listening *has* evolved as a principle constituent of what it means to drive a car, then as Hawkins presupposes, people who reminisce about driving or simply picture themselves sitting in the driver's seat, may actually *hear* a music soundtrack playing above the imagined engine revs. Finally, it should be pointed out that unfortunately a few traffic studies (for example, Matthews, Quinn, & Mitchell, 1998) have employed tracks from these collections *as if* the pieces themselves, or the complete CD programme, were reliable or valid materials for use in scientific investigations.

Few studies have endeavoured to place drivers behind the wheel on real-world roadways while listening to music they have actually brought from home (as their preferred driving tracks), and then documented the authentic music playlists of these pieces. One exception is Brodsky and Slor (2013) who implemented a two-year investigation for the Israel National Road Safety Authority (described in Chapter 6). Accordingly, 85 young novice drivers between 17–18 years of age were recruited to drive in instrumented learners vehicles with a driving instructor. Each driver first created their own playlist from their home music collections and brought the CDs along for two 45-minute trips. Unal et al. (2012) described this strategy as advantageous because it increases *ecological validity* as participants make their selections based on what they perceive they usually listen to while driving; this methodology assures that participants are *familiar* with and *prefer* the heard music in the background while driving in the experiment. In total, 1,035 music tracks were brought to the study, representing roughly 12 tracks per driver. The majority of the music pieces (70%) were international tracks in the English language, while the remaining selections were local songs in Hebrew. Further musicological analysis indicated that six overriding music genres accounted for more than 90% of all music tracks heard in the study. These were: Dance/Trance/House, Hebrew-Middle Eastern, Rock/Hard Rock/Punk, Pop, Hebrew-Pop, and Hip-Hop/Rap styles (see Figure 6.4). From the playlists themselves, 20 international English-language song tracks were found to be highly popular between 25–50% of the drivers. These pieces are listed below (in alphabetical order):

'Bass Down Low' (Dev/The Cataracs)
'Champagne Showers' (LMFAO)
'Club Rocker' (Inna)
'Danza Kuduro' (Don Omar)
'Dirty Situation' (Mohombi/Akon)
'Don't Want Go Home' (Jason Derulo)
'Give Me Everything' (Pitbull)
'Glad You Came' (The Wanted)
'Hey Baby' (T-Pain/Pitbull)
'I Like How It Feels' (Enrique Iglesias/Pitbull)
'I'm Into You' (Jennifer Lopez)
'Little Bad Girl' (David Guetta/Taio Cruz)
'Miami To Ibiza' (Swedish House Mafia)
'Mr Saxobeat' (Alexandra Stan)
'On The Floor' (Jennifer Lopez/Pitbull)
'Party Rock Anthem' (LMFAO)
'Rain Over Me' (Pitbull)
'Tonight' (Enrique Iglesias/Pitbull)
'Where Them Girls At' (David Guetta)
'Yeah 3x' (Chris Brown)

Finally, in 2012 *SEAT Motors* reported a rather new on-road phenomenon they refer to as *car-aoke* (Barari, 2012; Barretts, 2010). Accordingly, this most popular activity is one whereby drivers turn their car into a personal karaoke booth. Conducting their own in-house research on driving songs among almost 3,000 British motorists, *SEAT* reported that male drivers most happily swap their stereotypical driving music for 'chick hits' when they are on their own in the car. Perhaps this has to do with gender-biased social expectations that are somewhat shed when driving alone in the car. Specifically, one-in-five (20%) men confessed that when they are alone in their vehicle they enjoy nothing more than to sing along with Shirley Bassey, WHAM!, or Take That; 71per cent of male drivers even admitted they enjoy a bit of falsetto to 'Stayin' Alive' (Bee Gees). The study also showed that 17% of women drivers love to belt out 'macho classics' from the likes of The Clash or Thin Lizzy. Indeed as Lezotte (2013) points out, men and women perceive and experience cars differently because of gendered expectations about driving behaviours and automobile use. These variances also extend to the employment of music inside the vehicle. The *SEAT* study was commissioned alongside the 2010 launch of a TV reality show called 'SEAT Sex Drive', which aimed once and for all to settle the debate over which of the sexes are better equipped behind the wheel. Two main findings of the study are paramount. First, more than half of the participating British motorists viewed their car as a much-needed 'personal sanctuary'. Second, 13% of the drivers admit to having 'embarrassing *closet* musical tastes' that they indulge *only* in their car. This finding is similar to that of Dibben and Williamson (2007) who found that for 17% of drivers in age groups 18–29 and 30–50 years old report that driving is the only time and/or place where they are able to listen to the specific music styles and songs that they wanted. *SEAT*'s research team claimed that these findings are hardly surprising, especially when considering that 32% of men drivers say they usually have to listen to what their wife or girlfriend insists on hearing, and that 22% of women drivers say they are forced to listen to the song choices of their children. But everything changes when a driver is alone in the car – 61% of men and 58% of women said that they listen to different music when motoring solo; they clearly admit to liking music that their friends, family, and partners can not stand to hear. The overall UK *car-akoe* favourite was Robbie Williams (formerly of boy band Take That). According to *SEAT*, when alone, male drivers reported to enjoy singing to 'Stayin' Alive' by the Bee Gees (71%), 'Girls Just Wanna Have Fun' by Cindy Lauper (15%), 'Flying without Wings' by Westlife (12%), 'Don't Cry for Me, Argentina' (12%), and 'YMCA' by the Village People (11%). Moreover, male drivers reported to enjoy singing an assortment of songs by Robbie Williams (29%), Abba (26%), Take That (25%), Shirley Bassey (20%), WHAM! (20%), Kylie Minogue (15%), Spice Girls (14%), Erasure (13%), and Britney Spears (12%). Similarly, women drivers will abandon Robbie Williams or Abba for something much rockier when behind the wheel with no one else around. For example, female drivers reported singing classics from The Clash (17%), Thin Lizzy (17%), The Jam (17%), as well as a host of heavy metal groups such as AC/DC (10%) and Motorhead (10%). Finally, the study found that the majority of both men (65%) and women (79%) own-up to singing out loud (and very loudly) in the cabin when driving alone. Ironically, they also reported to have felt very embarrassed when caught in the act by a motorist who pulls up along side them at a traffic light. Vanderbilt (2008) explains this latter feeling: because traffic is a place where no one knows the driver's name, and driving alone essentially means that no one is watching (or as in the current case, no one is listening), then drivers also believe that no one will see (or hear) them. Hence, when another driver pulls up, and looks directly into the cabin, or even worse makes eye contact, many drivers will not only feel embarrassment but that their privacy was violated.

3–1.2.1. In-car music subcultures – Boom cars

The experience of in-car listening is in some part due to the fusion between the technologies of car-audio and the music itself (Williams, 2010). Accordingly, there are even particular music styles, specifically Rap and Hip-Hop, that not only can be best understood *through* such compatibilities, but that from a historical vantage these were explicitly conceived as a form of musical expression *for* such compatibilities. A case in point: Rap music was explicitly created as a soundtrack to be heard by people while seated in a moving car. That is, the car was conceived as the optimal listening space for an entirely new distinctive musical expression – one that encompassed a range of visual, aural, and tactile components as no other musical art form before it.

Some time ago, Urry (1999) asserted that our sensing of the world through a 'screen' had become the most dominating mode of human experience in contemporary society. Although initially he referred to the screen as we experience with computer monitors (including desktops, hand-held PDAs, and laptops[4]), he extended his theoretical concept to circumstances involving automobility – conceiving the automobile windshield as one does a computer monitor screen. Yet, unlike all other flat screens that seemingly display an array of text and graphic information from afar, Urry claimed that the car windshield is essentially the portal of a capsule in which technologies are employed to guarantee a highly stable and relatively protected cabin, with sophisticated automated systems to monitor both internal and external environments, as well as deliver sounds – especially music – of the utmost highest quality in real time and space. It is interesting to note that the autosphere is similar in function to other more conventional spaces designed and constructed purposely for the transfer of sound and music – branded by Blesser and Salter (2007) as *aural architecture*. Blesser and Salter conclude that the audible attributes of physical space have always contributed to the fabric of human culture. Such a disposition can be seen in the descriptions of prehistoric artefacts and cave paintings, classical Greek open-air theatres, Gothic cathedrals, the acoustic geography of French villages, modern concert halls, and virtual spaces in home theatres. Accounting for our modern-day lifetime of experiencing environments that provide 'sonic envelopment', Williams (2010) feels that people have acquired a wide range of perceptual knowledge about what kinds of music are most typically fitting to distinct listening spaces. Williams perceives that such interpretations are inevitably linked to cultural, historical, and spatial cues. Accordingly, sound bites that consist of a muted trumpet, walking acoustic bass, and coarse-textured saxophone would most certainly prompt us to visualize the intimate middle-class smoky enclaves of a jazz club; or the sounds of airy legato brightly orchestrated strings would cause us to generate imagery of a wide-open up-market cushioned-seated concert hall; or the sounds of heavy pounding synthesized electronic drums reproduced through a laptop computer we would simply envisage a massive communal hanger filled with the rave; or the sounds of distorted metal guitars with a tight backbeat drum-groove played through a large PA system would suggest a large fan-based stadium. But, what about Rap music?

Rap and Hip-Hop styles are clearly unique cases – as far as in-car listening is concerned. Williams (2010, 2014) presents a case study of *Dr. Dre* (aka: Andre Romelle Young), who envisioned that from the millennium onwards the primary mode of music listening would be through car stereo systems. Already in the early 1990s he originated a music style labelled *G-Funk*. This 'gangsta pop' was created and mixed specifically for automobile drivers. In fact, Dr. Dre saw automotive listening space as representing an idealized reference. He perceived the car as the venue and format of the people who were going to listen to his music:

4 It should be noted that 'smart' cellular phones (*iPhone*) and tablets (*iPad*) were not yet available to everyday consumers at the time.

> I make the shit for people to bump in their cars. I don't make it for clubs; if you play it, cool. I don't make it for radio, I don't give a fuck about the radio, TV, nothing like that, I make it for people to play in their cars. The reason being is that you listen to music in your car more than anything. You in your car all the time, the first thing you do is turn on the radio, so that's how I figure. When I do a mix, the first thing I do is go down and see how it sounds in the car (Cross *In* Williams, 2010, p. 168).

Another musician described by Williams is rapper *Ice-T* (aka: Tracy Marrow), who classified the ideal listener as a driver with a car stereo sound system. Ice-T claimed that his *Power* CD, recorded in an L.A. studio, employed a custom-built system resembling a car stereo system in order to assess how it would sound to the people who would buy the CD. Finally, Price-Styles (2012) examined 'jeep culture' as reflected in the music of Hip-Hop artist and rapper *Masta Ace* (aka: Duval Clear); this music was originally seen as bass-heavy musical anthems of the low-rider sound movement.

For many years music producers have been saying that the automotive listening space is *the* ideal listening environment. As the car began serving music producers as a listening reference, they became more conscious of the idea that a recording was intended to "fill a particular space rather than to reproduce a performance accurately" (Williams, 2010, p. 166). Moreover, as producers considered the idiosyncrasies of the vehicle cabin interior, they also began mixing elements accordingly; they produced tracks with dynamic compression, the equalization of frequencies, and most specifically the sound quality of low frequencies (i.e., aural and tactile sensations produced by subwoofers). Both producers and artists came to see the car as a *cocoon* where not only is there a really big sound in an enclosed environment, but also where the listener experiences the optic flow of scenery moving by. Hence, the driving experience most certainly includes the way people view the world from the windscreen, which, when we add in the music accompaniment, is basically a soundtrack that articulates our similar experiences such as when we watch television or view the cinema screen (Paterson, 2007). Therefore, new musical styles after the millennium attempted to tap the greatest potential of the in-car listening experience, which is unquestionably through multi-sensory input and cross-modal perception involving aural, visual, and tactile (i.e., vibratory) sensations. Such, then, is the impending impact of 'boom cars' – as a vehicle of presence – to be seen, heard, and felt as one cruises the streets of the neighbourhood. For a time, some automotive manufacturers even fixated on the audiences of such innovative music styles, targeting their associated characteristic urban behaviours through marketing messages that emphasize the car as being more than a convenient form of individualized transportation, but rather as a vehicle of self-expression. For example, Lutz and Fernandez (2010) point out that the marketing materials of Ford's *Fusion* model employed the tagline 'The street is your stage', which was intended to serve as a subconscious anthem and buying incentive for prospective consumers from the Rap and Hip-Hop culture.

One contemporary audiophile subculture is known as the *Boom Car*. These automobiles are customized with sound systems producing what seems to be a boundless low-end frequency by way of subwoofer speakers (Shawcross, 2008). Boom Cars function in a similar way as did Boomboxes of the 1980s.[5] Boomboxes were highly popular among urban society, particularly African American and Hispanic youth, whose use (and abuse) in urban communities led to these music systems being coined a 'Ghetto Blaster' – a designated label that was soon employed to ban the use of such equipment, as well as to rebuke the whole Hip-Hop culture. As many cities began to ban Boomboxes in public places, and subsequently they became highly unacceptable for use on city streets, the advent of the Sony *Walkman* contested the need to carry around such huge and hefty audio components, and triggered a total eclipse from the streets. Nevertheless, the final end to the Boombox in popular culture may not simply be because listeners modified their sound reproduction preferences for a newer immensely popular fixed-field listening experienced via headphones, but rather because alongside of all of these developments was an evident impassioned and almost zealous escalation of the car-audio culture that arose from the streets themselves.

Williams (2010, 2014) claims that like the Boomboxes before them, *Boom Cars* have been a source of both intense competition and neighbourhood frustration. On the one hand, "some car audiophiles hold competitions for the loudest and highest quality automotive sound systems referred to as 'sound-offs', 'crank-it-up competitions' or 'dB Drag Racing'" (2010, p. 164).[6] There is even an international auto sound challenge association (IASCA, iasca.com) dedicated to the growth of the mobile electronic industry, with chapters in 34 countries around the world. However, for some this preoccupation may seem far from constructive. Yet, as Blesser (2007) points out, excessively loud music does serve a function by making the listener functionally deaf to everything except the music – transporting them into *aural space*, moving them from their social space of people to the musical space of the performer. When the reality of the neighbourhood is nothing but

5 The *Boombox* was a portable music player with two or more loudspeakers. It reproduced AM/FM radio, pre-recorded cassettes tapes, and CDs at a relatively high volume. Designed for portability, it featured dual power (direct line current and batteries). Later versions had input and output jacks to interface with a microphone and turntable. A desire for louder and heavier bass frequencies led to a bigger and heavier box. By the 1990s most models resembled a suitcase requiring more than 10 'D' size batteries making it extremely heavy and bulky. Throughout the 1990s Boomboxes grew to accommodate bass output featuring heavy metal casings to handle the bass-generated vibrations.

6 In the interests of safety, the 'racer' does not sit in the vehicle, but rather the car is sealed tight and the audio system is operated by remote control, with a decibel meter running inside the cabin. The winner is plain and simply the owner with the loudest system. Williams reports that in 2008 the winning record was 180.8dBl. For comparison, normal conversation (60dBl), a rock concert or jet airplane (120dBl), and a gunshot (140dBl) are all far less than the prise-winning car system.

heart-rending tragic disaster, who wouldn't want to be projected into an exciting aural space full of intense activity that generates energy. Loudness cues attention, and therefore perhaps also a sweet seduction among drivers wanting to continue to engage in such behaviour. For example, in some drivers' minds there may be great social reward of being recognized, admired, and remembered by others in the community. When a driver raises the volume of music, they are indeed transported out of the car on to the road and beyond to the urban district – and from the listener's vantage – from being a driver to commanding the centre stage of artistic performance. Moreover, as loud music changes the mood and behaviours of listeners outside the vehicle, the operator is 'driving' pleasure to the masses. Sheller (2004) claims that both artists and listeners of this music embrace the 'feel' of the car (both the external body shell and the interior cabin) as sensuous shapes and materials that not only project direct emotions upon the car, but also reflect something *about* themselves and *within* themselves. Both the lyrics of the tracks, and their associated videos, portray scenes in which the fantasy world of the driver is heightened through touching the metal bodywork, fingering the upholstery, caressing its curves, and miming driving with every possible part of the human body. Hence, Williams points out that like the Hip-Hop recordings themselves, which no doubt mix human elements with technology, the cinematographic clips ultimately suggest the conjoining of human-flesh and metal-mechanical bodies that depict the car as a prosthetic extension of the driver.

However, the unfortunate truth is that many drivers, who cruise the streets in vehicles customized with subwoofers, simply bombard their streets and neighbourhoods with extreme high-volume low-frequency music (Schwartz & Fouts, 2003; Williams, 2008; Williams, 2010). Blesser (2007) concludes that as with all other forms of pleasure, excess may produce damage, and therefore we must weigh the risks versus the rewards. Many people and organizations such as the National Alliance Against Loud Car Stereo Assault, see *Boom Cars* and the music associated with them as sonic weaponry that support 'audio terrorism'.[7] In the very least, *Boom Cars* are often linked to urban decay (McCown et al., 1997). In 2007, a New York City noise code came in to effect whereby the use of a car stereo was prohibited 'if audible from 25 feet'; the legislation gave municipal policemen a format to clean up the soundscape. Further, the Greater London Authority set up an assembly to study noise nuisance from car stereos; a published report has allowed authorities to seek initiatives for a public referendum towards changing legislation (Shawcross, 2008).

3–1.3. Drivers' Attitudes Towards Driving with Music

Despite the huge variation in the background music employed by drivers worldwide and the debate over the uses and abuses of car-audio, popular beliefs about in-car music listening are widespread. Attitudes explicitly expressed through survey

7 www.lowertheboom.org.

studies indicate that in general, drivers are confident that listening to music is a constructive helpful activity that all of us have engaged in for more than 50 years. For example, a study commissioned by Quicken Insurance SM (Quicken, 2000) found that playing the radio loudly was viewed by American drivers as a far less objectionable practice in the car than reading the newspaper, putting on makeup, and cell phone use, but was seen as a more cavalier activity than combing hair – albeit all of these were perceived as reckless. White, Eiser and Harris (2004) reported that listening to radio and singing to yourself were perceived by 200 British drivers from Sheffield as causing *little-to-no-risk* in comparison to another 13 activities that could potentially increase the chances of an automobile accident (see Table 3.3). Further, Patel, Ball, and Jones (2008) found that listening to music was accepted by 40 London drivers as the most valid activity one can engage in while on the road, perceiving the activity as providing the *lowest-risk-factor* among 14 factors identified as causing driver distraction. Similarly, Titchener, White, and Kaye (2009) studied 113 Australian women drivers who reported that listening to music was the most frequent propulsion-unrelated activity initiated by drivers (more than communicating with a passenger, adjusting climate controls, tuning radio/searching CD, listening to talk radio, or eating and drinking), and was perceived at the *lowest-level-of-risk* among 19 other activities that might lead to in-vehicle distractions and crashes. Further still, Young and Lenne (2010) reported that 287 Victorian drivers from southeastern Australia considered listening to music via radio, CD, tapes, or portable music player as the *least dangerous* of 23 driving situations that could be considered distracting risky activities. Finally, Unal, Steg, and Epstude (2012) reported that 69 undergraduate drivers from The Netherlands perceived music as a *non-distracting auditory stimulus* expressing an overall positive attitude for listening to music while driving.

3–2. Benefits of Driving with Music

As was pointed out in Chapter 2, when people choose to hear music they do so in different places and for many different reasons (North, Hargreaves, & Hargreaves, 2004). Accordingly, each reason for listening to music is significantly associated with variations in the listening situation (i.e., contexts), and hence peoples' motivations for music listening may be 'situation-dependent'. Namely, when we are doing an intellectually demanding task, music seems to help us to concentrate and think; in a pub or nightclub music may help create the right atmosphere; when we are deliberately listening to music at home, or when on a bus/train, we specifically choose to listen to music because we enjoy it, or because it helps to pass the time; when at a friend's house we might hear music simply because someone else liked it; while conducting routine everyday activities such as eating or doing housework music listening may simply be out of habit. In each case, not only are the motivations unique, but also the nature and level of benefit are quite distinctive. Along these same lines, there are various

Table 3.3 Drivers' perception of driving with music as compared to other possible driver distractions

Rank*	White et al., 2004	Patel et al., 2008	Titchener et al., 2009	Young & Lenne, 2010
1	Shave/Put On Makeup	Mobile Phone Use	Read	BAC Over 0.05
2	Dial Hand-Held Mobile	Look At Map/Book	Write	Send SMS Text
3	Look At Map	Grooming	Text SMS Message	Read/Write
4	Ans Hand-Held Mobile	Look For Object	Hand-Held Mobile	Read SMS Text
5	**Look For Tapes**	Eat/Drink	Look At Map	Dial Hand-Held Mobile
6	Adjust Seat	Look At Billboard Ad	Reach For Object	Fatigue
7	Put On Seatbelt	Smoke	Data Entry SAT NAV	Speed
8	Eat/Drink	**Adjust Device (MP3)**	Use Device (PDA)	Take Hand-Held Mobile
9	Sneeze	Look At Landscape	Grooming	Daydream
10	Dial Hands-Free Mobile	Hands-Free Mobile	**Search Radio CD/MP3**	Argue With Passenger
11	Ans Hands-Free Mobile	SAT NAV	TV/DVD In Rear Seats	Ans Mobile Phone
12	Smoke	Look For Road Signs	Eat/Drink	Deal With Children
13	Talk To Passenger	Talk To Passenger	Hands-Free Mobile	Look At Video Ad
14	**LISTEN TO RADIO**	**LISTEN TO MUSIC**	Adjust AC	**Adjust Portable MP3**
15	**SING**		Listen To SAT NAV	Look At Billboard Ad
16	Suck On Sweet		Adjust Seat Belt	TV/DVD In Rear Seats
17			Talk To Passenger	Smoke
18			**Listen To Talk Radio**	Take Hands-Free Mobile
19			**LISTEN TO MUSIC**	Eat/Drink
20				**Adjust Radio CD**
21				Talk To Passenger
22				Adjust AC/Mirrors
23				**LISTEN TO MUSIC**

* Rank = Highest To Lowest Perception Of Risk.

reasons for choosing to drive with music. In-car music background essentially offers drivers exclusive advantages and benefits that no other in-cabin secondary activity can offer.

North, Hargreaves, and Hargreaves (2004) implemented a study with roughly 320 British participants aged 17–78 years old, investigating various episodes in which ordinary people came into contact with music. In general, when we are in a position to choose the music genre or tracks heard in the aural background, we perceived there to be far more benefits of listening to music than in situations when we are not able to choose the music. This was true for the participants in all episodes involving background music, for all settings and undertakings, especially driving (see Table 3.4). By isolating the episodes of driving with music, the data seems to demonstrate several benefits above and beyond all other contexts. For example, in-car listening was found with higher levels of enjoyment, that ease the passing of time, and the fulfilment of habits – in comparison to all other contexts in which music was present. On the other hand, driving with music was less advantageous as a background to create the right atmosphere or image, to intensify an emotional state, or to learn more about music. 'Enjoyment' from hearing music significantly decreased when listening to another person's preferred music if not liked by the listener. Hence, driving with background music is an inferior activity that lacks much benefit when drivers are forced to hear passenger-preferred music. This circumstance is quite unique as in almost every other kind of pursuit and whereabouts, we half-heartedly appreciate listening to music we ourselves have not chosen; in most contexts we feel that the music background slightly helps to enhance the right atmosphere, and perceive the music as somewhat aiding the activity. Having said that, North, Hargreaves, and Hargreaves's study underlines the fact that we are far more tolerant of hearing music we have not chosen, and for a longer period of time, when we are in an automobile than when elsewhere – such as at home (doing housework, schoolwork, or eating), in a restaurant, in an exercise gym, at a shopping mall, at prayer in a church, travelling by bus/train, at a nightclub or pub, in a waiting room, visiting a friend, or even seated in a concert hall.

To date, only one survey has been implemented investigating the listening habits of drivers. Dibben and Williamson (2007) documented in-car listening from both AM/FM radio and pre-recorded tapes and CDs. They found music listening to be the preferred in-vehicle activity among 1,780 British drivers. The majority of drivers reported that they perceived driving with music to offer many benefits on the level of the overall driving experience, as well as directly related to improved driving performance. Roughly a quarter of the drivers claimed that the music advanced improvements to both concentration and relaxation. To the same extent that drivers reported to be actively engaged by singing, music seems to be an active force in minimizing boredom by heightening arousal and positive mood. A large proportion of the drivers (62%) reported that music soothes them; with music they were calmer drivers and hence felt that reduced stress subsequently leads to more considerate on-the-road driving behaviour. This last point was also reported by Wiesenthal, Hennessy, and Totten (2000, 2003), who investigated the use of music while driving to reduce mild roadway aggression.

Table 3.4 Benefits for driving with music background

Benefits	Benefits When Choosing Music		Benefits When Not Choosing Music	
	%Car[1]	%Tot[2]	%Car[1]	%Tot[2]
I enjoy it	56.5	48.6	15.2	26.5
It helps me pass the time	51.1	38.3		
Habit	47.7	23.2		
It helps me concentrate/think	19.6	19.3		
It helps to create the right atmosphere	17.5	32.3	23.9	32.5
Helps create or accentuate an emotion	13.9	19.9		
Someone else with me likes it	10.9	14.5		
It brings back certain memories	09.4	09.5		
Other	03.3	03.7		
Helps create an 'image' for me	02.1	08.9		
I want to learn more about music	01.5	03.5		
None at all			13.0	17.3
I want to hear music longer			08.7	06.0
Aids my performance			06.5	13.6
It annoys me			04.3	15.1
Want to turn off music as fast as possible			02.2	02.5
Hinders my performance			02.2	06.3
It helps to create the wrong atmosphere			02.2	01.7
It makes me look stupid			00.0	00.0
It makes me look good			00.0	00.3
Other			00.0	00.0

[1] = Percentage of episodes involving background music when driving.
[2] = Percentage of total episodes involving background music at home (housework, schoolwork, eating, listening to music, doing something), in a restaurant, gym/exercise, shopping (mall), place of worship, bus/train, pub/nightclub, waiting room, friend's house, concert hall.

Source: Adapted from North, Hargreaves, and Hargreaves (2004, Tables 16–19).

3–2.1. Music for Driver Fatigue

Certainly drivers listen to music while driving as it delivers a more entertaining and enjoyable experience (Dibben & Williamson, 2007). Moreover, drivers listen to music especially under more challenging circumstances, such as when driving alone, when traffic conditions are moderate or heavy, and in evening/night times (Stutts et al., 2003). However, Clarke, Dibben, and Pitts (2010) claim that beyond the advantage of providing enjoyment, drivers tend to use background music to alleviate boredom as in-cabin background music causes physiological arousal as well as elicits cognitive engagement. Namely, in a more active listening mode in which the driver is aware of the pieces, anticipates musical events, seeks to understand the lyrics, and perhaps even reminisces about where the music was heard before, a whole range of feelings and emotions surface that increase levels of physical and mental stimulation.

Nevertheless, perhaps the major benefit of driving with music is as a countermeasure for driver drowsiness, sleepiness, and fatigue. Drowsiness or sleepiness refers to the urge to fall asleep (a physical state of the body) whereas fatigue refers to an inability to perform the required task (Vanlaar, Simpson, Mayhew, & Robertson, 2008). Regarding fatigue, some people also differentiate between physical and mental/psychological fatigue; the former is directly related to muscular work followed by loss of strength as well as discomfort or pain, whereas the latter is a more subjective experience of incompetence or failure to maintain/complete on-task performance (Oron-Gilad, Ronen, & Shinar, 2008). Despite the semantic or structural differences between these distinctions, in all cases the result is the same – abilities to drive safely are compromised. According to Vanlaar et al., an American survey from 2002 found that 51% of drivers admitted to driving while drowsy, and 17% even admitted to dozing off (with 1% of USA drivers involved in a crash due to dozing off or fatigue), while a Canadian survey from 2005 reports that one-in-five drivers admitted to nodding-off at least once while driving during the previous year. American sources believe that up to 20% of serious crashes may be due to drowsy driving and estimate as many as 79,000 to 103,000 collisions (and 1,500 fatalities) per annum as a result of fatigue. In their own study, Vanlaar et al. found that 60% of the sample admitted that in the past year they drove while drowsy or fatigued (representing > 5 million Canadian drivers), and 15% said they nodded off to sleep while driving (representing > 1 million Canadian drivers). Anecdotal evidence suggests that after feeling drowsiness or fatigue (such as when we yawn or feel our eyes heavy), it is most common for us to open a window for fresh air, and then perhaps turn up the volume of the radio – both reported as techniques to offset falling asleep at the wheel (Stutts, Wilkins, & Vaughn, 1999). In general, drivers do believe they can overcome fatigue or drowsiness using certain prevention tactics. But beyond anecdotes, some research has shown stimulation to be beneficial. For example, in a very early study by Fagerstrom and Lisper (1977) 12 participants between 22–30 years were recruited to drive > 3 hours in a 1970 manual transmission Volvo *Express* estate car across 370 kilometres of motorway without pause. The driving consisted of three conditions (110 kilometres each): listening to two pre-recorded Swedish radio broadcasts (talk-only and music-only programming), as well as silence (for empirical control). The broadcasts were reproduced from a tape player and loudspeaker in the back seat. Although no differences surfaced between listening to the talk radio versus music radio, there were clear effects of stimulation (i.e., talk or music) compared to silence on fatigue or vehicular performance. Yet, doubts have been raised. For example, Reyner and Horne (1998) explored two countermeasures in an instrumented vehicle among 16 drivers who employed the use of cold air (i.e., air blown on face through AC air-conditioning) and/or listening to music (i.e., driver-preferred radio stations or driver-preferred music tapes brought from home) while driving under dull and monotonous road conditions throughout a 2.5-hour drive. The study found that neither action significantly reduced sleepiness – albeit listening to radio/music tapes demonstrated a positive trend for improvement. The researchers concluded that the use of AC or music seemed to be inferior to either caffeine (drinking coffee) or to stopping for a 15-minute nap. Further, Oron-Gilad et al. also report that drivers employ various behaviours to maintain alertness, such as: listening to the radio and increasing the volume of the radio, opening a window, following the lane markers, talking to a passenger, and drinking coffee. Yet, as far as music is concerned, the research team bring out that little empirical evidence demonstrates efficacy, and unfortunately, there have been relatively high rates of drivers who fell asleep with their radios turned on. In this connection Dibben and Williamson (2007) astutely comment that while many studies seem to have demonstrated that drivers' self-ratings targeting their level of sleepiness were indeed significantly lower when listening to the radio, methodologies employing EEG measurements do not demonstrate listening to radio as having a significant causal effect on sleepiness, and those studies targeting vehicle performance measures do not demonstrate significant improvement of driving. Namely, these studies raise the possibility that listening to the radio may ironically bias drivers' self-perception of drowsiness; as music is assumed to be a countermeasure for the effects of fatigue, such beliefs might actually sanction circumstances by which drivers continuing to drive in a heavy-eyed lethargic drowsy state while listening to their favourite ballads. Dibben and Williamson feel that this absurdity is particularly pertinent among young drivers who report using in-car entertainment as a deterrent to a greater extent, but fatefully it is specifically young drivers who are also most at risk for sleep-related crashes.

Nevertheless, there are many studies (Anund, Kecklund, Peters, & Akerstedt, 2008; Cummings, Koepsell, Moffat, & Rivara, 2001; Ho & Spence, 2008; Oron-Gilad et al., 2008; Vanlaar et al., 2008) that do point to music as a beneficial method for maintaining alertness to offset monotony and sleepiness while driving. For example, Cummings et al. interviewed 200 crash cases and 200 matched controls above 18 years old from Washington County; they explored the factors that might counteract drivers falling asleep while driving. The study found that breaks and highway rest stops were rather common (40% of drivers) and decreased the risk of crashing, while napping during breaks (2%) were rather seldom; that 28% of drivers open windows for a refreshing cool breeze, but that no correlation surfaced between fresh air and a decreased crash risk; that drinking caffeine (coffee, tea, soda, energy drinks) was common among 30% of drivers, and this action lowered the risk of crash by 50%; and that 68% of drivers who had a sound system playing were at lower risk of crashing. Subsequently, Hasegawa and Oguri (2006, Experiment 2) analysed drivers' heart rate (HR), pulse beat, and EEG brainwaves among ten undergraduate students during a 10-minute simulator drive with a music background of two well-known Japanese popular music artists. The music exemplars were classified as high-preference fast-paced tracks versus low-preference slow-paced tracks. The results show that high-preference fast-paced music made stimulating effects on drivers' levels of arousal and vigilance as demonstrated by increased HR and EEG with increased steering stability, while the low-preference slow-paced music caused sedative decreasing effects including steering instability. Hasegawa and Oguri concluded that maintaining levels of arousal and vigilance through music selections is a valid countermeasure for preventing drowsiness. A later study by Anund et al. (2008) employed a questionnaire survey among a random sample of 1,880 Swedish drivers between 18–65 years of age; the findings show that the most common countermeasures for fatigue were to stop and take a walk, to turn on the radio, to open the window, to drink some coffee, and talk to a passenger. Out of 22 different countermeasures delineated by the survey, 'music activity' was indicated by more than half of the drivers, including turning on the radio/stereo (52%), singing or whistling (31%), dancing/body movements or tapping to the music (27%), and turning up the radio/stereo volume (26%). Anund et al. conclude that music activity may be especially possible to carry out because it is less intrusive than stopping or taking a nap, as well as being a bit easier to apply. Moreover, Vanlaar et al. (2008), who polled 750 Ontario drivers aged 16 to 93 years old, found that among an inventory of 14 possible tactics to cope with drowsiness and fatigue, close to half of the drivers (44%) indicated that they open the windows for air or turn on AC – albeit they themselves perceived these manoeuvres to be only 62% effective. The two strategies perceived to be most effective by 83% of the sample were stopping to nap or suspending themselves as the designated driver – but yet these two were practised by just 22% of the drivers. It is interesting to note that music engagement (including turning on music at loud volume, changing radio stations or CDs, singing to the music, or tapping/moving to the music) was only employed by 29% of the drivers who assessed that music would decrease vulnerability by more than 53% (see Table 3.5). Finally, in an experimental programme Oron-Gilad et al. (2008) conducted a study exploring alertness maintaining tasks (AMTs) while driving; twelve professional truck drivers between the ages of 32 and 55 years old drove five 2-hour sessions with a fixed-base simulator. The study implemented three sessions, each with two 10-minute sequences of a cognitive distractor task (i.e., choice reaction time, working memory task, or long-term memory task); in addition there was a fourth session involving listening to driver-preferred music brought from home, and a fifth session of driving-alone (i.e., without AMTs). The study found that all drivers felt that music had a positive effect on their driving; when music was used to counter monotony there were no indications of deterioration, nor were there costs to feeling fatigue, physiological efforts, or cognitive resources. The researcher team concluded that "driving with music is a good method for maintaining alertness, or at least definitely better than driving without music" (p. 858).

Given the above, it should be of no surprise that an intervention was developed in which the drivers' state of mind and cognitive focus is captured (via brainwaves from the frontal region of the brain using a portable EEG device), and subsequently cued a *personalized-music-recommendation-mechanism* to play music through the in-car stereo system as the tool to 'refresh' the driver upon the detection of drowsiness from the driver's brainwave patterns. Sizable music databases – such as the *Pandora*'s 'Music Genome Project' that can propose individual songs by name, genre, lyrics, melody, harmony, rhythm, and instrumentation – have existed for some time but without such neurobiological applications. That is, the recommended music relies on a content-based filter, whereupon the music items can be tagged for textual, social, and cultural attributes, as well as for features within the music itself. Tags are then employed to compare and rank songs accordingly to each driver's own musical preferences along with their personal profiles. The brainwaves released thereafter by the driver, in response to various types of music and its content, continue to re-select appropriate music based on newly updated neural data. Liu, Chiang, and Hsu (2013), who implemented the intervention, point out that although a host of accidents are caused by driver fatigue, if warning signals initially occur when drivers become drowsy, then the incidence could decrease. But thus far, the detection of driver fatigue through sensors that monitor lane drift, accelerator activity, and eye movements, have all been somewhat tainted by a number of false events due to driver habit, road

Table 3.5 Tactics used as countermeasures for drowsiness and fatigue while driving

	Tactics	Drivers Use of Tactic %	Perceived Level of Efficiency %
1	Open Windows or Turn on AC/Fan	44	62
2	Talk to Passengers	34	65
3	Stop (Eat, Exercise, Relax) without Napping	31	67
4	Change Radio Station or CDs (Or Songs)	30	50
5	Turned Radio or CD on at Loud Volume	30	56
6	Drink Caffeine (or Take Caffeine Pills)	30	61
7	Sing Along to The Music	29	57
8	Eat/Drink (Not Caffeine)	28	56
9	Ask a Passenger to Take Over Driving	28	84
10	Move Around or Shake Head (to Music)	26	50
11	Stop to Nap or Sleep	15	82
12	Talk on Cell Phone	12	43
13	Pour Water on Face or Neck, Slap/Hit/Pinch Face	07	50
14	Take Stimulant (e.g. No-Doze)	02	51

Source: Adapted from Vanlaar et al. (2008).

conditions, and optic glare/blinking. Liu et al. feel that neuro-based detection methods of drowsiness that determine mental fatigue (as measured by EEG physiology), are more robust as they reveal the driver's cognitive state. In their study, ten undergraduates were recruited to drive a simulator over a period of six weeks. One hundred music selections in MIDI-format were employed, and these were classified as pertaining to one of two music types: *refreshing* (decreasing drowsiness) or *non-refreshing* (having no effect on drowsiness). The findings thus far demonstrate that the mental states of all drivers are effectively detected, and these produce positive outcomes by selecting a functional set of refreshing music that decreases driver drowsiness.

While this chapter has mainly appraised the incidence and attitudes of drivers towards driving with music, as well as explored the benefits of music as an accompaniment to driving and as a countermeasure for fatigue and drowsiness, the following two chapters explore the contraindications of in-car music and the detailed *ill-effects* of music on the driver and vehicle performance.

Chapter 4
Contraindications to In-Car Music Listening

This chapter focusses on the setting of driving a car targeting those circumstances in which the additional exposure of background music might be counterproductive. As seen in Chapter 1, the automobile has become a vehicle of daily life, and car-audio has developed as an integral feature component of the machinery. So much so have the two become embedded in the social system of Western culture that *cars-&-music* have merged into a single subculture itself – one that is re-experienced on a daily basis as an iconic emblem of civilization – that is represented in the industries of theatre and film, the recording arts, video and computer gaming, as well as among all mass-media communications. Obviously, music itself is a personal experience. Nevertheless, as Chapter 2 underlined, listening to music as background for non-musical activities has occupied a more central position in our general everyday life, and certainly this has much to do with mobile technology that allows us to take preferred music along *as if* in an effort to control the soundtrack of life itself. Chapter 2 notes that different musics offer a range of effects to people in a host of situations and contexts, and yet while an wide array of reactions is expected, there are also clear commonalities depending on the stereotypical characteristics of the listener, the context, and the activity in which music is employed as a background. Since the turn of the millennium, the context and activity that has been consistently reported to be the most popular location in which people can (and do) engage in music listening, is while they are in a car. Chapter 3 accounts for the frequency of in-car listening and outlines the central beliefs of drivers worldwide that background music is as much a natural and fundamental constituent of driving as is accelerating, looking ahead, steering, and braking. Moreover, as pointed out in Chapter 3, there seem to be many benefits of driving with music. However, adding music to a milieu consisting of driver performance and vehicular control within a highly dynamic and potentially hazardous traffic-based road environment, may also have some shortcomings as far as personal safety is concerned. The consequences may even be fatal. With this as a backdrop, the current chapter looks at inattention and distraction, as well as the contribution of devices that provide drivers with the opportunity to engage in music listening including the radio, cassette tape player, CD player, and MP3 digital music players. Further, the chapter outlines four specific contraindications to the use of background music while driving. It should be pointed out that there is a very limited pool of information concerning the interaction of music and cars to begin with, while data concerning the subsequent effects of driving with music is even scarcer. Having said that, much can be gained with evidence that has surfaced from empirical investigations on mobile devices such as cell phones –

which in the end take up drivers' attention, as they are also preoccupied with tasks secondary to driving. As with musical activity, the use of a cell phone also engages the driver in mechanical, perceptual, and cognitive processes, as well as in verbal activity. These may be executed when either alone in the cabin or when the driver is accompanied by other passengers.

Driving is probably one of the most difficult complex tasks we do in everyday life. For the most part, driving "does not take place in ideal conditions, in which a well-rested, well-trained and well-behaved individual interacts with a simple, undemanding road environment" (Horberry, Anderson, Regan, Triggs, & Brown, 2006, p. 185). Even on a normal everyday road, we should consider driving as a form of complex multi-tasking. Driving requires the effective orchestration of several executive functions, selective attention, and integration of information from a host of visual inputs as well as from additional sensory inputs, and the coordination of numerous compound behaviours such as divided attention, working memory, prospective memory, behaviour selection, and behaviour inhibition (Jancke, Brunner, & Esslen, 2008; Just et al., 2008). That is, we view the road scene in front of us through the windshield and the road scene behind us via the rear-view and side-view mirrors, we monitor the gauges on the instrument panel, we hear auditory sounds of our own vehicle as well as from those surrounding us, experience the proprioceptive qualia of our own vehicle's balance and stability, and execute a host of behavioural skills involving navigational steering, accelerating, and braking. Vanderbilt (2008) claims that driving consists of roughly 1,500 sub-skills; among those previously outlined by Strayer and Drews (2004) are proficiencies relating to on-line navigation, judging and maintaining speed, maintaining lane position, following distances, scanning for hazards and information, evaluating risks, anticipating future actions of others and reacting to unexpected events, adjusting instruments, and making about 20 decisions per mile – and all of this while we are involved with a host of other *irrelevant activities* such as drinking coffee, eating a sandwich, lighting a cigarette, thinking about what happened last night, talking on the phone, having a conversation with a passenger, quieting a child in the rear seat, checking SMSs and e-mails, shaving, putting on makeup, adjusting the radio, changing CDs or scrolling through MP3 files, and perhaps even singing a song or tapping out the rhythm of the greatest guitar riff or solo of all time. Because we seem to do all of these with an apparent ease, we assume that like breathing, driving is no more than an involuntary reflexive behaviour – not to be taken as seriously as other activities that require intense mental effort. But yet, the ill-effects of performing dual tasks have been observed since the mid-1960s when Brown, Tickner, and Simmonds (1969) recruited 24 men between 21–57 years of age, with a licence for at least three years, to drive a manual gearshift *Austin A40* estate car on an airfield set up as a test track. During the runs, each driver was presented with a telephoning task, which in the 1960s was a verbal task heard from a loudspeaker with driver responses via a telephonist's headset employing a *radiophone* with commercial band (CB) broadcast. Accordingly, the participants drove under conditions of divided attention for half of the runs; while completing phone tasks they had to judge which gaps on the track they could drive through (some of which were smaller than the 5-foot width of the car). The study found an overall 7% reduction in speed when engaged in the radiophone task but did not affect lateral position or longitudinal accelerations. Nevertheless, judgments of clearance were obstructed, and both response time and accuracy of the verbal task were significantly corrupted when speaking on the radiophone. From these findings, Brown et al. concluded that concurrent dual tasks impair perceptual-motor skills leading to errors. This study offered prima facie evidence suggesting that under increased workload drivers will decrease their speed. However, one might also infer from the data that driving while singing to songs (like engaging in verbal conversation) might also obstruct drivers' judgements concerning the visual field.

The National Highway Traffic Safety Administration (NHTSA, USA) has estimated that driver inattention is a contributing factor in 25 to 30% of crashes, which is 1.2 million crashes per year in the United States (Stutts et al., 2005). As a matter of fact, the official statistics (NHTSA, 2010) for year 2009 indicate that 5,474 people were killed and 515,000 injured in motor vehicle crashes attributed to distracted driving. The National Safety Council (NSC, 2012) points out that driver distraction has joined alcohol and speeding as leading factors in serious and fatal crashes; 21% of all crashes in year 2010 involved a cell phone (i.e., talking and/or texting) reflecting 1.1 million vehicle crashes. Certainly, a driver looking in the wrong direction or taking their mind off driving at a critical moment can lead to disastrous consequences. Indicators going back as far as 1977 acknowledge that *recognition errors* are *the* overriding marker of human factors involved in 93% of accidents (Stutts et al., 2001b; Victor, Harbluk, & Engstrom, 2005).

Driver distraction is understood as the delayed recognition of information needed to safely undertake the driving task that results from a behaviour or mind-set triggered by any object or event that draws attention away from driving activity or control over the vehicle. Some researchers differentiate driver distraction from other forms of driver inattention (such as absent-mindedness resulting from unrelated incidental thoughts), while all researchers view inattention and distraction otherwise from physical and mental states involving fatigue and drowsiness. Stutts et al. (2001b; Stutts et al., 2005) list eleven specific types of driver distraction that can be found among as many as 51% of all drivers. The most prevalent distractions were those resulting from objects, people, or events outside of the vehicle (29%), followed by adjusting radio/ cassette/CD (11%), passengers (11%), a moving object inside the vehicle (4%), using a device or object brought into vehicle (3%), adjusting the air-conditioner or other controls (3%), eating and drinking (2%), using a cell phone (2%), and smoking (1%). However, NHTSA collapsed these and other sources into four types of activity-related provocations that contribute to driver distraction. The British Royal Society for the Prevention of Accidents (RoSPA, 2007) describes these as follows:

1. *Visual distraction.* Occurs when a driver sees objects or events that impairs observation of the road environment.
2. *Auditory distraction.* Occurs when sounds prevent drivers from making the best use of their hearing because their attention has been drawn to whatever caused the sound.
3. *Biomechanical distraction.* Occurs when a driver is doing something physical that is not related to driving such as reaching for something, holding an item, or being out of the driving position.
4. *Cognitive distraction.* Occurs when a driver is thinking about something not related to driving the vehicle.

Recarte and Nunes (2009) provide another view of the four factors causing distraction:

1. *Visual factors.* Tasks involving visual demands (like searching MP3 files) that give rise to direct conflict at the level of visual input.
2. *Cognitive factors.* Tasks involving cognitive processing that do not explicitly require looking (such as listening to the radio) which give rise to direct conflict from the cognitive effort derived from input.
3. *Activation factors.* Tasks involving attentional dysfunction (such as singing lyrics of a song) that give rise to energy aspects of attention hampering the availability of attentional resources.
4. *Anticipation factors.* Tasks related to expertize or expectancy of explicit knowledge (like skilfully recounting the track details including the names of the artist or band) that cause the drivers to miss relevant information on the road due to a lack of training (especially among young novice drivers).

Although these above two classification systems may be useful for a greater empirical focus among researchers in their efforts to target specific potential harm associated with various driver behaviours, the truth may simply be that in-vehicle activities such as music listening and singing actually involve several of these categories in parallel. Hence, it seems cogent to view in-vehicle music behaviour via its *utility* or function (as well as 'dysfunction') of driver actions. Using the latter taxonomy as scaffolding, this chapter outlines four contraindications to background music while driving. These are: *structural mechanical interference* related to the ergonomics and/or mechanical configurations of the sound system found within the vehicle cabin; *perceptual masking* related to the capability of music to cover self-monitoring feedback sounds and external warning signals; *capacity interference* to central attention subsequent to overtaxed cognitive faculties; and *social diversion* related to recreational driving in the presence of passengers and music along with illicit activity.

4–1. Driver Inattention/Distraction and In-Vehicle Devices

Several empirical reviews of the literature mention driver inattention and distraction. For example: Neale et al. (2005) found driver inattention as contributing to as much as 68% of near-crashes and 78% of crashes; Lee (2007) found that distraction or inattention are attributed to 13–50% of crashes; Shinar (2007) concluded that the incidence of crashes due to some form of inattention is between 20–80%; and Recarte and Nunes (2009) claimed that the weight of distraction on accidents is estimated between 25–50%. Overall, the literature highlights the fact that when drivers are preoccupied with either an internal or external stimulus or event that attracts attention, typically there is a detriment of driving performance. For some time now, there has been a growing concern over the detrimental attributes of new technologies, and the effects of driver-device interactions within vehicles on cognitive functions (such as attention, expectation, and processing demands). RoSPA (2007) submits that drivers are distracted when they pay attention to a second activity while driving because the additional task places a further load on the primary driving task, and subsequently this reduces one's typical driving standard of controlling the vehicle, as well as causes one to be less attentive to the roadway, and impairs judgments resulting in failure to anticipate hazards leading to accidents. Yannis (2012) claims that driver distraction may have an impact on driver attention (hands-off-the-wheel and eyes-off-the-road) as well as on driver behaviours (vehicle speed, headway, lateral positions, and reaction times). Certainly, the most obvious distraction is looking away from the driving scene while engaged in another activity. Gazing at objects incites the opportunity for potential risk that can escalate depending on the amount of time the driver spends looking away from the traffic scene. Shinar asserts that, in reality inattentiveness is "the end result of our own needs to seek stimulation when the driving task is not very demanding" (p. 519). Accordingly, while we drive we find ourselves calling a friend, reading an SMS text, eating, drinking, smoking, searching for items in the vehicle, looking at the scenery passing by, turning on the radio, listening to songs, singing, drumming on the steering wheel, and even dancing in our seat – all of which have absolutely no relevance to monitoring the roadway or controlling a moving vehicle. Perhaps we engage in all of these, and other non-driving-related activities, simply because we do not consider driving an automobile as a challenge to begin with. Hence, distraction can occur willingly (Regan, Young, & Lee, 2009), seen as *driver-initiated* action in a second attention captivating activity, for example when we decide to scroll through a playlist to find our favourite driving tune. And, we assume: *Why not? If everyone does it, then it can't be that dangerous*. Vanderbilt (2008) reiterates that we are not driving machines, but once we feel like we have everything under control, we begin to act indifferently. Yet, much of the trouble comes from the fact that we *do* have perceptual limitations, and we cannot pay attention to everything all the time.

Lutz and Fernandez (2010) bring out the irony in the fact that as roads have become more secure, and safety measures have been added to automobiles, people tend to drive more recklessly. This phenomenon is referred to as *risk compensation*. While it may be absurd, the irrational truth is that straighter stretches of better lit road with wider lanes and shoulders, automobiles that support airbags and anti-lock brakes, and car systems that feature cutting edge technologies (such as electronic stability control, collision warning systems, and intelligent speed adaptation), make us all seem to be confident that we can drive faster, leave shorter distances between cars, change lanes more often, drive when fatigued, talk on the phone, and operate a host of infotainment devices. Ho and Spence (2008) state this very clearly:

> Researchers have found that drivers of cars equipped with safety features tend to be more aggressive, pushing their cars closer to the limit (that is, they exhibit a form of risk compensation)... As a result, it would appear that drivers offset the original positive intention associated with safety innovation and impose additional danger or risk to their road safety instead (p. 96).

Beyond the overconfidence that drivers may have in today's vehicles, there is another element of such a compensatory mind-set. Drivers are quite capable of convincing themselves that they are adept at prioritizing their skills and proficiencies in a way that enables them to apply the required amount of attention to the vital information and activities needed for driving – despite their involvement in other undertakings.

The above is eloquently illustrated by several surveys reported in both scientific literature and social media. For example, Royal (2003) conducted a USA nationally representative survey of 4,010 drivers in which nearly all participants reported to engage in distracting activities. Yet, most conceded that they were only 'somewhat aware' that such preoccupations do in fact impair their personal driving safety. Accordingly, the five most cited activities that drivers partake in (by percentage of engagement, and percentage of awareness that these are hazardous), are: talking with other passengers (81%, 4%); changing radio stations, tapes, or CDs (66%, 18%); eating or drinking (49%, 17%); using a cell phone (26%, 46%); and dealing with children or passengers in the backseat (24%, 40%). One cannot but note the high frequency of music listening. Further, other activities were acknowledged as causing greater impairment, and hence, drivers reported to partake in them less frequently, were: map reading (12%, 55%); grooming (8%, 61%); reading printed material (4%, 80%); responding to a beeper or pager (3%, 43%); and using the Internet (2%, 63%). Titchener, White, and Kaye (2009) reported another set of data highlighting the engagement of in-car activities while driving. Accordingly, the top-ten activities cited by 113 Australian women drivers (by percentage of engagement, with percentage of risk perception), were: listening to music (92%, 25%); communicating with a passenger (68%, 32%); adjusting climate controls (66%, 44%); tuning radio or searching CD tracks (65%, 54%); listening to talk radio (63%, 30%); eating or drinking (48%, 52%); reaching for an object (47%, 80%); adjusting the seat belt (46%, 42%); a text message (36%, 86%); and using a hand-held mobile phone (36%, 86%). Again, the extremely high frequency of listening to music while driving comes to the surface. Then, in 2010 the teen-oriented glossy magazine *Seventeen* teamed-up with the American Automobile Association (*AAA*) to conduct an online survey among 2,000 teenagers between 16 and 19 years of age (AAA, 2010a, 2010b; AASHTO, 2010; Copeland, 2010; DavisLawGroup, 2010, PRNewswire, 2010). This study found that 86% of the participants engaged in some form of distracted driving behaviour. Accordingly, the secondary tasks most undertaken by both male and female teen drivers, were: adjusting a radio, CD, or MP3 player (73%); eating food (61%); talking on the cell phone (60%); and text messaging (28%). Regarding the latter, the participants revealed that during the month prior to the survey every driver had sent an average 23 texts to friends while driving. Other behaviours reported as particularly distracting and hence undertaken less frequently, were: applying makeup and driving with four or more passengers. These findings were confirmed by an observation study outside 40 Canadian high schools in the cities of Vancouver, Barrie, and Halifax. Accordingly, Oda et al. (2010) found that driver distraction was as high as 30% among the 470 young drivers they documented; the most common distraction in Vancouver was the cell phone, while in Barrie and Halifax the most common distraction was loud music. It is interesting to note that 84% of the *Seventeen* survey were indeed aware that their actions increased their risk for being in a crash, and yet, they still persisted in these driving behaviours as they perceived their actions would take just a split second (41%), that they themselves would not get hurt (35%), and that like all members of Generation Y they are highly experienced masters of multitasking (34%).[1] The *Seventeen* survey illustrates previous findings by Lee (2007), which demonstrated that even when young drivers *do* perceive driving hazards, two behaviours contribute to them underestimating appropriate levels of risk. Foremost, they tend to overestimate their ability, and then they tend to accept greater risk. Such brashness and daring is especially dangerous for young drivers who do not recognize specific dangers that could possibly overwhelm their ability to control the vehicle.

Teenagers, such as those that participated in the collaborative venture between *Seventeen* magazine and the American *AAA* are of primary interest because many previous studies (for example, Harbluck et al., 2007) have shown that this sample

1 Generation Y (approximately 71 million) were born between 1980–2000 (hence also called the Millennial Generation). Members of Gen Y are racially and ethnically diverse, incredibly sophisticated and technologically wise. Most often, they have been raised in dual income or single parent families. They not only grew up with it all, they've seen it all, and have been exposed to it all – aided by the rapid expansion of the Internet, cable TV, satellite radio, and a host of mobile devices labeled 'i-' (such as *iPod*, *iPhone*, *iPad*, *iTunes*) or 'smart' (such as *smartphone*, smart cars, smart devices), with futuristic sci-fi associations (such as *Android*, *Galaxy*), with platforms that run *apps* and *sync* through *Wi-Fi* and *Bluetooth*. Gen Y is known for its intensive use of social media and virtual communications that engage anywhere up to 88 texts a day, as well as a willingness to conduct their lives through their *smartphones*, *iPads*, and laptops (Gratton, 2013; Schroer, n.d.).

has fewer visual scans toward the road scene, more occurrences involving hard braking when cognitively distracted, and demonstrate the greatest predisposition to use technology in their vehicles. In light of the *Seventeen* survey data pointing to tainted (and perhaps even utterly impaired) decision-making processes of young drivers who seem engrossed in secondary tasks while driving, one might question if young drivers do even engage in internal mental processes of *risk analysis* (involving reason, logic, and careful reflection about the consequences of driver behaviours), or alternatively do such young drivers just simply perceive *risk as feelings* (in the sense that more irrational motivations may be behind their driver behaviour). In this connection, Vanderbilt (2008) maintains that young drivers routinely calibrate the level of risk they are willing to take, and then when they drive justify their behaviour in relation to the benefit they estimate will be gained. For example, the *Seventeen* survey highlights the fact that while 86% of 16–18 year old teen drivers unremittingly partake in a host of secondary activities as well as operate many innovative fashionable (but non-essential) devices while driving, more than a third (38%) recollect frightening (and even 'terrifying') experiences while they themselves were a passenger of a distracted driver. Further, more than a third (36%) account for already having been involved in at least one near-crash because of distracted driving. However, a third (31%) still express their belief that nothing bad will happen to them while on the road. Arnett (2002) relates to this phenomenon as an *optimistic bias* found among teenagers in general. It would seem that young drivers have a tendency to see the likelihood of negative events as more probable occurrences for others than for themselves. The truth is, that the more drivers assume they are in control of both driving the car and the secondary tasks (i.e., devices employed while driving), the less they perceive there is any risk involved at all. Vanderbilt claims that "the relative ease of most driving, lures us into thinking that we can get away with doing other things… like listening to the radio… [and clearly] we buy into the myth of multi tasking…" (p. 77).

In 2012 Allianz Insurance surveyed 1,000 people between the ages of 18 to 60 examining the 'Top In-Car Distractions' (Allianz Insurance, 2014). The participants reported that the most prevalent secondary sources of distraction were: passengers talking (48%), children (44%), mobile phones (41%), food and drink (7%), and music (7%). In general, 70% claimed that most of time (60–100%) they are distracted by music, while 24% claimed they are distracted by music in roughly half of the time it is playing in the car; 10% admitted to nearly having an accident due to being distracted by music, and 7% revealed that they had already been in an accident because of music-generated distraction. (A partial data set is presented in Table 5.5). In an important study, Westlake and Boyle (2012) surveyed 1,603 drivers between 14–19 years old from Iowa (USA). The researchers examined driver distraction among teenagers, as well as the activities they themselves consider to be distracting. The survey targeted 13 different activities, including: dialling and talking on a cell phone, text messaging, eating or drinking, changing a cassette or CD, tuning the radio, adjusting the climate controls, reading, looking for an item in one's purse/wallet/

backpack, using an *iPod* or laptop, doing homework, daydreaming, and thinking about something difficult. One highly unique aspect of this study was that based on responses, every participant was classified in one of three driver profiles regarding their level of their engagement with secondary tasks:

1. *Infrequent engagers* (n = 790, 49%) often tune the radio and adjust climate controls while driving, but rarely text message, dial, or talk on the phone while driving. These drivers carry the highest proportion of adults as passengers (37%), as well as have the lowest percentage of crashes (27%).
2. *Moderate engagers* (n = 491, 31%) perform more distracting activities including increased cell phone use, daydreaming, eating and drinking, tuning the radio, and adjusting the climate controls.
3. *Frequent engagers* (n = 322, 20%) habitually perform a host of distracting activities. These drivers regularly engage in dialling a cell phone, texting messages, changing CDs or cassettes, using a mobile device (such as an *iPod* or laptop), looking for items in their wallet/purse/backpack, as well as contemplate the complexities of their problematic teenage life and relationships. They carry the highest proportion of friends as passengers (over 80%), engage in distracting activities while driving in nighttime traffic, and have the highest percentage of crashes (43%).

These characteristic features outlined among the last above noted profile have been previously described by Williams (2003) albeit in an unrelated study. Westlake and Boyle also found that although driving with some activities was considered moderately distracting, most of the participants still chose to engage in these often – as was the case for text messaging (86%), dialling a phone (72%), and talking on the phone (64%). Even activities considered to be highly distracting were still performed while driving, such as doing homework (80%), looking for an item in wallet/purse/backpack (77%), and reading (77%). Specifically regarding music-related driving behaviours, Westlake and Boyle found that 61% of the sample used audio devices brought into the car; 56% repeatedly switched between CDs and tapes, and 42% frequently tuned the radio. While these figures are far below the incidences presented in many other studies (as previously described in Chapter 3), the investigation offers a vantage targeting characteristic behaviours per driver profile, which itself is a distinctive contribution never before seen in the literature (see Figure 4.1). As can be seen in the figure, the study put in place an intensity-based response scale (1–7). Frequent engagers most regularly coupled *iPod*s and other devices to the in-car systems (Panel-A), most consistently switched cassette tapes (between sides A/B and between tapes) or swapped between several CDs (Panel-B), and endlessly sought-out radio channels of music broadcasts (Panel-C). Finally, based on a model projecting the associated probability of causing an accident, Westlake and Boyle found that drivers who were frequent engagers in distracting activities were projected to be 1.45 (i.e., one-and-a-half or 50%) times more likely to be involved in an accident than either drivers who were moderate or infrequent engagers.

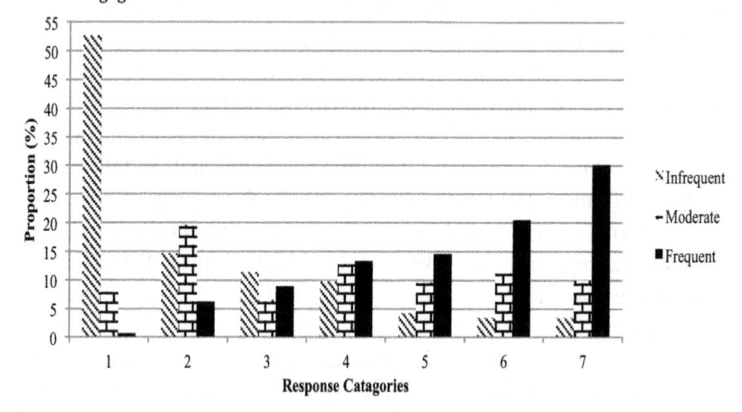

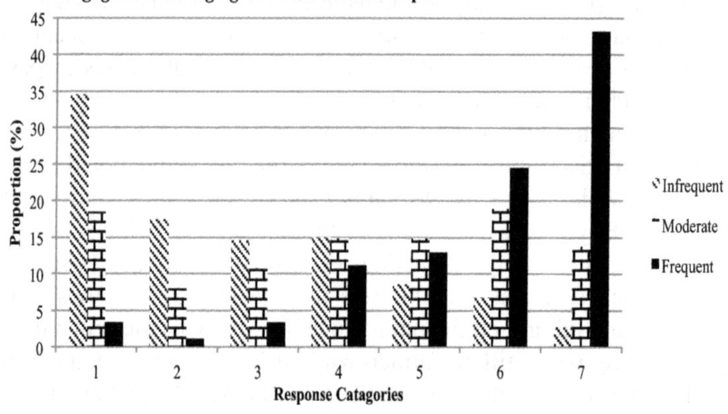

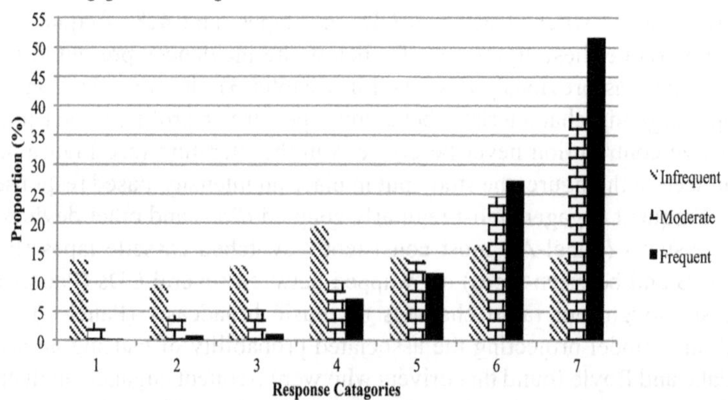

Figure 4.1 Proportion of drivers engaging in music listening devices
Source: Adapted from Westlake & Boyle (2012).

Undeniably, some of the activities we engage in during driving are internal diversions inside the cabin (caused by directing attention to objects, devices, events, and passengers). However, there are also deviations of attention from external incidents (such as billboards, street signs, sirens, pedestrians, and landscapes) that cause a highway visual clutter – all of which are involuntary and command the driver's attention (Horberry et al., 2006; Horberry & Edquist, 2009; Regan, Young, & Lee, 2009). Hence, besides driver-initiated engagement in various actions (in the colloquial: 'I WAS DISTRACTED'), clearly *non-driver initiated* distraction also occurs when a random event involving something or someone else simply happens ('I WAS LOOKING BUT JUST DIDN'T SEE IT'). Finally, there are many cases in which we may simply be 'absent-minded' because of being preoccupied with thoughts, daydreams, memories, and connotations that enter our mind ('I HAD MY MIND ELSEWHERE').

Among all previously published investigations and analyses of driver behaviour, RoSPA (2007) claims that the largest and most comprehensive examination is the American *100-Car Naturalistic Study*. This 2000–2001 study produced a public access database containing many cases of driving behaviour and performance including severe fatigue, impairment, judgement error, risk taking, willingness to engage, aggressive driving, and traffic violations. For many years (and even today a decade later) this database has served researchers worldwide; for example, see Klauer, Sudweeks, Hickman, and Neale (2006b). Neale (2011; Neale, Dingus, Klauer, Sudweeks, & Goodman, 2005) provide a clear outline of the *100-Car Naturalistic Study*: 241 drivers participated in 42,300 hours of on-the-road driving across 13 months covering over two million miles. The data was collected for 100 vehicles (78 driver-owned cars and 22 leased cars), employing an instrumentation system including a five-channel video feed (four strategic vantages of the automobile, two inside and two outside views, and the driver's face), as well as from vehicle kinematics (i.e., sensors that interact with the vehicle network), and GPS positioning. Accordingly, there were over 9,000 events reported during the investigation: 82 *crashes* (i.e., contacts between the subject vehicle and other vehicles, objects, or persons), 761 *near-crashes* (i.e., conflict situations requiring a rapid severe evasive manoeuvre to avoid a crash), and 8,295 *incidents* (i.e., conflicts requiring an evasive manoeuvre of a lesser magnitude than a near-crash). The data provided evidence that some form of inattention, distraction, or non-specific eye glance away from the forward road caused 64 (78%) crashes and 495 (65%) near-crashes. Moreover, the number of events induced by a secondary source, were: 10 crashes (12%), 116 near-crashes (15%), and 1,308 incidents (16%). The secondary sources attributed to these incidents (in order of frequency), were: wireless devices (such as a cell phone, Palm-Pilot, hand-held PDA computer), passengers, internal distractions, the vehicle itself, engaging in personal hygiene, dining (eating and drinking), external distractions, talking or singing (without a passenger being present), smoking, and daydreaming. In a reanalyses of the data, Hanowski (2006; Hanowski et al., 2006) targeted 'at-fault' incidents. The researchers found that the most frequent contributing factors were:

driving techniques (70%), distraction (22.5%), and aggressive driving (22.5%). The reasons for distraction as a contributing factor were talking or listening on a cell phone (22%), the adjacent passenger (13%), dialling a hand-held cell phone (9%), looking at the centre mirror (7%), looking out left window (7%), an external distraction (4%), adjusting the radio (4%), cognitive distraction (4%), grooming (4%), lost in thought (4%), smoking (4%), talking/singing/dancing (4%), eating and drinking (2%), lighting a cigarette (2%), operating a PDA hand-held device (2%), reaching for an object (2%), reading (2%), and looking out the right window (2%). It should be pointed out that Hanowski et al. are perhaps the first (and only) research team to differentiate between 'adjusting the radio' versus 'listening, singing, or dancing' to music during at-fault accidents. Moreover, such a stance was certainly without any a priori conceptual reasoning as they offer no explanation nor mention the implications of such analyses in their discussion. Further, they point to two other important findings of the *100-Car Naturalistic Study*. First, 'fatigue' was identified as the cause of 20% crashes and 12% near-crashes. Second, the rate of inattention-related events was seen to be four-times higher among drivers aged 18 to 20 year old than for more experienced older drivers.

The *100-Car Naturalistic Study* paved the way for a host of investigations in the 2000s. Among the studies that were modelled on it is an investigation by Goodwin and colleagues in 2010; that study was subsequently reanalysed and then published by the *AAA* Foundation for Safety two years later (Goodwin, Foss, Harrel, & O'Brian, 2012). In the study 50 families residing in North Carolina (USA) were recruited. Fifty-two drivers partook as active participants; 38 participants were newly licensed teens, and 14 participants were their high-school aged siblings. In every car a data recorder was installed to solicit mechanical parameters of the vehicle, and both video and audio data from the cabin were sampled. The study sought to depict teen driving behaviour in a naturalistic environment, including detailed information about their use of electronic devices, behaviours leading to driver distraction, and specifically to document circumstances that occur when other teen passengers are present (described in Section 4–2.4 below). The study covered 228 months (6 months x 38 vehicles), in which a total 24,085 driving clips were recorded for all 52 teens. There were 19,384 clips of the target teens and 4,701 clips of their sibling drivers; on average there were 463 clips per teen driver. In the final analyses, Goodwin et al. selected a sample of 7,858 clips covering all 52 drivers. The clips indicate that among 27 drivers (52%) there were no serious incidents, but among the other 25 drivers there were 52 clips that document a serious incident: 3 collisions, 30 near-crashes, and 19 events in which a driver lost control or weaved off the roadway (albeit, returning to the lane without further incident). Of these 52 serious incidents, 26 (50%) involved high g-forces of extreme acceleration, deceleration, left, or right turns. Elevated g-force events have been associated with reduced time intervals by which drivers respond to dangers, increased crash risk, and a higher probability of losing control over the vehicle (Simons-Morton et al., 2011). In general, Goodwin et al. found 86 clips with events triggered by acceleration and 166 clips by deceleration (both forms of longitudinal g-forces), as well as 214 events triggered by left turns and 226 clips by right turns (both forms of lateral g-forces). The study also points out that electronic music devices were used in 523 (6.7%) of driving clips – a figure indicating that nearly twice as many teen drivers operated electronic music devices than cell phones. Other irrelevant behaviours found in the driving clips included: adjusting controls (471 clips, 6.2%), activities of personal hygiene and grooming (287 clips, 3.8%), eating and drinking (211 clips, 2.8%), reaching for an object (191 clips, 2.5%), communicating with someone outside the vehicle (113 clips, 1.5%), turning around to view something in the back seat (71 clips, 0.9%), and reading (8 clips, 0.1%). It should be pointed out that although drivers were three times more likely to look away from the roadway when using an electronic music device, these were only weakly related to high g-forces or to serious incidents. Altogether, Goodwin et al. found that drivers engaged in at-least one distracting behaviour in 1,186 (15%) of in-cabin video clips.

Beyond naturalistic driving studies, there are also investigations that present data from accident crash reports of police officers (for example, Peek-Asa et al., 2010; Stevens & Minton, 2001; Stutts, 2001; Stutts et al., 2001a, 2001b). This method illuminates the incident as witnessed by a policeman arriving on the scene shortly after the event. Yet, for the most part, these records employ data solely from fatal crashes, and therefore provide very limited information about driver distraction – simply because it may be highly difficult for an officer to determine any specific behaviours that occurred prior to fatal crashes (Westlake & Boyle, 2012). However, there are also a few studies that have explored non-fatal crashes among young drivers; these are especially illuminating because they target traffic-related mishaps concerning distraction and inattention, and have been documented as occurring ten times more among young drivers than among adult drivers. Of this category, two studies that supplemented law enforcement reports with driver narratives (from either the accident reports themselves or from post-accident interviews) are highlighted below. The first is an investigation by McKnight and McKnight (2003) who examined 2,128 American police report narratives of teen drivers (60% male) in California (57%) and Maryland (43%). The analyses led McKnight and McKnight to identify 11 categories of accident-related driver deficiencies. The eleven deficient driver behaviours, in descending order of associated cause to accidents, are: attention (23%), adjusting speed (21%), searching ahead (19%), searching to side (14%), maintaining space (10%), searching to rear (9%), handling emergencies (9%), basic control (8%), driver-vehicle (6%), traffic controls (6%), and signals (1%). McKnight and McKnight found that new drivers had a significantly greater proportion of accidents than drivers with just one to two years more experience, and these accidents resulted from lack of attention, reduced visual search prior to left turns, not watching the car ahead, driving too fast for conditions, and failure to adjust to wet roads.[2]

2 This measure was adopted by Brodsky and Slor (2013) for their on-road high-dose double-exposure within-subjects clinical trial (described in Chapter 6).

Having said that, young(er) drivers in general were found to be highly prone to distraction and less efficient in processing visual information – especially when they were involved in a secondary task. Moreover, the study found that male drivers were more likely to lose control of the vehicle and crash when speeding, while female drivers were more likely to violate the right-of-way as they mostly failed to see traffic controls or other vehicles (inadequate search before left turns and at intersections). McKnight and McKnight conclude that the overriding number of non-fatal accidents among young novice drivers seems to be from a deficient level of operating the vehicle in a safe manner, as well as from a failure to appreciate hazards that might result from thrill-seeking and risk-taking. In the second study, Braitman et al. (2008) interviewed 260 young novice drivers (51% male) who were involved in non-fatal crashes (that for the most part did not involve injuries) occurring in Connecticut (USA) between 2005–2006. The interviews took place roughly one-to-four months after each incident. The incidences included multiple-vehicle (69%) and single-vehicle (31%) crashes, in which 198 drivers (76%) were considered to be 'at-fault'. At-fault crashes were classified as: ran-off-road (39%), rear end (31%), and violated right-of-way (20%) crashes. Among these incidents, 42 (33%) involved speeding violations, while 35 (83%) concluded with losing control of the vehicle. The five most common reasons for non-fatal crashes were: search and detection (39%), speeding (38%), loosing control (38%), road conditions (30%), misjudged evaluation (19%), and course (10%). The study highlights that the most common distractions causing search and detection violations leading to crashes, were: adjusting the radio and CD player (25%), interacting with friends or pets (12%), talking on cell phones (4%), and other manual diversions (33%). In general, Braitman et al. found that distraction and inattention triggered 60% of all right-of-way crashes and 55% of all rear-end crashes, while speeding was linked to 77% of all ran-off-road crashes.

In conclusion, inattention and distraction is a serious issue that is accountable for a high incidence of near-crashes, and crashes – some of which are fatal. The contribution of secondary tasks to inattention and distraction is well documented, and among these the engagement with music-related devices is noted. But yet, in-car music behaviour, especially among young(er) drivers, has not been taken seriously.

4–1.1. Problem of the Current Research Literature with Music

The literature of traffic psychology targeting driver behaviour has scarcely investigated *music* as a source of inattention or distraction. Foremost, there is great confusion regarding what is *music*, especially as the difference between 'music' and 'auditory' stimuli is not very clear. Hence, for the most part, researchers have employed most any aural stimuli – as if anything 'heard' is the same. Although some studies have claimed not to find effects (i.e., null-effects) for auditory stimuli (using exemplars that were solely speech and/or language content, or exemplars that simulate radio broadcast), others then cite the same findings *as if* relevant and transferable to stimuli based on tonal and rhythmic content (i.e., music). It is no wonder, then, that researchers have typically inferred that 'music causes little, if any, effects'. In one of the earliest studies, Brown, Tickner, and Simmonds (1969) stated "Steering skills are probably so automatized among trained drivers that they are minimally degraded when attention has to be diverted intermittently to an auditory stimulus" (p. 423); the robust findings were so prevailing that few if any studies after Brown et al. explored *music* as a first-level factor. That is, after Brown et al. traffic research studies in the main have employed music as a control task for comparative analyses *between* groups, whereby different groups of drivers perform a specific task under unique environmental conditions (one of which is auditory). In addition, few traffic studies employed music as a control task for comparative analyses *within* the group, whereby the same drivers perform a host of consecutive tasks under unique environmental conditions (one of which is auditory). Nonetheless, it is unfortunate that even traffic researchers employing *music* in their investigations have had little knowledge about musical structures (i.e., the actual complex of sound, rhythm, harmony) and further exhibit a total disregard for the level of rigor necessary to incorporate music stimuli within their empirical framework. For example, in the majority of music-related investigations the selection of exemplars are contaminated (that is, the songs or pieces are chosen inappropriately and are not suitable for experimentation), and the conditions of exposure are flawed (that is, participants in driving simulators are instructed to listen to music via sterile laboratory headphones or PC satellite speakers rather than from floorboard mounted speakers that are ecologically valid *as if* in a car cabin[3]). Moreover, if some form of hypotheses about *in-car music listening* does appear in the published report, these are generally based on intuition without any scientific grounding. Finally, the majority of studies in the literature do not even recount details about the music stimuli selected – as if to convey that music is music and anything heard is ultimately similar. Clearly, research reports should include particulars such as name of the music selections (and perhaps even the names of the artists, description of the genre, level of complexity, and average tempo of the item set); further they should describe the hardware used to reproduce music within the empirical setting (for example, details about audio player type, brand name, model number, type and number of speakers used, estimated distance between sound source and the subject, decibel level of volume intensity, etc.). As a result of this lower than acceptable scientific thoroughness, it is no wonder that for the most part traffic studies report to have found *null effects* – or claim that the effects of music are statistically insignificant. Even the best of researchers in the traffic domain have come to ill-fated conclusions. For example, in their exceptional 2008 book titled *The Multisensory Driver*, prominent researchers Ho and Spence collate the sparse literature about the effects of music on driver behaviour presented in a chapter titled 'Driven to Listen'. From the onset they state that one of the

3 Two popular models for comparison: PC Speakers: *Creative A35 2.0*, 2" drivers @2w, wide-range 200–20000Hz; Car Stereo Speakers: *Pioneer TS-A1773R*, 6.5" drivers, @35w, 3-way full-range 30–28000Hz.

most common non-essential things people do when driving is to listen to the radio, and "the most obvious prediction would be that listening to radio should impair performance under at least certain conditions" (p. 35). They themselves conclude the chapter with the statement:

> It is not uncommon for researchers to assume low distraction caused by radio primarily because so few of the available accident records suggest the use of the radio as a direct cause of car accidents. This may render a false impression when researchers attempt to compare the degree of distraction caused by mobile phone use to that of a radio use (p. 40).

Thus, Ho and Spence themselves raise the problem of ecological validity and the issue of a *False Positive Type II Error* that runs freely throughout the traffic literature (concerning the effects of radio and/or music listening). However, while stressing the lack of hard evidence, and having stated that only limited efforts have been made to understand whether or not listening to the radio affects driver performance, the authors nonetheless perpetuate the error by maintaining that "the majority of studies appear to show limited interference from listening to the radio on driving performance" (p. 40). One can only wonder how many studies represent 'the majority' and on what basis was 'limited interference' measured. Ironically, just one year later, and based on the same corpus of scientific literature, Spence and Ho (2009) bring forth a more definitive deduction: "Listening to an in-car entertainment system is typically less taxing (and hence less dangerous) than engaging in a conversation with a passenger or with someone on the mobile phone" (p. 192). Such an axiom is of course not only misinformed but largely cited in further published works.

In reviewing their own research, Engstrom, Johansson, and Ostuland (2005) reflect on the methodological procedures put in place when they investigated visual versus cognitive load in both real and simulated driving. The key issue they raise in their discussion is the 'artificial' nature of the tasks they employed referred to as *surrogate* in-vehicle information system (S-IVIS). Outright, they claim "experimental design is always a trade-off between experimental control and ecological validity" (p. 115), while pointing out that in the interest of systematic manipulation, research often pays a price of reduced ecologic validity. Having said that, when the former overshadows the latter, then one has to ask two questions: What exactly is under investigation? And on what basis are conclusions about *effects* on driving performance being made? To illustrate the problematic nature with the current state of the scientific traffic literature concerning music, the chapter brings forward four examples of studies that are highly cited by others.

1. Nelson and Nilsson (1990) compared headphones and speaker effects on simulated driving. In the mid-1990s driving with headphones was becoming highly fashionable as drivers began bringing mobile music devices into the cabin for more pleasurable driving. In a laboratory study consisting of two 3-hour driving sessions, each participant underwent a number of tasks (including steering manoeuvres in highway traffic, 13 speed changes, 6 gear shifts, and 5 random hazards) while listening to either Classical music or Modern music. In one session music was presented via fixed-field exposure

employing Sennheiser headphones, while in the other session music was presented via free-field exposure employing a compact 3-way speaker placed under the dashboard; a Sansui stereo cassette system with Dolby noise reduction was used to reproduce the music. The results indicate that headphone listening caused increased reaction times when adjusting speed, shifting gears, and to hazards, than when listening to music via speakers – but only differences of shifting gears were found to be statistically significant. The authors deduce that wearing headphones "produced a detriment in performance of a complex driving task that is dependent on auditory cues… [Hence, it is] likely that stereo headphones sometimes hamper motorists' ability to respond to highway incidents" (p. 528). Further, they conclude that lack of significant effects on steering accuracy, accelerator control, and speed changes may be because these tasks are visual by nature – as opposed to gear changes that are based on auditory cues. Therefore, "headphones reduce primarily auditory attention [that] may be critical in traffic" (p. 528). In short, the researchers found effects of *masking*. Given such conclusions, one could only wonder how much of the effect (or lack of effect) is actually attributable to the *music* that was presented – especially as listening to music with headphones may reduce attention to important auditory cues necessary for driving. But no attempt to control the music as a variable ever entered into the analyses. It is difficult to fully appreciate the findings as no details of the music selections were ever provided. It should be pointed out that 'Classical' music is a label used for Western art music spanning 18 centuries of music history ranging from Gregorian Chant through to Stravinsky. On the other hand, 'Modern music' is a label that could have several meanings, and could indicate any music such as twentieth-century Serialism (based on 12-pitched tones), Electroacoustic compositional practices, or perhaps simply mean any music that ranges from Blues through Pop to Rock or Heavy Metal styles. Modern music might be one of the 34 styles presented in Table 2.1. If as Nelsen and Nilsson state, there were equal number of participants who chose each music style, and if we substitute two highly popular tunes as representative exemplars for the two genres suggested (for the sake of the argument 'The Four Seasons' composed by Antonio Vivaldi in 1723 performed by the Manchester Halle Orchestra versus 'Brown Sugar' composed by Mick Jagger and Keith Richards in 1971 performed by the Rolling Stones), and reiterate that both were played at similar volume (i.e., 63dBls), then perhaps we might infer that the effects reported might have surfaced because of an interaction between the music (Classical versus Modern) and exposure conditions (headphones versus speakers). Namely, that drivers wearing headphones were more distracted when listening to the Rolling Stones than Vivaldi but not when listening via speakers. As only one variable (i.e., gear shifting) was sensitive to the exposure method (i.e., fixed-field versus free-field), one might assume that this finding was no more than statistical noise (pun intended).

2. Consiglio, Driscol, Witte, and Berg (2003) studied the effects of phone conversations and 'other potential interference factors' on reaction times (RTs) in a braking response. Their study somewhat replicated a previous experiment by Strayer and Johnston (2001, Experiment #1). In the earlier experiment, Strayer and Johnston recruited 45 undergraduates between the ages of 18 and 30 from the University of Utah (USA) to perform a simulated driving task. The students were randomly assigned to three groups in an effort to compare between hand-held versus hands-free cell phone conversation, along with a control group that listened to the radio. It should be pointed out that the students chose the radio programming, but no details are provided as far as 'content' (i.e., percentage of talk versus music), 'broadcast genre' (that is, sport, news, easy listening, classical music, pop chart radio), or 'exposure format' (headphones, speakers, intensity volume). In the experiment the students performed a pursuit-tracking task using a joystick to manoeuvre a curser on the computer display as close as possible to a moving target. There were two phases in each condition: a single-task (pursuit tracking alone) and a dual-task (pursuit tracking with either conversation or radio listening). Occasionally, red and green lights were flashed on the screen requiring the drivers to brake on the red lights by pressing a 'brake button' located in the thumb position of the hand-manipulated controller. The experiment found that: (1) when engaged in conversation braking reaction times (RTs) were one-and-a-half times slower, and drivers missed twice as many traffic lights; (2) deficits are similar for both hand-held and hands-free phones, which is an indication that interference is caused by averting attention to the cognitive aspects of conversation and not by holding a mobile cell phone device; and (3) there is no indication of a dual-task decrement for radio listening.

   Then, in the study by Consiglio et al. (2003) undergraduates in Miami, Florida (USA), were recruited for an experiment set up to emulate driving an automatic-transmission vehicle with an accelerator and brake pedal; the students sat in front of a laboratory station computer monitor with two foot-pedals on the floor as triggers. Using only their right foot on the right pedal (accelerator), the participants were instructed to depress the left pedal (brake) when a red lamp (simulating a brake light of the car ahead) was activated. There were five conditions: No manipulation (as a control); Radio (as a control); Conversation-only (with a mock passenger); Conversation via hand-held mobile phone (with a mock caller); and Conversation via a hands-free mobile phone (with a mock caller). The authors expected the highest increase for RTs in the conversation-only condition, no differences of RTs between the phone types, and little to no effect for radio. The procedure mandated five blocks of five trials (25 items in total), which were presented in a counter-balanced fashion across the sample. In the Radio condition a different contemporary song was played at a moderate volume for each of the five trials. Accordingly, the results indicated that RTs were significantly longer for all conditions involving conversation than for either No-manipulation or Radio control conditions; but no significant differences surfaced between the three conversation conditions. Finally, as expected, there was no effect of the radio on braking RTs. In fact, RTs for radio exposure were similar to those of the No-manipulation control condition. The authors concluded: "being in the presence of an audible signal such as music does not necessitate the allocation of attention to the extent that engaging in a conversation does" (p. 498). This is the axiom! But certainly those who cite this finding do not also indicate the authors' own cautionary warning: "Because of the simulated nature of the task, the results of the present study are only indirectly related to driving" (p. 497). Nor are other researchers mindful to cite Consiglio et al.'s final deduction: "It is recognized that the attentional demand placed on a driver could differ markedly depending on the type of radio broadcast listened to (music versus talk) or the type of radio operation being performed (e.g. listening versus tuning)" (p. 498). That is, they themselves acknowledge that perhaps the findings they report might need to be reconsidered. For example, undeniably their simulated radio condition did not in the least emulate a form of 'broadcast', but rather must be seen only as a form of 'listening to music' – and in this latter case we would expect a full description of *what* music selections were heard by the participants. In fact Consiglio et al. do report exposing the participants to five (unidentified) exemplars designated only as 'contemporary' pieces. One can also assume that these selections were personally chosen by someone without formal music training who designated them as 'contemporary' music genres based on intuitive and implicit musicological knowledge of musical styles. Clearly, it should be pointed out that developing music for use as stimuli in empirical investigations is not to be taken lightly. Hence, any serious reader would simply be confused as to what exactly did the participants hear? In addition, as no details are offered as to *how* the music was heard, we can only assume that the participants were exposed to music through speakers rather than headphones. Further still, there are no details 'how' the music exposure was controlled at a 'moderate' reproduction volume level. We might again presume that a comfortable volume was chosen by an experimenter based on their own personal aural sensitivities. Finally, the study does not in the least account for the fact that in real-life driving with background music, drivers do not sit in a catatonic statuesque-like fashion (as was the case in Consiglio et al.'s table + chair lab-station environment), but rather that most drivers participate in the music performance by singing, tapping along, playing air guitar riff solos, and even dancing in their seat. Hence, as a condition that might shed some insight on brake response times when driving with music, Consiglio et al.'s study is far from providing any convincing evidence, and therefore, the results should not be seen as a reliable empirical demonstration leading to their ultimate conclusion: "engaging in a paced conversation generates significantly more capacity interference than does listening to music" (p. 498).

3. Ho, Reed, and Spence (2007) who investigated multisensory in-car warning signals for collision avoidance, recruited 15 males between the ages of 17 and 41 with a valid UK driver's licence, to participate in a 60-minute simulated driving experiment. A car-following procedure was implemented in a fixed-base Honda *Civic* family hatchback (5-speed manual gearbox) with simulated engine, road, and traffic sounds heard via a stereo system (albeit, it is not clear if the system was the car's on-board in-cabin stereo or an external unit). The trips were on two-way roadways that passed though rural neighbourhoods at an average speed of 80 kph, while drivers experienced 42 randomly placed events that concurred with one of three warning signal types: *Vibratory* (vibrotactile), *Auditory* (car-horn sound), and *Audiotactile* (vibration + car-horn sound). Vibrotactile warning signals were presented through a vibration generator (i.e., vibrator) placed on a stomach belt worn over the driver's clothing, while the Auditory warning signals were presented though a speaker mounted above the dashboard instrument panel. In addition, Ho et al. assessed the effectiveness of these three warning types in the presence/absence of background radio. In total, there were three blocks of 20 minutes driving per warning signal condition, and these were further subdivided by two 10-minute segments with or without background radio. It must be pointed out that the rationale behind 'enhancing' (or perhaps we must unfortunately view a more appropriate 'contaminating') the study with background radio was two-fold: (1) "to better understand whether concurrent auditory stimulation may distract a driver and hence possibly reduce any beneficial effect of in-car warning systems" (p. 1,108); and (2) "to provide a somewhat more ecologically valid driving setting, given that the majority of drivers report listening to the car stereo while driving" (p. 1,108). Background radio was heard from two loudspeakers positioned on the rear passenger seat; drivers heard a 20-minute clip of the *Chris Moyles Show* (an original broadcast of UK's 'BBC Radio 1') consisting of three songs, informal conversation among various guests, and news/sports reports. In general, the study found that responses were significantly decreased when Audiotactile signals were presented (in comparison to either single Auditory or Vibrotactile warnings). Further, Ho et al. report no effect of listening to background radio – as listening to radio did not facilitate nor hamper reactions times. Regarding the latter finding, they conclude: "The effects of music on driving-related performance have been studied by many researchers over the past few years… however, the presentation of the radio programme in the background in the present study had no effect on the speed or accuracy of the participants' braking in response to the sudden deceleration of the lead vehicle" (p. 1,112). As this conclusion reflects a critical point than can have repercussions on future investigators, then perhaps it is wise to take a more in-depth look at

the operational methods that led the researchers to such a finding: The study employed warning sounds @85dBA presented from a frontal dashboard position; engine, road, and traffic sounds @75dBA were presented from side door panels; and background radio broadcast @65–78dBA were presented from a Boombox in the back seat. We might question, then, if there is any possibility that the reported effects (or lack of effects) are no more than artefacts of *perceptual masking* or *interference of spatially conflicting sound locations*. (For more information on the latter concept, see section 4–3.2 below.) Moreover, Ho et al. seem to be alluding to their use of 'radio' *as if* equivalent to 'music', and hence present their findings in a fashion that readers might see these results as a viable contribution to understanding the interaction between driver performance and listening to background music. Clearly, no information was offered as how many or what kind of songs were played within the *Chris Moyles Show*. Simply for the purposes of replication it would have been important to access details such as song titles, name of performers, or simply musicological music genres. However, when attempting to account for the procedures employed, the reader is faced with other dilemmas: given the duration length of an average pop song (three minutes), and with details presented stating that there were three songs in every 10-minute radio programme, then either the radio broadcast consisted of three 30-second links in between the songs or that the background radio programme was actually talk radio with a few intermittent musical interludes (of an average 60-seconds each). It is vital to know how much (the percentage) of each ten minute driving segment was actually covered by music stimuli. Hence, one can only wonder what do these findings indicate about music effects on multisensory in-car warning signals.

4. Oron-Gilad, Ronen, and Shinar (2008) evaluated the effectiveness of alertness maintaining tasks (AMTs) on driver performance among professional truck drivers. The drivers participated in five simulated driving sessions, whereby each session contained two road sections (labelled '1' and '2') that appear repeatedly in five segments (1a, 2a, 1b, 2b, 1c). In Session #1 a 'driving only' condition for empirical control and baseline was implemented. In Sessions #2–4 participants drove with one of three AMTs presented in segment 1c (in a counterbalanced fashion): a vigilance choice reaction time task to visual cues; a working memory test resembling the 'Simon' game; and a long-term memory 'Trivia' verbal task. In Session #5 participants drove while listening to driver-selected music cassette tapes brought from home (also seen as a control condition). The results indicated that: (1) AMTs were not equally effective in their function to maintain driver alertness (i.e., speed was significantly lower in the working memory task than the other two AMTs); (2) that drivers maintained a better vehicular performance when engaged in AMTs (i.e., steering wheel control was worse in the no-AMT control condition than all three AMTs); and (3) there was no difference between driving with music and driving-alone without

music (see Figure 4.2). As can be seen in the figure, the analyses present comparisons between segments 1b (No-AMT) versus 1c (AMT). Yet, unlike all three AMT conditions, comparing between segments 1b versus 1c for either No-AMT condition or for the music condition is neither effective nor valid because the empirical procedure mandated music exposure throughout all five segments (1a–2a–1b–2b–1c) of the procedure. Hence, why should there be differences between identical segments 1b versus 1c to begin with? In fact, the authors reported this exact occurrence: "[there was a] lack of significant changes in any of the measures while listening to music" (p. 858). Yet, when faced with *null effects* concerning the music condition, Oron-Gilad et al. deduce that "the lack of significant changes in any of the measures while listening to music indicates that driving with music was a good method for maintaining alertness, or at least better than driving without music" (p. 858). Such an interpretation about music and driving seems to be premature as no analyses indicating significant differences between driving with music versus driving without music were ever undertaken – or at least none were presented in the research report. However, it is not exactly *what was said* about music that misinforms future researchers, but ironically *what wasn't said.* Unbeknownst to the authors, a truly important result about driving with music surfaced in their study. Perhaps, such an oversight stems from the fact that traffic researchers themselves are unmindful to the characteristics of music effects on driver behaviour. Specifically, the data in Figure 4.2 clearly shows that the average acceleration speed of drivers in segment 1c is higher (i.e., faster speed) when participants listened to music than when they drove in all other conditions. This fact that driver-preferred music brought from home significantly increases driving speed has only recently been documented by Brodsky and Slor (2013). Nevertheless, inklings of this driver behaviour appear four years earlier in Oron-Gilad et al.'s study. If only details about the music selections drivers brought to the study, and the methods employed for music reproduction were reported, more could have been learned from such a find.

Driving performance measures for the (1b) and (1c) road segments in each driving conditions in Experiment 1

Measure/condition		No-AMT	CRT	Memory	Trivia	Music
Average lane position [cm]	(1b)	139 (8)	140 (15)	144 (8)	146 (12)	140 (14)
	(1c)	140 (13)	140 (7)	139 (8)	144 (14)	151 (13)
S.D. lane position [cm]	(1b)	24 (5)	24 (4)	24 (4)	27 (7)	27 (5)
	(1c)	25 (7)	26 (6)	24 (8)	24 (5)	28 (4)
S.D. steering wheel rate [deg/s]	(1b)	5.22 (4.70)[a]	2.93 (1.03)	3.35 (1.62)	3.72 (255)	3.48 (1.20)
	(1c)	8.48 (8.42)	4.64 (5.16)	4.52 (3.06)	3.77 (2.48)	3.86 (1.81)
Average speed [km/h]	(1b)	70 (13)	72 (14)	76 (14)[a]	77 (9)	78 (8)
	(1c)	66 (15)	72 (16)	70 (15)	75 (11)	80 (8)

[a] Duncan post-hoc test significant ($p < .05$) for the difference between (1b) and (1c).

Figure 4.2 Alertness maintaining tasks (AMTs) while driving
Source: Oron-Gilad, Ronen, and Shinar (2008). Used with permission of Elsevier.

Given what is known about driver inattention and distraction, and in an effort to clarify more precisely what are the specific demonstrated ill-effects of music on driver behaviour (presented in Chapter 5), four wide-ranging *contraindications* for listening to music while operating an automobile are outlined below. These are: structural mechanical interference, perceptual masking, capacity interference to central attention, and social diversion in the presence of passengers.

4–2. Contraindications to In-Car Music Listening

The *100-Car Naturalistic Study* described above illustrates that everyday driving is not a simple task. In fact, the whole body of transportation research literature highlights the fact that driving is a highly complex skill requiring the sustained monitoring of combined perceptual and cognitive inputs, from a host of competing sensory informants via gauges and devices that are both on-board and peripherally linked, that limit and divide our attention. In a personal reflection, one highly esteemed traffic researcher captured the dynamics of the matter most elegantly:

> Today there is an abundance of new devices that we can look at or interact with while driving. Some of these (and this number is growing) are embedded in the vehicle while many others we bring into the vehicle ourselves. Although these devices are a great resource, offering information, connectivity, and entertainment, we must handle them appropriately least we fail in our driving duties. Vision is a limited resource and, since we have difficulty looking at two places at the same time, we must allocate our attention in a manner that ensures a successful interaction. I learned first-hand how these in-vehicle devices might act as 'attention sinks' when I bought my first satellite radio. While I enjoy countless hours of entertainment, I was amazed at how compelling it was to look down at the display for prolonged periods of time, waiting for the artist or album name to finish its crawl across the small screen (Horrey, 2009, p. 151).

It certainly is difficult to influence how people drive in a safer manner, especially as each trip further reinforces the image of a conventional 'safe trip' (Vanderbilt, 2008). As most of us get through a lifetime of driving without ever being involved in a fatal car crash, it then seems of no consequence that as we drive we also spend much time talking on the phone and texting, adjusting the radio, switching between CDs, searching through playlists, and singing happily along with the music as we shimmy in our seats. By all accounts, drivers (especially young novice drivers) may miss a critical traffic signal (stop sign, set of lights), fail to detect the sudden halt of a car ahead, or apply a braking response in a slower fashion because of structural interference from in-vehicle technologies, competing demands that cause perceptual masking and limited cognitive resources, or even because of involvement with passengers. All of these can cause a collision with the vehicle in front (Drews, Pasupathi, & Strayer, 2008;

Hancock et al., 2003; Ho & Spence, 2008; Peek-Asa et al., 2010; Spence & Reed, 2003; Strayer & Drews, 2004). A full description of these contraindications to in-car music listening follows.

4–2.1. Structural Mechanical Interference

As cars elicit a range of feelings from the pleasure of driving to the thrill of speed (Sheller, 2004), and drivers envisage feeling secure and protected by their automobile, the last thing any driver would ever think about is: *How safe is it to turn on the radio, toggle a channel knob, adjust the volume, flip a cassette tape, or swap a CD?* (Power, 2009). In the initial days of automobile use, the legislatures in the State of Massachusetts and in the City of St. Louis proposed regulations to ban listening to the radio while driving in public. Automotive historian Michael Lamm (in DeMain, 2012) claims that back then in the 1930s the opponents of car radios argued that the device would cause accidents. They claimed that when tuning to different frequencies drivers would not be heeding attention to the road; that drivers might often be distracted by the content of the programme; and that music broadcasts could lull drivers to sleep. Accordingly, the Auto Club of New York followed pursuit, and subsequently conducted a 1934 poll finding that 56% of respondents considered the car radio a 'dangerous distraction'. In opposition, of course, was the American Radio Manufacturers Association, who adamantly professed that car radios could be used to warn drivers of stormy weather and bad road conditions – as well as to keep drivers awake when they get drowsy.

Today, over 80 years later and well into the next millennium, we might continue to ponder: Are car radios safe? If digital music systems are indeed installed as features of 'intelligent vehicles' that the auto-industry has referred to as *smart cars* (including: cruise control, voice activated entry and ignition, GPS navigation, front and rear distance-monitoring sensors, automatic braking systems, Bluetooth technology, hands-free mobile phones, telemetric devices for cellular Internet and e-mail, and high-end entertainment systems with on-board HD flat screens), then: *How unsafe could it be to simply listen or sing to music while driving an automobile?* The answer, although seemingly quite intuitive, is one that reflects years of evidence-based research: Structural mechanical interference through physical manipulation of in-car audio equipment causes driver distraction and road accidents. When drivers use their hands to tune the radio receiver, seek an Internet radio station, change a tape/CD, or search for their favourite driving tune, the action can interfere with driving ability and performance – such as instigating a slight difficulty in steering the car or changing gears (Dibben & Williamson, 2007; Titchener, White, & Kaye, 2009). Structural interference results when physical mechanical manipulations are a source of performance decrement, simply because a hand can only be one place at a time, and the eyes can only be focussed on one target source at a time (Consiglio et al., 2003).

4–2.1.1. Eye glances, operational steps, and response times

Fine-tuning the radio controls, swapping sides of cassette tapes, inserting CDs, and scrolling through MP3 playlists, have all been linked to two significantly high-level risk behaviours: eyes directed inward away from the road and one hand on the steering wheel (Horberry et al., 2006; Lee, Roberts, Hoffman, & Angell, 2012; Stutts et al., 2005; Wikman, Nieminen, & Summala, 1998). Originally, car radios were tuned manually with controller knobs (potentiometers) and then later with mechanical push-button pre-set radio stations – both of which obliged the driver to take several glances to find the right station. Only towards the millennium did high-tech digital car radios focus the driver immediately on the exact frequency region – which seemingly cuts down the number of glances required to locate one's favourite stations. Vanderbilt (2008) claims that these operations, whether the driver employs yesteryear's receiver of AM radio transmission or today's satellite terminal for digital audio broadcast, seem to cause drivers to take their eyes off the road for at least 1.5 seconds. The norm for most driving activities is between 0.6–1.6 seconds (Wikman et al., 1998), while glances above two seconds are associated with higher crash risks, minor incidents in real traffic, as well as increased frequency of near-crash situations. Further, the frequency of a very short < 0.4 seconds glance is an indicator of driver uncertainty, which then requires further subsequent glances for the secondary activity (Kujala & Saariluoma, 2011). Moreover, as car audio infotainment technology incorporates central displays loaded with information and features, such as color-coded animations or details outlining the performers names and song titles, glances off the road are considerably longer. Especially when we search for a function option menu that has not been used in a long while (like speaker controllers for left-right balance or front-rear spatial fader), or simply to scroll through playlists searching for a specific tune, our off-the-road glances will be roughly 10% longer in duration. Having said that, most of us are generally aware of the risks in visual distraction, and as a result we do try to keep our in-vehicle off-the-road glance lengths below 1.6 seconds. For the most part, drivers understand that although they themselves rationalize certain occurrences to have been 'random' circumstances (that 'just happened'), everyday driving experiences do confirm that increased visual demands of secondary tasks lead to occasional errors in vehicular control as a result of diminished visual sampling of the traffic environment (Horrey & Wickens, 2006). Hence, in addition to the physical manipulation itself, there is also visual (in)attention associated with in-car music listening that stems from the driver concentrating on the in-cabin entertainment system, and then while temporarily distracted, their eyes are taken away from the road.

The level of structural mechanical interference on driving performance was initially explored by Wikman et al. (1998) who recruited a total of 47 drivers of which 24 were 18–24-year-old novice drivers and 23 were 29–44-year-old experienced drivers. Each participant drove for a 3-hour session covering 126 kilometres (78 miles), at a maximum of 100 kph (62 mph) in an instrumented manual-gear compact car. The tasks Wikman et al. investigated were: (1) Cassette tape – taking a cassette from the side doorbox and putting it in the cassette player, then removing it from the player and replacing it in its box; (2) AM/FM radio – searching for a station with soft relaxing music; and (3) Mobile phone – dialling participants' own number, and dialling a number dictated by a research assistant. The study found a main effect of experience on glance durations. That is, there were significantly more short-normal glance durations (in the range between 0.5–2 seconds) for experienced drivers, while inexperienced drivers demonstrated 46% more >2 seconds and 29% more >3 seconds glance durations. In addition, there were increased lengths of glances when drivers engaged with in-car tasks (that were associated with increased lateral deviation swaying). Wikman et al. conclude that inexperienced young(er) drivers are potentially more at risk because they look away from the road for longer periods of time and hence increase the possibility of running off the road. Most specifically, the study is perhaps the first to demonstrate variances between diverse in-car secondary tasks, and these suggest a divergent level of structural mechanical difficulty via glance durations. For example, glance durations for the radio were on average 1.02 seconds, for the mobile phone 0.96 seconds, and for the cassette tape 0.91 seconds. Regarding variances between the two music devices, it would seem logical that searching for a station is both manually and visually more demanding than inserting (and/or removing) a tape cassette. From a safety point of view, evidence points to the fact that as the driving task becomes more difficult, drivers need to devote more attention to the control of the vehicle by looking more frequently and for longer periods of time at the road. Victor, Harbluk, and Engstrom (2005) feel that when in-vehicle secondary tasks require vision, driver behaviour involving timesharing occurs with the eyes being continuously shifted back and forth between the road and the secondary task. Namely, visual tasks dramatically reduce viewing time of the road centre area. However, it was found that secondary tasks of an auditory nature increase viewing time of the road centre area – but at the expense of glances to the peripheral field. Research has indeed demonstrated strong associations between glance-frequency/glance-duration and lane-keeping.

A second more comprehensive investigation that explored differences between in-vehicle tasks including several devices linked to car-audio was implemented by Angell et al. (2006). The tasks investigated, were: (1) Cassette tape – taking a cassette tape from the case, and inserting it into the tape player on either A/B side as specified; (2) Compact disk – taking a specific colour coded CD from a visor wallet, inserting it to the CD player, and selecting track #7; (3) AM/FM radio – tuning to a specific frequency; (4) Hand-held mobile phone – entering an area code and phone number on hand-held flip phone; (5) Audio book – listening to a 2-minute story; (6) Conversation – listening to and answering questions about the driver's biographical background; and (7) Driving alone without undertaking any task (for empirical baseline control). Angell et al. found that driver manipulation of devices was much more detrimental than aural perceptual processes (see Figure 4.3).

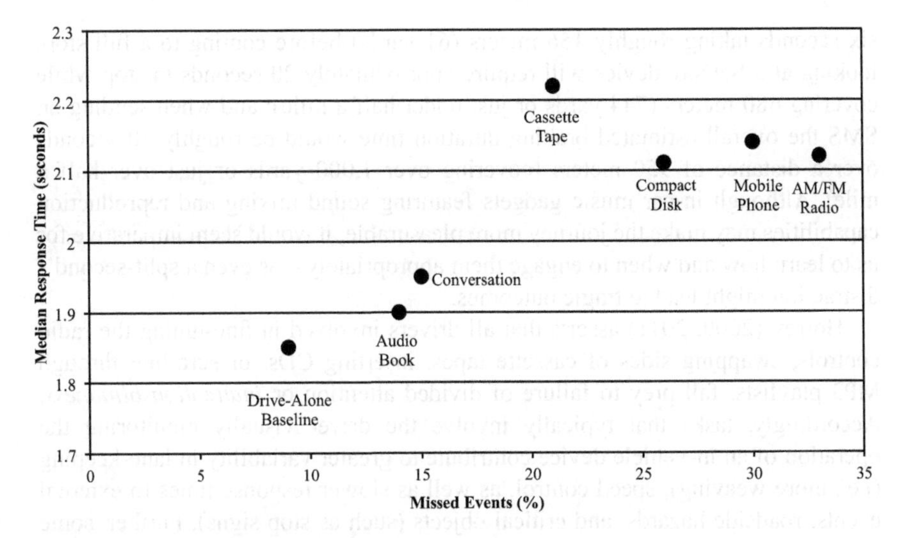

Figure 4.3 The effects of in-vehicle tasks on driving performance
Source: Adapted from Angell et al. (2006).

As can be seen in the figure, structural mechanical tasks caused the highest percentage of violations (i.e., missed traffic directions from signs and lights, as well as the longest reaction times for required braking). In a later report, Ho and Spence (2008) confirmed these findings and concluded that driver manipulation of car devices such as cell phones and car radios impair drivers' executive decision-making and vehicular control.

Lee (2007) claims that using an infotainment system can easily interfere with vehicle control on operational, tactical, and strategic levels. Consequently, drivers' ability to keep the car in the lane and to respond to the red brake lights of the vehicle in front is considerably reduced when they look away from the road while engaged in using peripheral devices. Even when our preoccupation does not involve looking away from the road (for example, singing along with a song playing in the background), our involvement may still hamper perceptual judgements, and reaction times may increase by approximately 300 milliseconds (Horrey, 2009). In the 2009 Annual Report of the British Royal Auto Club (RAC), a survey of 1,109 UK motorists was disclosed in which 433 (39%) stated that they become seriously distracted when driving (RAC, 2009). Accordingly, the top five driver-perceived most distracting in-car technologies, were: car audio devices (57%); SatNav systems (47%); mobile phones (32%); AC controls (31%); and dashboard warning lights (21%). However, most certainly such driver insights are purely subjective, and they do not reflect the reality of the dangers involved. Ultimately, the way to quantify the real threat of manipulating in-vehicle devices while driving is to calculate the *time* and *distance* a car will have travelled when attempting to halt the car at 70 mph. In this connection the RAC estimated that when changing a CD we will be faced with a braking duration time requiring about six seconds taking roughly 156 meters (61 yards) before coming to a full stop; looking at a SatNav device will require approximately 20 seconds to stop while covering 680 meters (744 yards or just under half a mile); and when sending an SMS the overall estimated braking duration time would be roughly 30 seconds over a distance of 950 meters (covering over 1,000 yards or just over half a mile). Although in-car music gadgets featuring sound mixing and reproduction capabilities may make the journey more pleasurable, it would seem imperative for us to learn how and when to engage them appropriately – as even a split-second's distraction might lead to tragic outcomes.

Horrey (2009, 2011) asserts that all drivers involved in fine-tuning the radio controls, swapping sides of cassette tapes, inserting CDs, or scrolling through MP3 playlists, fall prey to failure of divided attention or *inattention blindness.* Accordingly, tasks that typically involve the driver visually monitoring the operation of an in-vehicle device contribute to greater variability in lane-keeping (i.e., more weaving), speed control, as well as slower response times to external events, roadside hazards, and critical objects (such as stop signs). Further, some in-vehicle devices require us to scan back-and-forth from the road to the device repeatedly, and certainly as we can only gaze at one location at a time we are then more vulnerable and at greater risk. In this connection Horrey found that when engaged with in-cabin devices, our average response times becomes 130–210 milliseconds longer. Accordingly, this is clear evidence that in-vehicle secondary tasks interfere with our ability to detect, identify, and respond to distinct road events. Undoubtedly, some in-vehicle activities demand only one short glance. For example, when we monitor instrument gauges such as the speedometer our average glance duration is 0.78 seconds, but while checking the left side mirror we require an average of 1.10 seconds glance duration (Bach, Jaeger, Skov, & Thomassen, 2009). Yet, there are many other in-vehicle tasks that require us to manage quite sophisticated devices, with more complex operating systems and functional menu structures demanding multiple interactions. For example, Horrey found that most in-car entertainment activities necessitate between two to six glances, each requiring a glance duration between 0.5–1.6 seconds, that together average four seconds of total glance duration of time with our eyes off the road.

The truth is that *scrolling* music playlists is of great concern as it may affect driving performance and visual control. In a study by reported by Young, Mitsopoulos-Rubins, Rudin-Brown, and Lenne (2011), 37 drivers aged between 18–48 years (M = 30, SD = 4.13) participated in a 90-minute simulated driving session at Monash University (Melbourne, Australia). The participants drove while performing music selection tasks (such as searching for a song on a short 250-item playlist versus long 860-item playlist) employing an *iPod Touch.* Of the sample, 72% had previous experience with a digital music player featuring a touch screen interface. The study used a PC-based simulator, three 17" monitors, and a *MOMO Racing Force Feedback* (Logitech) steering wheel with accelerator and brake pedals. The results indicated that regardless of list length, engaging in music-search tasks on a touch screen interface significantly hampered driving performance and vehicular control as demonstrated by increased lane excursions and significantly increased eyes-off-the-road glance durations. Further, the study found that short playlist searches required an average 28 seconds to complete, while long playlist searches required an average of 52 seconds to complete. Therefore, although drivers attempted to regulate driving when distracted by decreasing the speed, there was an association between list size and the degradation of driving performance. These results were also confirmed by Lee, Roberts, Hoffman, and Angel (2012) who recruited 50 young drivers between 18–25 years old ($M = 21$, $SD = 2$, 58% male) to drive in a fixed-base simulator consisting of a full 1992 Mercury *Sable* vehicle at the University of Wisconsin–Madison (USA). During the trips the drivers were required to find different target songs among a short 20-item playlist, a medium 75-item playlist, and long 580-item playlist using an *iPod* featuring momentum scrolling. Perhaps searching through music playlists on MP3 players during driving *should* be condemned (and prohibited by law such as SMS texting), especially as studies have shown that repeated usage, familiarity, and practice do not improve driving performance (Chisholm, Caird, & Lockhart, 2008). To illustrate the point, while listening to music on our *iPhone*, we would need three discrete operations if we were using a random 'shuffle' order. Alternatively, we would need to proceed through four to five separate menus (including scrolling) if we wanted to listen to a more personalized playlist (for example, a playlist titled 'Cool Tracks for Driving' that we previously programmed for driving prior to the trip). But, we would have to engage in five to six individual manoeuvres (including searching by name, genre, or scrolling) had we wanted to listen to our favourite preferred specific album (for example, Frank Sinatra's 1984 *LA is My Lady*, arranged and conducted by Quincy Jones). It is important to point out that each interaction (i.e., step, operation, manoeuvre, or menu) that preoccupies the driver while completing an in-vehicle task represents only one single period in which a driver gazes at the device display, but there may actually be multiple fixations. In the end, it is the cumulative time that indicates total eyes-off-the-road duration (Horrey, 2009).

Different interface systems require distinct hierarchical structures for music retrieval – and the most popular are not necessarily the most efficient. For example, Garay-Vega et al. (2010) compared manual- to voice-command interface systems. The findings indicate that although the most widespread visual-manual interaction enables us to find music quicker, it also causes us to spend more time with our eyes off the road. Hence, the effect of *iPod* manual interaction use on driver distraction has been of some interest. Mouloua et al. (2011) examined driving errors (lane deviations) and physiological measures during a driving simulation task among 30 undergraduates between 18–25 years old (60% female) from the University of Central Florida (USA). Data were collected when driving while searching for specific songs (by name of artists, albums, or title as pre-determined by the researchers), as well as when driving without scrolling activity (i.e., trials of driving alone pre- and post-*iPod* music searches). The results demonstrate increased theta brain activity (i.e., attention) during the music search as well as a significantly greater number of lane deviations ($M_{Pre\text{-}iPod} = 3.27$; $M_{iPod} = 6.93$; $M_{Post\text{-}iPod} = 3.4$; $F_{(2, 28)} = 37.24$, $p < 0.001$). In another timely investigation of in-vehicle infotainment devices, Kujala and Saariluoma (2011) explored differences between menu structures (i.e., browsers) and scrolling styles on driver behaviour, including frequency and duration of off-the-road glances and vehicular performance (i.e., speed maintenance errors and lane excursions). This study must be seen for its primary stride in the process of educating drivers (as well as engineers, designers, and manufacturers) about the risks that most in-vehicle devices place on drivers. Namely, most in-vehicle devices specifically require drivers to take their eyes off the road to inspect items, icons, and texts appearing on the device display. Therefore, by engaging in such operations, drivers interrupt their primary task involving the more functional scanning of the traffic environment. In two driving simulation experiments, Kujala and Saariluoma investigated the effects of alternative widely used display features of operating systems in mobile devices: the visual *list* versus *grid* menu structures found in software, and the arrow *button-press* versus kinetic-motion *finger-swipe* method of scrolling touch screen displays. The menu structures examine the number of items and font size on visual display, while scrolling operation examined the effects of motion (whereby up- and down-arrow buttons support a row-by-row view, and kinetic touch screen scrolling affords a continuing variable movement of the items dependent on the amount of kinetic force applied to the movement). In each experiment, 20 participants of mixed sexes between 20–34 years old were recruited from the University of Jyvaskyla (Finland); all had a valid driving licence, with an average experience of on the road driving totalling 95,000 kilometres (59,000 miles). Each participant drove in a medium-fidelity fixed-base three-display driving simulation for 20 minutes in a rural environment resembling the Polish countryside, involving roads with varying curves. The experiment required two 10-minute segments in which the first was baseline single-task driving and the second was experimental dual-task driving. When comparing between *list* versus *grid* menu structures, Kujala and Saariluoma found that list structures required a decreased number of < 1.6 second glances ($M_{List} = 13$; $M_{Grid} = 25$) and less 1.6–2 second glances ($M_{List} = 5$; $M_{Grid} = 12$) for a lower total number of glances ($M_{List} = 93$; $M_{Grid} = 124$), as well as shorter average glance durations ($M_{List} = 1.06s$; $M_{Grid} = 1.08s$) for a lower total eyes-off-the-road duration ($M_{List} = 2.4s$; $M_{Grid} = 3.7s$). Simply said: when we use a list structure menu we are less at risk. The study found that neither dual-task condition caused a significant increase in speed maintenance errors ($M_{List} = 3$; $M_{Grid} = 4$), although differences of the number of lane excursions approached statistical significance ($M_{List} = 7$; $M_{Grid} = 19$; $p = 0.084$). When comparing between the *button-press* versus kinetic-motion *finger-swipe* scrolling, no significant differences surfaced for the number of < 1.6 second glances ($M_{ButtonPress} = 26$; $M_{FingerSwipe} = 23$) or number of glances between 1.6–2 seconds ($M_{ButtonPress} = 11$; $M_{FingerSwipe} = 10$); albeit, the total eyes-off-the-road duration approached statistical significance ($M_{ButtonPress} = 2.89s$; $M_{FingerSwipe} = 3.6s$; $p = 0.071$). Further, while button-press scrolling did not cause a significant reduction in lane excursions ($M_{ButtonPress} = 15$; $M_{FingerSwipe} = 16$), drivers significantly decreased the number of speed maintenance errors ($M_{ButtonPress} < 1$; $M_{FingerSwipe} = 5$). To summarize, mobile devices employing arrow-button scrolling on a list-wise presentation of items are more user-friendly and support increased driving safety than the more fashionable highly popular touch-screen smartphone devices that contain kinetic scrolling, panning/zooming features, fingertip manoeuvres, and a grid menu structure. Unfortunately, millions of us mostly use the latter on a daily basis – especially when we open smartphone applications and music operations.

Controlling the car on the road while regulating music listening devices seems to be perceived by all drivers as a natural event. For the most part, we all sense that we have a seasoned level of expertise enabling us to deal effectively with in-car entertainment devices. But the truth is that such mind-sets might not be entirely accurate. For example, in an early study employing an instrumented car, Jordan and Johnson (1993) found that the operation of high-end car stereos significantly affected driving performance. The study compared a number of pre-defined driving performance variables such as speed, direction of visual gaze, and subjective measures of mental workload in a dual task situation. Other studies such as McKnight and McKnight (1993) found that young novice drivers respond less frequently to traffic signals when tuning the radio than more veteran drivers. In addition, Wikman et al. (1998) found that younger drivers (with less than 15,000 km driving experience) were particularly more at risk than older drivers (with more than 50,000 km driving experience) when tuning the radio to a 'soft music' station or changing a music cassette tape in a dashboard-mounted tape player. Accordingly, less experienced drivers significantly increased the lateral displacement of their car (i.e., weaving) while engaged in operating the car-audio. Taken together, all of the above studies provide ample evidence that infotainment systems hamper driving safety, if for no other reason, by weakening abilities to anticipate roadway demands. Wikman et al. also found that when interacting with a radio or cassette player, not only did 29% of younger drivers glance away from the road for longer than three seconds, but there was a greater variance in their glances (including short ineffective scans). For these reasons, both Sagberg (2001) and Ho and Spence (2008) have suggested that radio and CD players cause more accidents than mobile phones. Looking at the broader picture, Lee (2007) and Horrey (2009) point out that younger drivers may be especially vulnerable than more experienced older drivers not only because they are less efficient in time-sharing between driving and non-driving demands but rather because they are more likely to adopt evolving technologies, bring devices into the car, use them frequently, and look at them more often for longer durations of time.

4–2.1.2. Rate of operation

Regulating in-vehicle entertainment devices, especially in-car audio equipment, is hardly a 'one-time incident'. Truth be told, such actions occur frequently during a quick run to the corner store and most certainly repeatedly while on extended road trips. For example, in an analysis study of in-cabin videos among 70 American drivers from North Carolina and Philadelphia during a seven day period of naturalistic driving capturing a selection of six random half-hour drives (totalling three hours of driving per participant), Stutts et al. (2003, 2005) found that all drivers were engaged in some form of potentially distracting activity for at least 16% of the total driving time, with some events and activities, such as eating and drinking and turning on cruise control occurring in parallel. Accordingly, operating vehicle controls (other than radio) occurs roughly ten times per driving hour, with each manoeuvre lasting for about 5 seconds.[4] Yet, most specifically, Stutts et al. found that 91% of drivers constantly manipulated audio-controls while driving, taking up to 2% of total trip time, with hand movements occurring roughly eight times per hour, lasting roughly 5.5 seconds for each operation. McEvoy, Stevenson, and Woodward (2006) found that among 1,347 drivers from New South Wales and Western Australia, all reported to engage in at least one distracting activity every six minutes (reflecting 15% of driving time). Further, it seems that drivers search for something (such as sunglasses, breath mints, small money change for the toll) at least 11 times per hour, with the average driver looking off the road for 0.6 seconds every 3.4 seconds (Vanderbilt, 2008). Other studies reported that at least 55% of drivers change radio stations once or more per journey, and that younger drivers change CDs several times per trip (Young & Lenne, 2010). The Arbitron/Edison (1999, 2011) research group who studied media trends among representative samples of Americans found that changing radio stations was a highly repetitive behaviour for drivers between the ages of 18 to 34. For example, in 1999 44% of young(er) drivers claimed they 'frequently' changed radio stations, and in 2011 34% of young(er) drivers reported the same behavioural pattern. The Arbitron/Edison studies point out that ironically among old(er) drivers the picture is different; in 2011 only 26% of middle-aged drivers (35–54 years old) frequently manipulated audio controls while 11% of old(er) drivers (above the age of 55 years) frequently manipulated audio controls. Perhaps one reason for the lower frequency of engagement among older drivers is that the visual demands on them are 15–50% greater (Green, 2001); this difference is quite substantial, and may also account for an overall poorer driving performance that occurs specifically while engaging in secondary tasks. For example, older drivers require roughly 40% longer time to respond to warnings, 33–100% longer time to read maps, and 80% longer time to enter destinations on SatNavs. Thus, older drivers may just feel the need to see the road much more frequently than younger drivers, and therefore have much less time to look away from the road at in-vehicle devices. Finally, Arbitron/ Edison survey data collected between 2003–2011 indicate that driver behaviours involving manual search for radio stations has become even more dynamic with the development of satellite radio. For example, 42% of drivers reported that they simply drive along leaving the stations alone during trips, while the majority 56% of drivers report they change stations often.

4 Later studies, such as Recarte and Nunes (2009), demonstrated distraction times over two seconds to be highly unacceptable for safe driving.

4–2.1.3. Inattention and distraction

Some time ago Brown (1994) claimed that skilled drivers generally learn when and for how long they can completely withdrawal their attention from the driving task – and they can therefore adapt their concentration such that their satisfactory performance can be maintained for long periods. While this may be true from a hypothetical vantage point, the reality is that a *general withdrawal of attention* from road and traffic demands that is associated with structural mechanical interference can impair vehicle control and subsequently disable drivers' collision-avoidance competencies. For example, Reeves and Stevens (1996) found that radios and in-vehicle information systems have greater potential for distraction than other conventional vehicle instrumentation. Accordingly, drivers spend more time gazing at the cassette tape player (3.5 seconds when rewinding the tape; 5.5 seconds when playing the tape) and the AM/FM radio receiver (13 seconds when searching for a specific frequency), than when looking at vehicle controls (0.5–1 second when glancing at the speedometer) or when monitoring the road itself (Stevens & Rai, 2000). Clearly, 'fine-tuning a radio' takes up much more visual attention than 'changing a cassette' or even 'using a mobile phone' (Wikman et al., 1998). Namely, adjusting the channels of an AM/FM receiver results in longer periods of eyes off the motorway and larger deviations in road position than similar occurrences such as when a driver answers a hands-free mobile phone. In a study published by the American *AAA* Foundation for Traffic Safety, Stutts et al. (2001a, 2001b) found that during a three-hour period of driving, 91% of the drivers were remiss by fiddling with music or audio controls taking up a total of 13.5% of driving time (which accounted for a total of 24 minutes throughout the trip). The *AAA* claimed that such manipulation is a form of inattention or distraction and must be seen as a contributing factor of near-crashes or crashes. In a later analysis, Stutts et al. (2005) revealed that when manipulating audio controls while moving, 85% of drivers spend an overall 2% of their time with no-hands on the steering wheel, and 23% of their time with eyes directed inward at the display of the radio centre – for a total 10% of events and violations per hour. These levels are just slightly less than distraction caused by 'reaching or looking for something' (which consists of 4% no-hands on steering wheel, 20% eyes directed inward, 18% total events/violations), or being 'distracted by a passenger' (which consists of 2% no-hands on steering wheel, 19% eyes directed inward, 20% total events/violations). Engström, Johansson, and Östlund (2005) point out that performing a visually demanding secondary task while driving – such as operating a radio – results in time-sharing subsequent to attention being diverted from the road. Hence, the driver is in a position of diminished tracking responses that results in extended intervals of fixed steering-wheel angle, heading errors, and reduced lane keeping (i.e., weaving). Horberry et al. (2006) found associations between operating an entertainment system with lower speeds and greater deviations from the posted speed limit – and these were statistically greater (i.e., worse) than when the same drivers used a cell phone. In their driving simulator study (at Monash University, Australia), a group of 31 drivers (variegated by three age groups < 25, 30–46, 60–75) participated in one driving session consisting of six short drives; both simple and complex versions of non-distracted driving, driving with mobile phone distraction, and driving with entertainment system. In the study, each drive was along a 6 kilometre (3.6 mile) single carriageway, employing the same number of curves and hazards, with speed limits posted between 60–80 kph (37–50 mph). The entertainment system task required the drivers to tune the radio, change the tone (bass-treble frequencies), change the speaker balance, and insert/eject cassette tapes. The study found that drivers perceived the car-audio tasks to contribute to significantly greater demands and were seen as severally greater in level of distraction ($M_{CarAudio} = 10.3$; $M_{MobilePhone} = 8.1$; $t = 4.978$, $df = 30$, $p < 0.0001$). Finally, several recent studies have found that involvement with music devices generate at least similar effects of driver distraction as do mobile phones (Brumby, Salvucci, Manowski, & Howes, 2007; Salvucci, Markley, Zuber, & Brumby, 2007; Young & Lenne, 2010). Unfortunately, many young drivers place themselves doubly at risk. For example, in an American study with 27 young-adult drivers in Miami, not only did every participant account for driving while listening to music, but they also reported to do so simultaneously while talking on the cell phone (Bellinger, Budde, Machida, Richardson, & Berg, 2009).

In a report titled *Driven To Distraction* (RAC, 2009), the British Royal Automobile Club claimed that modifying radio channels, regulating audio sources, or selecting music tracks, were not simply matters of physical manipulation with associated visual diversion per se. But rather, during each five-and-a-half seconds needed for every operation, the car will have travelled roughly 156 meters (which is the length of two soccer pitches or 512 feet) while drivers are essentially unaware of the road environment. Such claims confirmed early analyses pointing out that shifting attention to audio controls denotes neglecting primary tasks such as lane-keeping and looking for other vehicles that increased the possibility of running off the road (Wikman et al., 1998). Further, the RAC report revealed a questionnaire survey among 1,109 UK motorists that found 40% had claimed to become seriously distracted when driving, and that almost 60% who felt that adjusting radio controls or changing a CD was the 'most significant of all in-car distractions'. Moreover, the report disclosed that young drivers were the most likely to lose concentration behind the wheel, and that one-in-five were listening to music through headphones while driving on the road. Given that headphones seem to cause increased RTs that hamper driving tasks dependent on auditory cues (Nelson & Nilsson, 1990) as well as cause larger interference effects with visual stimuli (Fagioli & Ferlazzo, 2006; Ferlazzo et al., 2008), perhaps we should see stereo headphones as increasing risk factors of driver inability to respond to highway incidents.

As a result of the evidence that has surfaced, the British Royal Society for the Prevention of Accidents (RoSPA, 2007) engaged in a national campaign; they distributed an information letter to all UK certified drivers and registered car owners. In the leaflet, RoSPA conceded that the percentage of drivers absorbed in operating radio and music devices (96%) is by far a higher proportion than those who engage in any other potentially distracting activity, including: events external to the vehicle (86%), conversation (71%), events inside the vehicle (67%), eating/drinking (65%), grooming (45%), reading/writing (42%), mobile phone (24%), and passenger in backseat (15%). Subsequently, the UK-based Direct Line Insurance company (Direct Line Insurance, n.d.) released the following suggestions to their client base:

- Avoid adjusting equipment until safely parked.
- Load MP3 players with playlists that last the full duration of the journey so as not to induce file-searching behaviour.
- Invest in safer audio devices such as radio receivers that do not need constant tuning and auto-changing CD players.

4–2.1.4. Near-crashes and crashes

While the frequency of music-related automobile accidents is not known, and perhaps this statistic is too difficult to account for while investigating accidents, there are numerous anecdotes (with accompanying photos) that depict such collisions. Many can be found on public access Internet sites (for example, see 'Music-related Crashes' at Car-Accidents.com). Among the vignettes with pictures, are:

Case A Distracted driver from looking at the CD player in a 1988 Oldsmobile (*Olds 88*), subsequently rammed by a GM Chevy Van (USA).

Case B Distracted driver from reaching for a CD in a sun-visor wallet, subsequently wrapped around a light pole (USA).

Case C Distracted driver from looking for a CD in an *Audi TT*, subsequently drove under 18-wheeler trailer while braking at 140 kpm (Saudi Arabia).

Case D Distracted driver from looking down at stereo to increase volume while entering a curve at 140 kph, subsequently ran off to road shoulder, lost control, flipped and rolled 7–8 times side-to-side before landing in ditch (Canada).

Case E Distracted driver from looking down at passenger seat while reaching for favourite CD, subsequently veered off mountain road, rolled front-to-back continuously for 250 feet down mountain slope (USA).

Unfortunately, collisions linking structural mechanical interference with music devices have been known for some time. For example, Sagberg (2001) conducted a survey study with 6,306 Norwegian drivers who had reported an accident to their insurance company; the drivers were distributed by responsibility as decreed by liability between insurance companies towards the event, with 3,340 (53%) of the drivers pronounced as the guilty responsible party. Accordingly, the proportion of accidents caused by sources of distraction for the guilty drivers, were: conversation with passengers (8%), children in back seat (3%), searching for street name or house number (2%), use of in car equipment (2%), changing cassette or CD or tuning/adjusting radio (2%), and other sources < 1% (such as advertising/billboard signs, objects/insects inside car, cell phone [0.66%], smoking, eating/drinking, and map reading). Sagberg states that both radios and CD players caused more accidents than cell phones, in which the relative risk of being in an accident was increased by 120%. Sagberg concludes: "In terms of the number of accidents mobile phones seem to be a relatively small problem, compared to e.g. distraction by passenger, navigation, or use of radios and CD players" (p. 68). In another study, Stevens and Minton (2001) annotated 5,740 fatal accidents among UK and Wales Police accident reports covering the period between 1989–1995. Their data showed that in-vehicle distraction was a contributory factor in 2% of all fatal accidents. Accordingly, changing a radio channel, cassette tape, or CD was reported as the second most frequently cited cause of distraction leading to an accident. Music distraction via structural mechanical interference was cited more frequently than cellular phone use, map reading, fixation on dash-board clocks and gauges, eating, or smoking – surpassed only by interacting with other passengers. Cases mentioning entertainment were five times greater than those citing a telephone. It is interesting that in the appendix of their report, two out of four case vignettes (pp. 544–5) relate to structural interference of a car-audio device:

Case 9724: Driver collides with stationary vehicle. Front seat passenger killed. Driver admitted he was tuning the radio when the collision occurred.

Case 2263: Driver looses control on bend and collides other vehicle head on. Driver killed. Driver found clasping audiotape by emergency rescue team. Probable loss of control of vehicle whilst changing tape.

The American *AAA* Foundation for Traffic Safety also reported similar findings about incidence of crashes due to structural mechanical interference (Stutts, 2001; Stutts et al., 2001a, 2001b). Accordingly, police reports for North Carolina and Pennsylvania (between 1995–2000) covering more than 5,000 crashes pointed to adjusting in-vehicle audio equipment as the source of distraction for more than 11% of crashes. As a matter of fact, Stutts et al. (2001b) found that young drivers < 20 were most likely to be distracted more often by adjusting the radio and music cassette or CDs (especially at nighttime). Further, younger drivers were more likely than older drivers to be distracted at the time of the crash. Namely, that 12% of drivers < 20 years old reported to be distracted when an incident occurred compared to 8% of drivers > 65 years old. In the following year 2002, the US National Highway Traffic Safety Administration (NHTSA) blamed 66% of the previous year's 43,000 fatal car crashes on 'playing with the radio or CD' (DeMain, 2012). Finally, in a study that recruited 1,347 drivers from New South Wales and Western Australia, Stevenson and Woodward (2006) found that all drivers engaged in an average of ten distracting activities for every hour of on-road driving. Accordingly, adjusting the stereo unit was revealed as the fourth most distracting activity, and considered to be substantially more distracting than eating, drinking, smoking, using a mobile phone, seeking directions, or map reading.

Not only do individual studies point to the hazards of structural mechanical interference, but two recent meta-analytic comparisons of distraction sources found the contribution of structural mechanical interference to crashes (referred to as 'associated crash risk'). Young and Lenne (2010) and Young and Salmon

(2012) highlight the fact that adjusting the radio, cassette tape, CD, or MP3 player is much more complex than most drivers seem to comprehend. In both analyses, data pointing to mechanical manipulation and engagement with music devices were collated and evaluated. The reviewers came to view these activities as a distraction, with pooled cognitive-visual-physical (manual) processes that resulted in a near-crash or crash, with odds estimated between 0.6–2.3%.

4–2.1.5. Benefits versus costs

Certainly new-fangled novel and innovative portable devices are swiftly being brought to vehicles as peripheral add-ons to enhance the driving experience. Infotainment systems include a wide range of gadgets that facilitate our engagement in a host of tasks unrelated to driving, including making telephone calls, watching videos, managing e-mail, sending and reading messages, scrolling through music playlists, and listening or singing to music. Even the ordinary and more commonplace car radio receiver has drastically changed as a result of *SiriusXM* satellite radio and MP3 (*iPod/iPhone*) music players. Although many fresh technologies offer benefit, some precaution must be taken for their use. For example, engagement with new technology – that in the end not only represents know-how of an operating system but is also a form of hardware machinery – means that an entirely new set of tasks are embedded in the cabin milieu for us to perform while we are behind the wheel (Lee, 2007). Unfortunately, no matter how these have been programmed to function in a 'user friendly' or instinctive fashion, all of these devices provide tasks that, on one level or another, compete for our limited resources. If the overriding conceptual designation for driver distraction is "the diversion of attention away from activities critical for safe driving to a competing activity" (Lee, Young, & Regan, 2009), then, certainly we need only look at the number of actions and manoeuvres we initiate with car-audio systems and infotainment apparatuses while driving. It is interesting to note that a decade ago Featherstone (2004) prophetically reflected upon the development of technological advancements that have intruded and violated the autosphere:

> The automobile is one everyday object where human beings regularly encounter new technologies in their everyday lives and learn to 'inhabit technology'. More and more, aspects of everyday driving becomes a mediated process in which technology ceases to be a visible tool or technique, but becomes a world in which the boundaries and interfaces between humans and technological systems become blurred, refigured and difficult to disentangle. The automobile becomes a world in the sense that we not only use the mobility of motor technology… to travel to things and places [but] we also use micromotors and embedded chips in the driving environment to bring things and places to us. [While] we look out of the windscreen to see the world to be driven through… we also consult the instrument data screen with its increasingly sophisticated, but 'user friendly' graphics. We consult the screen of the geo-positioning system to find out where we are, where to go and how to get there. We keep a sidewise glance and ear

> attuned to the television screen positioned for passengers to watch… This increasing isolation from other traffic and the physical surroundings outside the car, which occurs as other multitasking activities take over, can lead not only to 'absenteeism', but also to accidents… [as] car collisions frequently occur while drivers are engaged in activities other than driving (dialling a telephone number, fiddling with the radio, television or CD deck, listening to music, surfing the web, etc.) (pp. 10–11).

Clearly, some vehicular activities seem to interfere with other driving tasks – and some do so more than others. Albeit, in most cases drivers tend to focus on the physical constraints, totally ignoring the fact that each activity also involves other senses or means of distraction that hamper driver behaviour (White et al., 2004). In this connection, Consiglio et al. (2003) point out that decrement of vehicular performance occurs from events and undertakings that combine structural interference subsequent to poor mechanical configurations with other sources such as the blocking of driver awareness and/or obstruction to central attention subsequent to overtaxed cognitive faculties.

4–2.2. Perceptual Masking

Most drivers take driving an automobile for granted. In some ways, we are oblivious of the fact that driving is challenging for both visual and aural intelligences as well as reflects a highly demanding cognitive activity. Driving requires us to perform multiple concurrent tasks, such as to scan the road and periphery for potential hazards, to maintain lane position, to regulate the appropriate speeds, and to monitor signals and signs – all of which compete for the same limited perceptual and cognitive resources (Sethmudhaven, 2011). Auditory distraction, as a form of *masking*, is rarely mentioned in the driving literature. Nevertheless, the presence of music in vehicles, along with road noise and passenger speech, does mask the sound of auditory feedback from the vehicle itself including RPMs and engine sounds that are critical for self-monitoring driving performance (Dibben & Williamson, 2007). As visual displays in automobiles have become instrumented control panels that allow drivers to monitor the vehicle performance, they have become overloaded with combinations of analogue and digital animated clocks, gauges, and LED spotlights of various sizes and colours. We find a car console cluttered with gauges, such as the speedometer, tachometer, ammeter, voltmeter, odometer, fuel gauge, oil pressure, battery charging, engine temperature – to mention but a few. Subsequently, cognitive overload, which can hamper drivers' decision-making performance, has become an area warranting increased mechanical and ergonomic development. In fact, Spence (2012) states that it is vital to introduce a variety of innovative in-car warning signals into automobiles that alert drivers about impending threats, and try to direct a distracted driver's attention back to the road, especially when they might be in danger of leaving it too late to brake. In this connection, it seems that there is a consensus that the best alternative means of interaction to convey information to the driver is via the auditory modality.

Long before the computer age, audio alarms and signals such as horns, whistles, bells, buzzers, chimes, and oscillators were widely used (Blattner, Sumikawa, & Greenberg, 1989). But, more so after the advent of the PC, audio messages in the form of *earcons* have become a highly viable method employed to a great extent by design engineers in the auto industry. Auditory-based iconic information works well because it takes up less cognitive space through its simplicity, and drivers seem to have the ability to understand the conveyed message quickly as they do inherently reflect ordinary sounds and forms of the real world. Blattner et al. point out that earcons may either be representational (synthetic figures of natural sounds) or abstract (conceptual sounds with theoretically implied meaning). Further, Gaver (1993a, 1993b) claims that earcons more closely resemble *auditory icons* in that they are essentially 'caricatures' of naturally occurring sounds ('aphoristic' in that they reflect the tangible physical environment, 'connotative' in that they rely on social convention for meaning, and 'allegorical' in that there are similarities between the auditory simulation and the real world). Ho and Spence (2005) claim that long ago research initiatives explored the use of ecologically valid auditory icons as a component in interface design, while in terms of capturing and redirecting a driver's attention, evidence demonstrated that auditory icons (such as, 'beep beep' of a car horn) work better than pure tones (i.e., 'buzzzz'). Further, that auditory spatial cues (such as 'Look Right') work better than generalized non-spatialized cues (Spence, 2012). Indeed, auditory warning signals may be effective when we are distracted – even when engaged in a mobile phone conversation – but for the most part, have not been easily incorporated within car systems. Perhaps, one reason is our rampant use of background music that often 'covers' most in-cabin auditory messages meant to indicate ordinary engine processes or cautionary warnings of operation. When in-car music listening eradicates such codes, as well as covers external signals such as sirens and car horns that are meant to keep the driver out of harms way (Slawinski & McNeil, 2002), then, there is certainly a case for accident risk through *perceptual masking*. In a recent paper that forwards a cognitive model describing how sonic content relates to indicators of safety versus danger, Andringa and Lanser (2013) warn:

> The omnidirectional sensitivity of audition makes it ideal to monitor the proximal environment and to warn when something unexpected happens, but the more the ambiance is masked, the less our situational awareness can be based on easily available sonic information about the environment (p. 1,442).

Finally, in an interesting vignette, Vanderbilt (2008) perceptively outlines a variety of studies showing that when drivers loose auditory cues masked by listening to the radio or music CDs at a high volume, they also loose track of how fast they are going – especially as we perceive acceleration and speed from the sound of road resistance and tyre friction.

4–2.2.1. Cross-modality processing

Today we engage in a host of unrelated tasks while we drive in traffic. We have developed an overall inclination of multitasking behaviours, such as eating, drinking, smoking, talking on a cell phone, interacting with a passenger, as well as tuning the radio, listening to music, and even singing along with the tunes (Sethumadhaven, 2011; Unal et al., 2013a). As computer technology has totally dominated our modern-day cars with various secondary systems (such as advanced infotainment and navigation structures), more and more demands have been placed on our attentional resources. Many drivers bring along their own portable appliances for use while on the road – more commonly known as *nomad* devices (Engström, Johansson, & Östlund, 2005). Such devices increase the number of tasks we perform, for example allowing us to text, check e-mail, and scroll through music libraries – all the while driving in a dynamic milieu of traffic. Although Horberry et al. (2006) demonstrated that both visual and manual distractors impose serious demands on drivers and subsequently hamper vehicular control (as many of these actions seem to rely on the same mental resources as driving itself), Unal et al. are fairly unique in their claim that auditory distractors can actually be handled quite well. Accordingly, they outrightly claim: 'auditory distraction is not detrimental to driving'. Nevertheless, it would stand to reason, that even if Unal et al. found that when required to handle increased task-demands some drivers are capable of paying less attention to audio sources (by *blocking-out* auditory interference), it must be pointed out that the mainstream of research efforts have indeed demonstrated that auditory distraction impairs performance for most drivers.

Traditionally, theorists have accounted for modality specific independent neural channels involving auditory and visual information processing. However, *cross-modal* links in spatial attention between vision and audition are often found. In such cases, auditory cues are highly effective in enhancing the performance of visual tasks. Therefore, one would assume that separate sensory modalities that interact with each other do influence on-task outcomes, and that such a process also occurs within the context of driving. Hence, while 83% of information used to drive an automobile is visual by nature, it is essential to note that input from other senses (such as audition, touch proprioception/kinesthesis, and vestibular) augment our perception by providing ample feedback when driving. In this connection, Spence and Ho (2009) point out that the remaining 17% of input is roughly split as follows: kinaesthetic (11%), tactile (6%), and auditory (1%). They illustrate with the following examples:

- Drivers employ their car horn as an auditory signal to alert other drivers about their presence.
- Drivers listen to engine revs (RPMs) as a means of determining when to manually change gears.
- Drivers listen to engine revs (RPMs) as a means of determining when to lift their foot from the accelerator pedal after overtaking another vehicle in an automatic gear car.

- Drivers perceive the need to slow down when approaching a roundabout or train crossing because of kinaesthetic exoskeleton suspension moment resultant from passing over a hump or series of rumble strips.
- Drivers are warned as to their position of having veered out of their lane (centreline rumble strips) or approaching the outer road shoulder (edge-line rumble strips) by combined audio-tactile input dynamic road markings.

Spence and colleagues (Ho & Spence, 2005; Spence & Read, 2003) delineated much evidence to support the view that performance of in-cabin tasks traditionally thought to be exclusively visual, are indeed facilitated by auditory cues – especially when sounds prime essential actions and are generated from a source spatially located in the same direction as the required visual information or operation. For example, Spence and Read found that experienced drivers *speech-shadow* more words correctly (i.e., higher percentage of repeating words heard aloud) when they were presented from a front dashboard-mounted loudspeaker than when presented from a loudspeaker mounted on the passengers' side door ($M_{Front} = 85\%$, $M_{Side} = 49\%$; $F_{(1, 7)} = 86.0, p < .001$). Such results seem to point to spatial advantages of attention that occur when the attentiveness to both visual and auditory involvement are synchronously directed toward the forward roadway. In another study, Ho, Reed, and Spence (2007) found that drivers initiated significantly faster braking responses to potential front-to-rear-end collision events when a cross-modal audio-tactile (vibration + car-horn sounds) warning was presented than when either uni-modal warning was presented ($M_{Audio\text{-}tactile} = 806$ milliseconds; $M_{AuditoryAlone} = 887$ ms; $M_{VibrationAlone} = 1001$ ms). Hence, for some time auto manufacturers have placed their utmost efforts in incorporating new features that integrate combined cross-modal signals designed especially for warning drivers of impending harm.

Considering the above, and accounting for the wealth of research denoting impairment of driver judgement and decision making, increased RTs to unexpected incidents, and the escalation of missed events that result from performing an auditory task while driving (Hancock, Lesch, & Simmons, 2003; Ho & Spence, 2005, 2008; McKnight & McKnight, 1993; Strayer, Drews, & Johnston, 2003; Strayer & Johnston, 2001), we might be inclined to wonder about the effect of conversations (that is an activity involving both auditory listening and verbal production). In this connection Crundall, Bains, Chapman, and Underwood (2005) claimed that secondary activities involving verbal activity (such as conversations and even perhaps singing) should also interfere with processing visual information during driving. One possible explanation they raise is that conversation causes a *bottleneck* in attentional operations by flooding perceptual capabilities of divergent aural sources of information that stem from two peripheral directions. That is, we seem to be obstructed by aural sources such as the cell phone or car-audio speakers that seem to be located otherwise than the actual direction of incoming visual information from the road ahead. Such incongruences make it harder to perceive what is happening on the road, as drivers have to direct both aural and visual attention to two different directions (Ho & Spence, 2005; Vanderbilt, 2008). We might, then, further contemplate the perceptual effort placed on drivers when a multi-channel sound system with multiple speakers of various configurations and spatial locations is installed in the car.

It should be pointed out that multi-channel music has been moving into the mainstream of popular listening for some time now. Along with the new generation of music performers and recording studios who explore the possibilities of this medium for their recordings, car-audio enthusiasts have already installed comparable hi-fi reproduction systems for their in-cabin listening pleasure. As a result, there is a commercial demand that studios reissue a host of yesteryear's albums remixed for the new format, especially when considering the popularity of automobile listening. According to the Crutchfield technical team (Crutchfield, n.d.), the audio advantages of multichannel surround recordings are many:

> If you're listening to a 'live' performance, a multichannel recording will put you in the middle of the audience. In-studio recordings can capture the acoustical ambiance of the studio or concert hall in which they are recorded, or put you right in the middle of the band itself.

It should be noted that the multi-channel surround sound system most commonly found in high-end cars (such as those manufactured by BMW, Mercedes, Lexus, and Infinity) is the *DTS 5.1*. This sound system is the most popular sought-after after-sales add-on car-audio system. The system incorporates six channels of which five are full-bandwidth outlets with a frequency range between 3–20,000 Hz for the surround sound (front L/R, centre, and back L/R), plus a low frequency effects subwoofer channel covering a frequency range of between 3–120 Hz (News Home, 2013). The automotive sound system is coupled to a speaker configuration typically consisting of two pairs of full-range speakers (installed in the dashboard and back) as surrounds, with mid-range speakers (installed in kick-panels under the dashboard, in the side doors, and rear deck). Moreover, a centre channel speaker is placed in front of the driver to establish a well-defined 'soundstage', and a subwoofer speaker is installed in the trunk to add fundamental low frequencies giving more depth and punch to the bass and percussion sounds. In 2013 the Chinese Disi Riu *Shakespeare* model B-class entered the automobile market specifically targeting passionate audiophile drivers by featuring a full theatre-like *DTS 5.1* surround sound experience with a standard 14-speaker configuration that had previously only been featured in the especially high-end Mercedes Benz *S600L* and BMW *530Li* models (News Home, 2013) (see Figure 4.4).

In light of such state-of-the-art in-car audio systems, we must contemplate the consequences on braking times when drivers listen to music reproduced through multiple speaker configurations. This question is raised on the coattails of Ho and Spence (2005), who found that roughly 190 milliseconds more time is required to halt a vehicle – beyond the average 700 millisecond response time for expected braking, and 1,500 milliseconds response time for unexpected braking – when drivers need to switch their attention between auditory and visual modes with

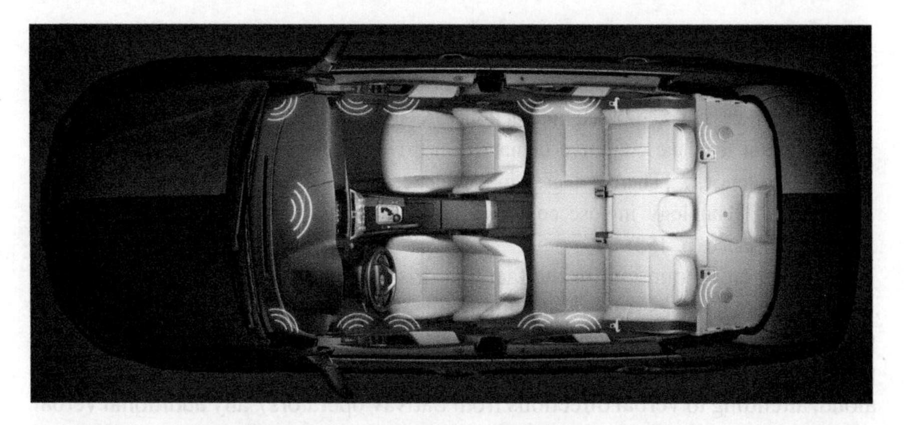

Figure 4.4 2013 *Shakespeare* Disi Rui with DTS 5.1 14-speaker configuration
Source: Used with permission of BYD America.

stimuli that are presented from different spatial positions in the car. Therefore, while cross-modal shifting from auditory spatial cues to those that are more predominantly visual may seem to be highly successful in aiding visual search tasks, and in fact are even more efficient than purely autonomous visual cues themselves, drivers employing surround-sound music systems may simply have to come to accept that listening to music reproduced through multiple speaker configurations is particularly demanding on their perceptive resources – and hence places them at a higher risk.

The idea brought forward above relates to spatial regions within the physical realms of the vehicle cabin. Yet, 'space' may also denote a conceptual perceptual realm as far as driver function is concerned. For example, Fagioli and Ferlazzo (2006; Ferlazzo, Fagioli, Nocera, & Sdoia, 2008) claim that drivers today are continuously forced to shift from a more *personal space* where conversation (or in-car music activity) takes place, to a more *extra-personal space* where driving tasks are executed. Namely, when speaking on the phone (or listening to music, or singing to songs) drivers repeatedly need to cross the threshold of functional utility from *listening* (i.e., an auditory mode of attention involving the voice and the language content) to *looking* (i.e., the visual mode of attention involving the traffic on the road ahead). Clearly, this shifting back-and-forth happens time and time again while we drive. Question: Are cross-modality shifts more of an advantage, or do they place drivers at risk? On the one hand, Ho and Spence (2005) see a significant advantage in detecting potential emergency events when drivers are cued by auditory signals; they claim that either non-semantic auditory signals coming from the relevant direction (such as, 'Beep Beep' of a car horn), or by drawing attention to the appropriate direction by semantically meaningful verbal warnings (such as, 'Look Ahead' or 'Look Right'), are of much benefit. On the other hand, Bach, Jaeger, Skov, and Thomassen (2009) maintain that speech technology for in-vehicle systems might be less of an advantage than perceived:

> ...speech-based interactions do not distract drivers, because drivers are not required to take their eyes off the road or their hands off the steering wheel... Nevertheless, voice-based interactions are not effortless, and they have the potential to place cognitive demands on drivers with mind-off-the-road as a consequence... [There is] growing evidence suggest[ing] that systems with speech technology impose cognitive load on drivers that can affect driving performance (p. 458).

Namely, when accounting for the fact that drivers typically preoccupy themselves with secondary tasks involving linguistic elements (such as conversing with passengers in the car, talking with callers via the cell phone, listening to radio programme announcers and news broadcasts, singing lyrics from songs heard aloud, attending to verbal directions from SatNav operators), any additional verbal warnings may unsurprisingly hamper driver perception leading to confusion and misinterpretation. Therefore, if cross-modality warning systems do have potential, then they might do better if containing only nonverbal auditory signals as these seem less susceptible to be corrupted by other concurrent linguistic elements in the autosphere. For this particular reason, several auto manufacturers have begun to feature a pro-active driver's seat that vibrates on the left or right side of the buttocks as a format of providing a warning signal for lane-departures (Spence, 2012).

4–2.2.2. Inattention and distraction

There are now more distractions in the car then ever before. Among them are cell phones, GPS systems, and entertainment systems. Most drivers feel that they are skilful in engaging these carefully, and that indeed they can adapt their concentration and attention such that they continue driving on the road with adequate control and safety. Nevertheless, long ago Brown et al. (1969) made a much different claim: "Although more automatized control skills may be affected minimally by this division of attention, some perceptual and decision skills may be critically impaired" (p. 423). In a later study Brown (1994) relates to this circumstance as *selective withdrawal* of attention whereby vehicle control may be left intact, but drivers' collision avoidance ability is highly compromised. Each and every new technology that enters the vehicular cabin truly seems to come with a proprietary control panel, set of slider buttons, touch screen, as well as a unique range of earcons that serve as aural feedback linking specific actions. As earcons are the aural equivalent of the visual icon (which can also be thought of as an 'eyecon'), these serve to convey messages and inform the user about device-related functions (Blattner et al., 1989; Gaver, 1993a, 1993b). Therefore, if visual scanning is critical, not only to control and manoeuvre the vehicle, but also for the important task of detecting and responding to hazards (Horrey, 2011), then we must consider if in-vehicle music-related tasks that carry additional demands not only affect how drivers take-in information about the environment, but also mask the necessary auditory perceptual cues of driving altogether. In the very least, we must consider if in-vehicle music-related tasks impede on drivers' abilities to distinguish the spatial location of auditory signals that warn of potential hazards and risks.

Although rarely documented in police reports, some anecdotes do provide evidence indicating how loud music played by drivers masks sirens of emergency vehicles coming from behind or from side streets. In one vignette, a Detroit (USA) fire truck en route to a scene crashed through a fence and into an empty house while trying to avoid an oncoming car in which the driver claimed not to have heard the sirens over his loud music; four fire-fighters were treated for minor injuries while the car driver was hospitalized in a more serious condition (Clickondetroit, 2006a, 2006b). For a long time lawyers have asked those involved in an accident if they had been listening to the radio at the time of the crash. But, no matter what the answer may be, or how a driver perceives music activity as a diminutive risk factor, the magistrate's inference is that most likely the driver wasn't attentive to the road because they were listening to music (Smith, 2006). A second example was an occurrence in Utah (USA) when an ambulance responded to an emergency 'Code 3' call and went through a red-light intersection with lights and sirens flashing, while the driver of a Suzuki sedan who entered the intersection on a green light failed to hear the sirens of the approaching ambulance because he was listening to loud music, and smashed into it; neither drivers were hurt, but an EMT riding in the back of the ambulance suffered injuries (Bruce, 2010; EMS1, 2010).[5] A third and tragic incident involved a Montana (USA) driver who was hit by a freight train because the music in the vehicle was so loud that the driver did not hear the train approaching – nor did he notice the flashing railroad signal (Montlick & Associates, 2011). According to accident attorneys in Atlanta Georgia: "This fatal train accident highlights an aspect of distracted driving that is fairly common but rarely addressed… Many serious car accidents can be averted by relying on one's sense of hearing as well as a driver's sense of sight." Accordingly, the attorneys point out that an approaching train is typically far from silent, besides being much larger than a passenger car. In this case, the train collided with the vehicle at a railroad crossing while the car was reportedly moving at a very slow rate of speed, with highly intensive loud music playing in the background that seems to have disrupted the driver's focus as well as his ability to hear what was going on around him. In their statement, the accident attorneys conclude:

> It is easy to understand how this could happen if you consider how often you have witnessed other drivers with loud music playing as they sing along, dance and gesture with the music or bang their hands on the wheel while paying no attention to the traffic around them.

Consequently, it is highly cogent to question the volume intensity of in-cabin music exposure: Do the actual sounds of in-cabin music obstruct drivers from monitoring driving-related mechanical and environmental 'clatter', that by all accounts are essential for efficient driver safety?

As safe and effective driving necessitates the detection of auditory information embedded in a background of continuously changing sounds (Slawinski & MacNeil, 2002), the presence of music in vehicles must be considered by drivers, traffic

5 See video: http://www.ksl.com/?nid=148&sid=12911899.

researchers, legislatures, and law-enforcement agencies. There are essentially two issues at hand: the volume or *perceptual intensity* of the exposure, and the complexity or the *cognitive intensity* of the music heard. Much research (Bellinger et al., 2009; Dibben & Williamson, 2007; Ho & Spence, 2005, 2008; RoSPA, 2007) has shown that a number of factors contribute to *if* and *how* music exposure (that is, the decibel level of background music) might cover vehicle-designed critical alerts (various 'beeps' and 'buzzes'), self-monitoring sounds (such as echoes from engine-revs, braking, the engine roar of an adjacent motorcycle), as well as external warning signals (such as car horns, skidding sounds, railroad bells, sirens of approaching emergency vehicles, screeching braking of nearby traffic). Acoustic clues may be the only warning a driver receives about road hazards before a fatal car accident occurs. After all, sounds are evident before visual cues, hence providing drivers with an extra moment that may be needed to safely avoid a serious car crash. Not only does loud music make it difficult to hear other potential sound sources that warn of approaching road hazards (whether they be from within the cabin or from external sources), but loud blaring music is distracting to the drivers themselves as they dominate awareness causing drivers unnecessary difficulty in focussing attention to the road (see 'Capacity Interference', Section 4–2.3 below).

To some extent mechanical and acoustic engineers can equate the level of perceptual masking: the degree of impairment depends on the size of the automobile cabin, the materials used as seat upholstery, the number of passengers, the transfer function of loudspeakers and their resonant frequency, and the relative state of the shell seal (i.e., if the windows are opened or closed). Yet, whether or not music listening is a contributing factor to inattention and distraction may still be relatively unknown (Eby & Kostyniuk, 2003; RSC, 2006; Smith, 2006, although see Brodsky & Slor, 2013). Nevertheless, we can only assume that any competing stimulus or activity that interferes with processes that have detrimental effects on driver awareness, road position, speed maintenance, control, reaction times, or negotiation of gaps in traffic, should be treated as a serious risk factor for inattention and distraction (NHTSA, 2010; Shinar, 2007; Young & Lenne, 2012; Young & Solomon, 2012).

4–2.2.3. High-intensity low-frequency 'boom' music

Much research has already shown that multisensory cues (i.e., audiotactile signals which are gestures that combine sound and vibration) are more effective than uni-sensory cues (i.e., a sovereign auditory or vibrotactile signal). For example, Ho, Reed and Spence (2007) and Spence (2012) proclaimed that both bimodal and multisensory signals break through and capture a driver's attention despite engagement in a secondary task, and are not countermanded by concurrent dual tasks. To support their declaration, they outlined several investigations in which experienced drivers reacted over 600 milliseconds faster to a sudden braking of the car in front of them when they sense an audiotactile spatial warning signal at the time (triggered by a sensor's input from the lead car's red brake light having flashed unexpectedly). It should be pointed out that an estimated savings of just 500 milliseconds could result in a 60% chance reduction of being involved in a front-to-rear-end collision. Furthermore, Spence proposes that insights from neuroscience research indicate multisensory signals to be exceptionally operative when they are timed effectively:

> If one is thinking about activating specific brain structures in order to elicit a particular behavioural response from the driver, then one needs to think about stimulus timing. It has been argued that humans are 'designed' to respond to audiovisual stimuli at a distance of 10 metres. Any further away and the sound lags the light; any closer and the visual stimulus arrives after the sound due to the physics of the situation and differences in transduction latencies… This means that inside the car, all multisensory warning signals are, in some sense, going to be presented sub-optimally (i.e., from too close to the driver), at least as far as the multisensory brain of the driver is concerned. Hence, one active line of research for a number of car companies is to see whether they can introduce a slight temporal asynchrony between the auditory and visual or auditory and tactile signals. By so doing, these car companies hope to ensure that the various unisensory elements of the multisensory warning signal reach the appropriate parts of the driver's brain (presumably those controlling orienting or response selection) at the right time (p. 666).

Therefore, we can assume that efforts have already been carried out on a variety of new non-visual *looming signals* that start out quietly with weak audiotactile sensations, but gradually grow in potency, and rapidly intensify to warn the driver of a pressing and perhaps critical need to perform a behavioural response. Such warning signals employ an *urgency mapping* principle (Wiese & Lee, 2004) in which the urgency of the situation is matched by the perceived urgency of the alert. That is the psychoacoustic properties, or envelope of the sound and vibration (including the attack, burst, timbre, intensity, and frequency), are employed in such a way as to depict the need for an immediate response.

Given the above, and considering the idiosyncrasies of the in-car listening experience, which is unquestionably through multi-sensory exposure involving all three aural-visual-tactile/vibratory senses, one can only question if technological innovations and advances in early warning systems based on multisensory input can indeed serve drivers in any effective manner at all. Let's consider for a moment the elements found in some modern music genres such as Drum&Bass, Hip-Hop, Rap, and Techno. Undoubtedly, these genres feature dynamic compression, the equalization of frequencies, and most specifically the sound quality of low frequencies. In addition, let's consider the architectural complexity of vehicle interior. Indisputably, the automobile cabin is functionally designed, upholstered, and equipped for sound, including multi-channel dispersement among the strategically placed tweeters, full- and mid-range speakers, and subwoofers, that collectively provide aural stimulation and bass-generated tactile sensations. Hence, there is every possibility that drivers' sensorial awareness and perceptual understanding of multisensory warning signals and cues may simply be completely distorted or even obliterated all together. Or, perhaps any supplementary information that is meant to reach drivers in a more systematic staggered timed fashion, may be totally eliminated from the acoustic environment. It is not just simply an issue of volume (which is hazardous because music exposure prevents drivers from hearing sirens or road traffic), but rather that the vibratory nature of many musics themselves can place drivers at high risk. In this connection, the World Health Organization (WHO) claims that in-vehicle 'boom' noise, resultant from music-generated high-intensity low-frequency (HI/LF) exposure, can cause physical damage at the cellular level. Accordingly, HI/LF music can damage human organs, induce sensory and mechanical hearing loss, evoke hypertension, cause physical and mental stress, and subsequently lead to sleep loss and depression. Furthermore, the WHO also warns of impaired cognitive development and thought processes as well as learning impediments (Lowertheboom, n.d.). Finally, HI/LF music exposure most specifically interferes with motorists' ability to hear and respond to emergency vehicles and has also been associated with aggressive driving behaviours. In a most recent highly published case that was judged in the Bristol Crown Court (UK), civil charges were brought up against a 25-year-old motorist who was caught driving erratically under the influence of Drum&Bass music (Bellamy, 2013, Farrier, 2013). As described in the UK (tabloid) press, the defendant pleaded guilty for being 'high' on the 'intoxicating effects' of his favourite tunes. Accordingly, this delivery driver was spotted by traffic police taking a sharp turn and then running two sets of red lights, before aggressively chasing and overtaking another motorist, and finally rounding a corner so sharply that the delivery van nearly tipped over. The plaintiff reported that the traffic officers who pulled over the driver could not substantiate DUI by either alcohol or drug tests. The defendant, who blamed HI/LF Drum&Bass music for driving erratically and dangerously, was sentenced to 80 hours of unpaid community service, banned from driving for a year, ordered to pass an extended driving test, and fined £60 (ca. $100). It is clear that although there might come a time when affordable automobiles are sold with manufacturer-installed looming warnings signals as standard on-board features, and these may present slight temporal asynchronies between auditory visual and tactile/vibratory inputs towards more effectively warning drivers of a required critical behaviour, there is no doubt that the vibratory nature of current and future music styles along with listeners' preference for a tonal spectrum that overemphasises the bottom bass frequencies, will cause all of these to be highly ineffective and even useless. When drivers are engrossed in activities involving listening, singing, drumming, and dancing to songs heard in the background, both auditory and audiotactile warning systems – and in fact all other external sounds of the traffic environment – seem to be completely eclipsed by in-car musics (especially HI/LV music genres such as Drum&Bass, Hip-Hop, Rap, and Techno).

4–2.2.4. Perceptual modes of listening

Gaver (1993a, 1993b) makes a unique argument for two distinct types of auditory involvement that are qualitatively different, and these certainly apply to the act of driving an automobile. Accordingly, the distinction is made between the two modes at perceptual resolution levels by which listeners experience music. But, the modes of listening are more psychoacoustic in nature than musical (i.e., the conceptions and models presented previously in Chapter 2, Section 2–1.1, that outlined theories forwarded by Meyer [concerning perceptual tenacity of listeners referred to as *Absolutists* versus *Referentialists*] and Kemp [concerning the cognitive-based perceptual styles, referred to as *Field-dependence* versus *Field-independence*]). Further, although using the terms 'Everyday Listening' and 'Musical Listening' as nomenclature to differentiate between the two distinct modes, these are not in the least related to musical notions (i.e., the conceptions and models presented in Chapter 2, Section 2–2.1 that outlined theories by Behne [concerning *everyday listening*] and Sloboda [concerning *music in everyday life*]).

Gaver (1993a, 1993b) delineates *everyday listening* as hearing the 'source of the events', while *musical listening* relates to the 'sensory quality of the sounds'. In the former, we listen to things going on around us and hear those things that are important to avoid, as well as those that offer possibilities for action. Namely, we listen to perceptual dimensions of auditory events in the environment – while not necessarily focussing on the sounds themselves. As an example for Gaver's first mode of listening, let us imagine that when walking down a street we hear an approaching car; we can judge how close it is and how quickly it is approaching, just from the perceptual dimensions of the auditory event. Moreover, by simply hearing an aural event we are capable of distinguishing an abundant amount of information from it. For example, we might discern if people are running upstairs or downstairs; the magnitude of a mass as it strikes a water surface; the speed of an object as it collides with another; the height of an object falling from the sky or the time it might take until hitting land; if pouring water has completely filled a cup; the size of a barking dog; and maybe the dimensions of an empty room from its echo. Gaver asserts:

> We can hear an approaching automobile, its size, and its speed. We can hear where it is, and how fast it is approaching. And we can hear the narrow echoing walls of the alley as it is driving along (1993a, p. 8)...We can listen to the sound of an approaching automobile, its engine, or perhaps one of its cylinders (especially salient when something has gone wrong) (1993a, p.17)... Most [automobiles] involve a number of rotting parts which can be expected to produce repetitive contributions to the overall sound... and hence a change in the regular (habituation) sound should alert the user to a malfunction (1993b, p. 306).

Gaver's second mode of listening is concerned more with the sound itself; we listen to the sensory dimensions of auditory attributes of the sound in the same way as a composer does when creating music. From a conceptual point of view, Gaver notes that *music sounds* do not necessarily represent the range of sounds we normally hear in the real-world environment, and therefore are relatively uninformative in that they reveal little about their source or acoustic transformations. Accordingly, musical sounds are harmonic, smooth, and feature a relatively simple temporal evolution. On the other hand, *everyday sounds* provide a great deal of information about their source, as well as reflect physical changes in nature. Yet, as far as artistic applications are concerned, everyday sounds are musically useless as they are inharmonic, more complex, and can even be considered 'noise'.

Certainly, it is imperative that we are able to hear and listen to any sound within the vehicle cabin itself or from the external roadway while we focus on both the events that caused the sound as well as on attributes of the sound itself – especially since both are ample resources of information that are vital for vehicular performance and driver safety. Nevertheless, we can only surmise that in-car listening to background music totally masks mechanical sound events from being heard by the sheer volume intensity of the music exposure. That is, when we jollily tap along and sing aloud with our favourite tunes blasting from the car stereo, we are not aurally aware of the acoustic nuances that accompany the proficient performance of a vehicle while travelling on the road. Furthermore, there is a possibility that when we are completely immersed in one particular mode of listening – for example, being totally absorbed in the sensorial dimensions and attributes of a James Jamerson bass guitar line in any one of a thousand great *Motown* hit songs he recorded – our ability to switch between Gaver's perceptual resolutions (i.e., modes of listening) will be compromised. Such a situation might represent a form of perceptual blindness (or perhaps, in the current context we should appropriately label the consequence as a form of *perceptual deafness*), in which the colloquial statement 'I WAS LISTENING BUT DIDN'T HEAR IT' might certainly be most warranted. This proposition is especially relevant for young drivers who have been found to be more susceptible to 'attentional capture' (Crundall & Underwood, 1998; Lee, 2007), and demonstrate the inability to reallocate their focus of attention for periods up to 400 milliseconds longer than more experienced drivers (Underwood, Chapman, & Crundall, 2009). Hence, it is particularly important that drivers (as well as traffic researchers) carefully deliberate the safety implications of music listening in the car, as to the extent by which the radio, CDs, MP3s, smartphones, and other peripheral devices coupled to the entertainment system potentially mask critical warning signals and sounds of the car itself.

4–2.2.5. Benefits versus costs

In Chapter 3, it was stated that automobile drivers seem to have an uncanny, extraordinary ability to cut themselves off from the world around them, albeit at the same time for safety reasons, they seem to be able to remain aware of every inch between them and the car in front of them (Bull, 2001; Miller, 2001). While this apparent behaviour was listed among the various *benefits* of driving with music, one must also look at such behaviour as a contributing *risk factor* that can lead to incidents, near-crashes, and crashes. Ho and Spence (2008) found evidence demonstrating inattention due to perceptual impairment caused by the division of attention between the different sensory modalities (eyes and ears), as well as by processes of switching between the input modes themselves. Therefore, while many drivers (and few traffic researchers) perceive the costs of operating the radio or MP3 players as negligible (at least on the more automatized driving skills such as steering), there is still the possibility that higher level decision-making abilities may in fact be compromised, and the impact could be life threatening. Hence, such a position that driving with music may be a risk factor for distraction must be considered seriously (Brodsky & Slor, 2013). If driving with loud and utterly absorbing music is seen as a positive activity because it can allow the listener to lose him or herself in the music even for only a second or two, then such inattention – even if just for a second or two – has to been seen as having the potential for fatal consequences. Interestingly, there is anecdotal evidence suggesting that when driving conditions become more difficult – such as driving in congested traffic or unfamiliar locations – drivers do turn the music down or shut it off altogether (BBC, 2004; Brown et al., 1969; Konecni, 1982; North & Hargreaves, 2000). The implication may be that listening to music can be both a diversion in that it energizes, as well as a hazard in that it masks critical information.

4–2.3. Capacity Interference

In-vehicle systems seem to have grown by a factor of ten during the last two decades. As a matter of fact, although these are highly dynamic and the levels of interactive behaviour differ from one device to another, the industry seems to have put in place standards and benchmarks concerning how manufacturers must create these peripherals to be user-friendly for automobile drivers. Today several brands not only include 'learning curves', 'task efficiency', and 'error handling', but also have extended their product lines to include criteria such that 'driver distraction' must be avoided. Yet, no matter what industrial partners claim, the honest to goodness truth is that either singularly, or in combination, in-vehicle systems tend to contribute to an increased load on attentional reserves and consequently to driver distraction (Bach et al., 2009; Tijerina, 2000). Therefore, while new applied technologies (including futuristic-like dashboard features, after-sales gadgets, and peripheral infotainment devices) all seem to be highly advantageous for drivers, their use raises concerns about safety risks – especially as evidence points out that those engaged in complex irrelevant secondary tasks while driving tend to be more involved in accidents. After all, we have only a limited supply of resources to process information, and when our eyes and mind become directed away from the roadway in front of us – along with our hands being located elsewhere other than being firmly planted on the steering wheel – concerns about in-vehicle systems as a safety risk to the driver have surfaced. The crux of the matter is that when secondary tasks deplete our attention (such as engaging in cell phone conversation, or listening to sports radio, or singing along with background music), we have reduced the assets available that can be dedicated to driving, and subsequently a decrement of vehicular performance occurs. This is referred to as *capacity interference* to central attention (Consiglio et al., 2003). As pointed out in the previous section, blindness to visual perceptual abilities is caused by auditory secondary tasks (Ho & Spence, 2008). Certainly, the extent of such impairment has much to do with the cognitive workload (or overload) that is associated with performing secondary auditory activities.

While Section 4–2.1 (above) described a more general withdrawal of attention in regard to structural mechanical interference (denoted as *eyes-off-the-road* distraction), and Section 4–2.2 (above) detailed a selective withdrawal of attention in regard to perceptual masking (consequent to active involvement with two tasks that either require the same channel of input or diverging cross-sensory inputs), the resultant driver mind-set subsequent to circumstances of capacity interference reflects a third manifestation of inattention referred to as *mind-off-the-road* distraction (Bach et al., 2009). Accordingly, in its basic form, this type of distraction indicates obstructed cognitive processes implicating diminished capacity for interpretation and decision-making. Long ago, Brown (1994) referred to a circumstance of *driving without awareness* that occurs by switching driver focus of attention to inner thought processes. In general, this kind of attention deficit is commonly associated with daydreaming. But yet, as drivers seem to be unaware of this impaired state, they are not in a position to remedy it. Further, there is evidence that talking on the phone or even conversing with other passengers can lead to this same level of mental interference. For example, Strayer and Drews (2004) implemented a simulated driving investigation in which 40 drivers (20 younger drivers aged 18–25, and 20 older drivers aged 65–75) drove on a multi-lane 24-mile freeway with on- and off-ramps, overpasses, and two- or three-lane traffic in each direction. The participants drove under two conditions: single-task (driving-only) and dual-task (driving with cell phone conversation on topics that were identified in advance as being of interest to each participant). To rule out manual components of phone use, and place an exclusive focus on the cognitive constraints of driver conversation, dialling and initial exchanges with a caller commenced before the onset of the dual task, and the study solely employed a hands-free cell phone. During the experiment, the drivers followed a lead car and were instructed to pace themselves appropriately; they were required to keep their distance, maintain speed, and brake when necessary. The study found that when drivers were engaged in cell phone conversation their reactions were 18% slower, the following distance was 12% greater, it took them 17% longer to recover the speed that was lost following braking, and there were twice as many rear-end collisions. No significant differences surfaced between the younger and older drivers. Strayer and Drews conclude that these findings undoubtedly demonstrate the distraction effects of cell phone conversations. One can only imagine that such effects might have surfaced also had the drivers been required to sing aloud with all their heart to their favourite tunes while driving.

'Mind-off-the-road' distraction has also been referred to as *inattention blindness* referring to the phenomenon whereby drivers fail to see an object in the environment although they were looking directly at it (Strayer, Drews, & Johnston, 2003). The US National Safety Council (NSC, 2010, 2012) indicates that when drivers use a hands-free phone they often have a tendency to 'look at' but do not 'see' objects caused by the cognitive diversion of a cell phone conversation. According to NSC estimates, drivers using cell phones look but fail to see up to 50% of the information in their driving environment. Conversations can indeed cause a 'withdrawal' of attention from the visual scene, even neglecting explicit memory recognition of objects, traffic signs, and roadside billboards (Ho & Spence, 2008; Strayer & Johnston, 2001; Strayer et al., 2003). Like 'tunnel vision' in which drivers may be looking out of their windshield but yet do not process everything in the roadway environment, cognitive distraction during conversation can draw the driver's attention causing them to be incapable of effectively monitoring their surroundings, identifying potential hazards, or responding to unexpected situations. The NSC concluded that being involved with listening and responding to a disembodied voice (i.e., paying attention to conversations) is in fact cognitive distraction that contributes to numerous driving impairments. It is interesting to note that anecdotal evidence points to the fact that most drivers recognize when they are visually or mechanically distracted, and subsequently, they seek to disengage from a secondary activity as quickly as possible. Moreover, many drivers become aware that they have been distracted by 'inner thoughts', especially when their attention is suddenly redirected to the road. Yet, as we typically do not realize we are cognitively distracted (such as, during a phone conversation or a chat with an accompanying passenger), then the actual risk lasts much longer. Ironically, we appear to have absolutely no recall whatsoever of events that occurred during the period of our unawareness (which in some aspects may be reminiscent of a trance-like state), and we can only assume that at the time our responsiveness to potential hazards and collision avoidance would have been less than adaptive. 'Cell phone-induced-inattention' suggests that the core of the problem lies at the central attentional level and is not due to structural mechanical interference or perceptual masking (Strayer & Johnston, 2001). Finally, He, Becic, Lee, and McCarley (2011) reveal yet another category of capacity interference that they refer to as *mind wandering behind the wheel*. 'Mind-wandering' is a particular pattern in which there seems to be an independent shift of attention away from the immediate driving task towards task-irrelevant thoughts – regardless of the fact that the driver is not involved in a secondary task. Accordingly, mind-wandering is accompanied by performance losses, including: slower and more shallow information processing, declines in vigilance and memory, narrowing of visual attention with wider peripheral gazing, larger fixation durations, and overall failure to scan/monitor the traffic environment. Stutts et al. (2001a) found that among accident analyses records of cases described as failures of attention, there were roughly 5% of drivers who were not necessarily engaged in a secondary task but rather described themselves as being 'lost in thought'.

Now, in reference to the *music* itself, little information is available. Does active listening to music, or singing along with the songs, or tapping in beat, or playing 'air-guitar', or dancing in the seat cause cognitive diversion and visual inattentiveness that can lead to diminished capacity for interpretation and decision execution? For the majority of us, there seems to be no answer! First, traffic researchers and accident investigators are not mindful of the risks associated with music (i.e., the sounds themselves). One reason for this gap is that many studies have totally 'side-blinded' readers (including advanced graduate students who are tomorrow's researchers) as well as current authorities on traffic safety, by disregarding music as a risk factor (Dingus et al., 2006; Klauer, Dingus, Neale, Sudweeks, & Ramsey, 2006a; Klauer, Sudweeks, Hickman, & Neale, 2006b; Stutts et al., 2001b). Yet, while occurrences may seem minor, data surfacing from studies such as by Stutts et al. (2005) do in fact indicate that driving with music (that is, exclusively listening to music while not manipulating audio controls) does affect driver behaviour and is accountable for incidences and events that can cause harm. For example, Stutts et al. (2005) found that 'listening to music' is accountable for 2% of incidences involving one hand on the steering wheel, 2.4% of incidences in which drivers' eyes were directed inwards inside the cabin (later corroborated by Horberry et al., 2006), and 8% of adverse vehicle events per hour (including lane swerving, crossing over lane boundary, and sudden hard braking). But, as the mainstream has disregarded in-car music as a risk factor, driver's education courses are not concerned with the effects that background music may have on vehicular control, and driving instructors do not even mention *music* in their educational programming. Second, many researchers have made erroneous assumptions that the presence of music "is not, in itself, distracting, [and therefore] entertainment systems only constitute a distraction when they need to be attended to, i.e., intermittently, to change a tape or re-tune a radio" (Stevens & Minton, 2001, p. 542). Further, some researchers have out-and-out wrongfully declared that *music* is 'not at all associated with negative driving performance' (Stutts et al., 2003, p. 205). Certainly, suppositions as these are internalized *as if* they were proven *truisms*. Third, the literature is full of reports by researchers who carelessly perpetuate dissemination of faulty deductions about music simply by citing findings from previously published studies without scrutinizing the data themselves. One example will suffice to illustrate the point. In a well-cited document by the US National Safety Council (NSC, 2010, 2012) titled *Understanding the Distracted Brain* the following axiom is presented:

> Listening to music does not result in lower response time, according to simulator studies. But when the same drivers talk on cell phones, they do have a slower response time. Researchers have concluded that voice communication influenced the allocation of visual attention, while low and moderate volume music did not (p. 8).

Those who read the above statement unfortunately perceive the following message: *Music is completely safe as far as there are no cognitive interference effects that cause increased braking responses*. However, had the writers of the NSC's document looked into the original source of the citation (i.e., Bellinger, Budde, Machida, Richardson, & Berg, 2009) they would have no doubt seen that the study did not implement a research structure that can boast to be an ecologically reliable resemblance of a driving environment – albeit, the study is an excellent cognitive psychology laboratory experiment highlighting isolated human perception and performance. Bellinger et al. recruited 27 undergraduates from Miami University to sit in front of a PC workstation (with two floor pedals as response triggers) with which they were to execute a braking manoeuvre in response to a table-mounted red-light lamp (emulating the rear brake light of a car in front). Certainly, this experimental driving task was far removed and isolated from the context of controlling a vehicle on a virtual road (as is the usual case within simulated driving studies). There were six conditions: one single-task (braking-only) and five dual-tasks (1 x braking + cell phone conversation; 2 x braking + music-only [with either low or medium volume]; 2 x braking + conversation + music [with either low or medium volume]). However, when examining the means used for music exposure, the experiment employed a home-movie surround-sound system of five speakers to reproduce the music at either 66dBA or 78dBA. Clearly, such a set-up is far from the typical car-audio speaker configuration (albeit highly impressive for a psychology laboratory). In addition, given that the volume intensity served as a dependent variable, we might wonder if such volume levels (defined as low/medium intensity) and the differences between them, are in fact valid to begin with. Therefore perhaps the lack of findings here are no more than artefacts of 'floor-effects'. Moreover, the music exemplars employed by Bellinger et al. were two tracks from the 2006 CD titled *Surrounded* by the urban gospel group Men of Standard; the two tracks were 'I Will' and 'Power'. There was no explanation why these particular selections (or for that matter the band, the album, or the tracks) were chosen. Moreover, the duration length of the two music exemplars are 2:41 and 3:09 minutes, while the blocks themselves are described to have been of a shorter duration (although the exact length is unspecified). In addition, as the two exemplars were used repeatedly in both music conditions, as well as perhaps in practice trials, we might wonder if there are effects of repetition (i.e., familiarity). Further still, it should be noted that both tracks chosen include excessive vocals (i.e., three-part harmonized texts as well as some 'rapping'), and hence these stimuli might just have provided the participants with perceptual qualities that are much more similar to listening to a conversation than to what Bellinger et al. describe as a pristine example of "soul rhythm-and-blues music that involves electronic instruments" (p. 444). Thus, one can only question the validity of the data to begin with – let alone the interpretations as conveyed in their published report. Finally, one could only assume that like most 19- to 23-year-old listeners, the participants may have been instinctively 'humming', or internally singing along, in at least one of the four exposures – especially during the music-only condition. Internal humming activity has been demonstrated to involve sub-vocal activity with phonological involvement (Brodsky, Henik, Rubinstein, & Zorman, 2003; Brodsky, Kessler, Rubinstein, Ginsborg, & Henik, 2008). Hence, if no facilitation or degradation effects of the music listening surfaced, it would have been warranted that Bellinger et al. look at their experimental methodology for imperfections and empirical errors. Having said all the above, and while the study was published in the scientific literature, readers of the NSC dispatch would have expected a USA funded safety agency to have inspected the facts of any study before citing it as having demonstrated that "listening to music does not result in lower response time..."

4–2.3.1. Neuroergonomics

Cognitive distraction consists of absorbing thoughts that take up the driver's attention. Albeit, it is highly difficult to assess how much attention is a contributory factor in real driving incidents. Yet, perhaps *neuroergonomic* studies that apply neurobiological explanations to issues of human factors in driving may shed light that can widen this inquiry. To clarify, neuroergonomics presents an outlook on human performance and ergonomic design based on neural limitations in information processing.

In an early study that attempted to map-out the 'driving brain', Walter et al. (2001) developed a driving simulation device that could be used within a functional magnetic resonance imaging (fMRI) environment. In that study, there were 12 German drivers who participated in two 10-minute sessions; one session was defined as 'passive driving' (i.e., looking at a video screen of the road scene *as if* from a passenger's perspective) while the other session was defined as 'active driving' (i.e., two simulated driving activities involving velocity/acceleration and road-tracking/steering tasks). The road environment was presented to the participants via LCD goggles, and steering was implemented via an fMRI-compatible joystick (replacing the more standard windshield and steering wheel). The study reported brain activations during active driving compared to the passive passenger control condition in the left sensorimotor cortex, the vermis, both cerebellar hemispheres, occipital areas, and in parietal cortex bilaterally. This finding suggests that simulated driving mainly engages areas concerned with perceptual-motor integration, rather than areas associated with higher cognitive functions. Hence, the authors concluded that driving actually *deactivates* a number of brain regions. Walter et al. infer that the activations of the left sensorimotor cortex and the cerebellar regions were essentially neural correlates of motor activity necessary for steering the car with the right hand. Finally, they presume that activation observed in the right occipital and parietal regions were probably a result of the perceptual processes associated with driving. In summary, the research team suggest that simulated driving requires the coordinated activity of occipito-parietal and motor brain areas while other regions, including motion sensitive areas, are less active compared to more passively viewing the road scene.

In a second neuroergonomic study, Jancke, Brunner, and Esslen (2008) specifically targeted one type of driver behaviour – fast impatient driving. The overriding foundation of the study accounted for the great strides that driving psychology had made during the eight-year interim since Walter et al. (2001) first published their findings. Namely, between 2000–2007 research scientists had come to view controlling a driving environment as a highly complex endeavour requiring the effective planning and implementation of several executive functions, among them selective attention, alertness, divided attention, working memory, prospective memory, behaviour selection, and behaviour inhibition. Jancke et al. hypothesized that if fast impatient driving places greater demands on the utilization of executive functions (such as inhibition of inappropriate behaviour, increased alertness, and explicit management of changing situations), then, there should be greater activation in the dorsolateral and the ventrolateral prefrontal cortex. But, Jancke et al. also foresaw the possibility that as driving speed increases, there may be less emphasis placed on using executive control processes subsequent to diminished activation of the lateral prefrontal cortex. In the study, 28 German males were recruited to drive a virtual vehicle (set up as an intact cabin of a Ford *Focus*), all the while undergoing electroencephalography (EEG) recordings with 30 scalp electrodes for the brain, plus four electro-oculogram conductors for eye movements. Simulated driving was implemented in two trips; one trip was a virtual highway with normal traffic, while the second trip passed through small village streets with less traffic. Both trips were undertaken during foggy weather conditions. Among the parameters collected were the actual cruising speeds and the number of near-accidents and accidents. The research team found increased α-band activity in the right lateral prefrontal cortex during trips with excessively fast driving compared to those in which the driving was performed in more appropriate speeds and when drivers stayed within the boundaries of other traffic rules and regulations. It should be pointed out that such an escalation in α-band activity correlates with reduced hemodynamic responses (i.e., activation). Further, the researchers also infer that as the right dorsolateral prefrontal cortex plays a prominent role in behaviour inhibition and control of risk taking behaviour, they presume it is quite possible that decreased neural activation in these areas during fast driving indicates diminished behavioural inhibition and capacity for control of risk-taking behaviour. Therefore, the decreased neural activation that surfaced in the study might be related more to the utilization of other executive functions than meets the eye (such as inhibition of inappropriate behaviour, increased alertness, and explicit management of changing situations). In light of their finding that fast driving is associated with a decreased neural activation in the lateral prefrontal cortex, Jancke et al. raise the possibility that increased driving speed is also associated with a reduced engagement of executive control processes in favour of more automatic control processes. And lastly, as this area is also known to be involved in the control of multitasking, neurophysiological deactivation might indeed indicate that the utilization of less neurophysiological effort in controlling multitasking behaviour *is* an adaptive behaviour.

A final third study that employed fMRI with drivers while engaged in a secondary task presents some highly convincing evidence about 'attention' as a contributory factor in real driving incidents. The study is positioned as the first investigation to use brain imaging in an effort to explore the effects of performing an auditory language comprehension task while simultaneously engaged in a simulated driving task – which are in essence, two tasks known to draw on different cortical networks. Just, Keller, and Cynkar (2008) recruited 29 American drivers (50% males) who were between 18–25 years of age. While undergoing fMRI brain scanning, each participant navigated a virtual vehicle in daytime driving, on streets with good visibility (without intersections, traffic, or hazards) along a curving winding computer-generated road cruising at a steady speed (43 mph). Two conditions were implemented in alternate orders across the sample: a single-task condition of undisturbed *driving-alone*, and a dual-task condition consisting of *driving while listening* to spoken sentences (presented over fixed-field high-fidelity MRI-compatible electrostatic headphones that mask scanner noise at 60dBA). The drivers were scanned at 3-Tesla with a blood-oxygenation level dependent fMRI acquisition sequence while they manoeuvred a virtual car using a computer mouse or trackball in the right hand. In the dual-task condition, the participants not only steered the car but also listened to general knowledge sentences, confirming each statement in a two-alternative forced-choice methodology. Namely, the participants employed two response buttons held in the left hand to respond to the verbal statements with 'True' or 'False' responses. Behavioural measures included reaction time and response accuracy, while simulated driving was assessed through performance measures such as percentage of hitting the berm (i.e., side edge of the road), and standard deviation of lane position. The central findings were: (1) listening tasks (i.e., processing of heard sentences) consistently hampered driving performance by producing a significant decline in driving accuracy; (2) listening tasks cut back anatomical brain activation in key regions that supported driving; and (3) listening tasks reduced signal intensity in anatomical regions of interest (which indicated levels of functional conduct for spatial, visual, motor, executive, and language areas). These findings are immensely important here (see Figure 4.5). As can be seen in Panel-A of the figure, driving-alone produced large areas of activation in the bilateral parietal and occipital cortex, motor cortex, and the cerebellum. When sentence listening was added to the driving (i.e., dual-task), the same driving-related areas were activated, but yet listening also triggered additional areas related to language comprehension – Panel-B – and these were the bilateral temporal and left inferior frontal regions. Moreover, Just et al. demonstrated that in fact a decrease of activation in the brain areas that underpin the driving task does occur, especially as processing spoken language draws on attentional and brain resources. When comparing Panel-A and Panel-B, driving related activation in the bilateral parietal lobe associated with spatial processing in the undisturbed driving task, can be seen as 37% less when participants were concurrently listening to sentences. Finally, Panel-C indicates reduced signal activity for all anatomical regions of interest. Namely, there were drops in the percentage change of signal activity for all four functional groupings (networks) of cortical areas, including spatial, visual, motor, and executive components; however, the visual, motor, and executive components did not reach levels of statistical significance. Yet, a statistically significant reduction in driving-related neural activity secondary to dual-tasking did surface for spatial processing components ($M_{\text{Single-task}} = 0.258$, $M_{\text{Dual-task}} = 0.163$; $F_{(1, 28)} = 29.38$, $p < 0.05$). Furthermore, the addition of listening significantly increased neural activity for the language areas ($M_{\text{Single-task}} = 0.070$, $M_{\text{Dual-task}} = 0.196$; $F_{(1, 28)} = 64.43$, $p < 0.05$); an increase most prominent in bilateral primary and secondary auditory areas of the temporal lobe, the pars triangularis region of Broca's area in the left hemisphere, and the hemologous region of the right hemisphere.

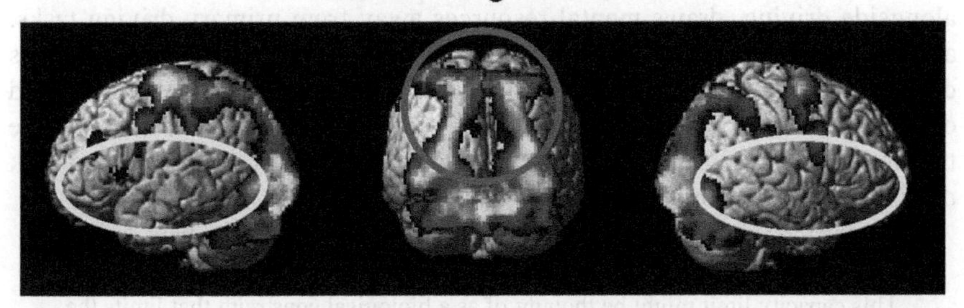

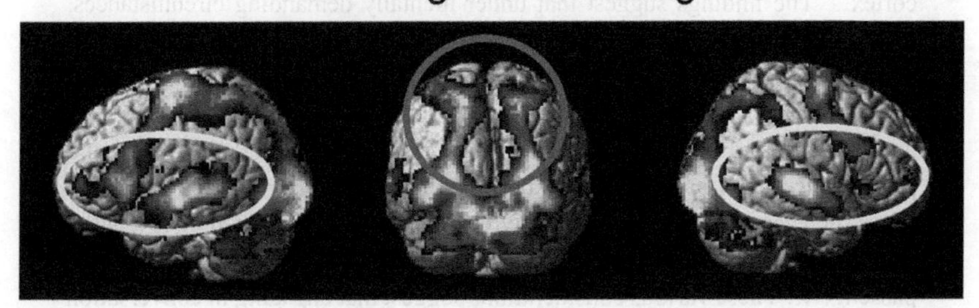

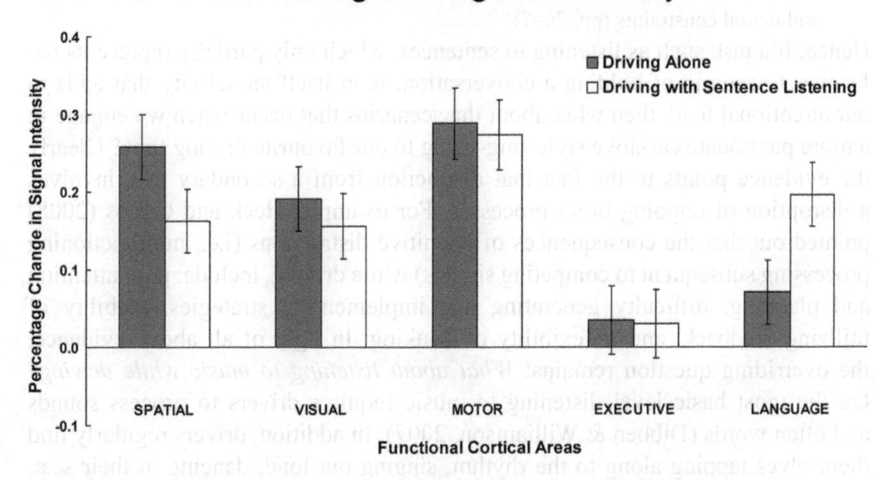

Figure 4.5 Brain activation while driving when listening to speech

Source: Just, Keeler, and Cynkar (2008). Used with permission of Elsevier.

Just et al.'s (2008) findings show that language comprehension performed alongside driving, draws mental resources away from primary driving tasks, and subsequently weakens driver performance and vehicular control as demonstrated by increased driving errors of hitting the berm and larger standard deviations of lane position. Certainly, the study is of particular interest because the findings show limitations of brain activation during concurrent dual-task driving. Just et al. conclude:

> We interpret this diversion of attention as reflecting a capacity limit on the amount of attention or resources that can be distributed across the two tasks. This capacity limit might be thought of as a biological constraint that limits the amount of systematic neural activity that can be distributed across parts of the cortex... The findings suggest that under mentally demanding circumstances, it may be dangerous to mindlessly combine the special human capability of processing spoken language with a more recent skill of controlling a large powerful vehicle that is moving rapidly among other objects... The new findings raise the obvious point that if listening to sentences degrades driving performance, then probably a number of other common driver activities also cause such degradation, including activities such as tuning or listening to a radio, eating and drinking, monitoring children or pets, or even conversing with a passenger... These various considerations suggest that engaging in conversation while concurrently driving can be a risky choice, not just for commonsense reasons, but because of the compromised performance imposed by cognitive and neural constraints (pp. 76–7).

Hence, if a task such as listening to sentences, which only partially represents the far-worst scenario of holding a conversation, is in itself an activity that adds to our attentional load, then what about the scenarios that occur when we engage in a more passionate car-aoke style sing-along to our favourite driving tune? Clearly the evidence points to the fact that distraction from a secondary task involves a disruption of ongoing brain processes. For example, Heck and Carlos (2008) pointed out that the consequences of cognitive distractions (i.e., malfunctioning processing subsequent to competing signals) while driving, include: poor attention and planning, difficulty generating and implementing strategies, inability of utilizing feedback, and inflexibility of thinking. In light of all above evidence, the overriding question remains: *What about listening to music while driving*? On the most basic level, listening to music requires drivers to process sounds and often words (Dibben & Williamson, 2007). In addition, drivers regularly find themselves tapping along to the rhythm, singing out loud, dancing in their seat, and even recalling memories triggered by the music. At the very least, the costs associated with this level of cognitive distraction might be observed in diminished driver response to unexpected events or hazards.

Considering the demands of the traffic environment, when we engage in secondary tasks involving music we might do well by regulating and modifying our driving performance to some extent for the added *cognitive workload*. There have been reports that some drivers do so by reducing the rate at which driving-related information is taken in and processed. Namely, we slow down. Brown, Tickner, and Simmonds (1969) were the first to document that drivers deliberately reduce their speed in order to handle additional cognitive loads. Thirty years thereafter, Wikman et al. (1998) explained that drivers react to more complicated driving situations by allocating more attention to the primary driving task – foremost by reducing their speed – then as attention is brought back to the driving task, the time used to consult the instrument panel decreases and peripheral glances to secondary tasks targets become minimal (especially when it is anticipated that the difficulty of the driving task will increase). Strayer and Drews (2004) confirmed such suppositions in further experimental demonstrations. Jamson and Merat (2005) asserted that this sort of manoeuvre is the result of developed strategies with intentions to "free-up resources for the secondary task by simplifying the primary task" (p. 93), and Engström et al. (2005) presumed that such a manoeuvre can only be interpreted as a "compensatory effect where the driver reduces the primary task load in order to maintain driving performance on an acceptable level" (p. 99). Horberry et al. (2006) concluded that the lower speed limit enforced by drivers when engaged with in-vehicle entertainment distraction is presumably a "compensatory mechanism adopted by participants to increase their margin for error when they were distracted" (p. 189). Shinar (2007) claimed that instinctually drivers slow down to increase the headway because "when we slow down we decrease the rate of information flow, and when we increase headways we increase the time to respond to sudden and unexpected or unplanned events, such as an unexpected stopping or slowing of the car ahead" (p. 536). Although it appears that many drivers do in fact employ such strategies when they perceive to be somewhat pre-occupied by a secondary activity, ironically listening to music involves quite unconscious processes, and therefore, it would stand to reason that since music influences drivers without them even being aware of it, there is little that they do to strategically remedy the situation.

4–2.3.2. Inattention and distraction

Given that most of the information used by drivers is visual, it has often been assumed (and some claim to have demonstrated) that combining driving with an auditory task such as listening to music over the car radio should cause little if any decrement in driving performance (Bellinger et al., 2009; Unal et al., 2012, 2013a, 2013b, 2013c). But yet, when looking at findings on mobile cell phone use, perhaps a different interpretation does surface. For example, many researchers (Just et al., 2008; Ho & Spence, 2008; Horrey & Wickens, 2006; Spence & Ho, 2009; Spence & Read, 2003) point to the fact that as no wide-ranging all-encompassing differences between hands-held and hands-free cell phone use have ever surfaced, and if there had been differences these would surely signify some level of manual limitation and motor deficit associated with holding or manipulating a phone while steering the car, then any difficulties encountered by drivers stem from cognitive impairment associated with trying to divide their attention between their eye and their ear. That is, the issue is more of a competition for mental resources at a central cognitive level. In this connection, Recarte and Nunes (2000) found spatial tasks with a verbal component (i.e., repeating letters) caused the reduction in eye gazing to specific spatial regions during driving. Further, Strayer and Johnston (2001) found that simulated driving while engaged in a phone conversation caused drivers to miss traffic signals, and that RTs for braking to critical events were increased (i.e., 30% slower) compared to listening to talk radio. Further still, Strayer, Drews, and Johnston (2003) found that driver conversation hampered drivers' capability to focus on foveal visual information resulting in missing information. So again, we might ask if these findings are applicable to driving while singing songs heard in the background (which no doubt is also phonological verbal activity).

Consiglio et al. (2003) demonstrated that engaging in a conversation decreases the efficiency of oculomotor search, subsequently reducing accuracy of driver performance in a wide range of different driving-related tasks. There seems to be a wealth of evidence confirming that: (1) drivers do have a limited pool of attentional resources; (2) at best these are shared between processing of information across different sensory modalities; and (3) there seems to be a fundamental competition going on in the brain for access to whatever limited-capacity central-response output does exist. Perhaps, human brains do not efficiently perform two simultaneous tasks in parallel but rather sequentially – albeit as our brain continuously juggles each task back-and-forth ever so rapidly by switching attention between the tasks we are led erroneously to believe we are 'doing' two things at the same time. Yet, it should be understood that during the 'doing', our cognitive resources select which information the brain will attend to, which information will be processed, which information will be encoded, which information needs to be stored in memory, which information previously stored needs to be retrieved from memory – and only then – executes responses based on the procedural operation. Neuropsychology research long ago demonstrated that different neural pathways, and hence different areas of the brain, are more engaged in some activities than in others; these seem to depend, to a certain degree, on what type of information is involved. Therefore, when we attempt to perform two cognitively different complex tasks – such as driving and talking over the phone, or conversing with a passenger, or listening to songs and singing along – our brain continually shifts focus, and then some important information about the road 'falls out of view' because we are mentally taking on too much elsewhere. In an interesting study by Atchley and Dressel (2004) it was hypothesized that performing a conversational task while driving would result in a decreased functional field of view. The researchers conjectured there would be an increased risk while maintaining a conversation. Certainly, impairments of driving performance while performing a concurrent verbal task could be attributable to a number of factors. For example, several notions bring forward conceptual distinctions between the visual field itself and to what we can actually process via the visual field. Atchley and Dressel claimed that such distinctions, if they exist, stem from the physical nature of the visual system. Namely, they presumed constraints on the driver's view imposed by limitations in the information capacity, reflecting a biological competition between what is *functional* versus what is *useful*. Accordingly, "...what people see is not limited by their eyes but by cognitive operations they must simultaneously perform" (p. 665). Ironically, the reduction of the functional visual field has been seen to result in right-of-way accidents (i.e., failures to yield at intersections, and failures to heed to a stop signal). Atchley and Dressel found that when a verbal conversation was added to a perceptual task, limitations to the functional field of view increased, and drivers ability to localize peripheral information accurately, did in fact slow down. They concluded that drivers engaged in verbal conversations have a more narrow window of spatial attention. We must understand that as conversation is a paced activity, and pacing is inherently and ultimately a temporal quality, then the same impairment would also surface if drivers listen to songs – as music with lyrics is based on semantic content much the same as is verbal conversation.

The cost of switching between tasks, that is the measurable time when the brain switches attention and focus from one task to another, might only take a few tenths of a second per event. Yet, in the end, each episode eventually amasses the losses that hamper the system, and might even cause a bottleneck hold-up, in which different regions of the brain are left to retract from the shared limited resources in order to engage in the secondary driving-irrelevant task. When our brain experiences increased workload, and information processing slows down, we are much less likely to be able to respond to unexpected hazards in a timely fashion towards avoiding an accident. Unfortunately, there is evidence that attentional limitations and inattention is co-mediated by driver experience. That is, young novice drivers seem to lack the spare attentional capacity found among experienced drivers, who respond more quickly to the peripheral targets (Crundall & Underwood, 2001; Lee, 2007; Underwood, Chapman, & Crundall, 2009). In this context, spare attentional capacity refers to the difference between the cognitive resources demanded by the driving task versus the cognitive resources drivers have available to invest in a secondary task like conversation or singing songs. Therefore, driving about with favourite music tunes playing in the background, whether listening, singing in unison, or tapping in syncopation, requires a mental investment that may consume all available excess attentional capacity from the required cognitive resources demanded by the driving task. Finally, it is possible that any preoccupation with music can easily overwhelm the driver's system all together causing them to be highly vulnerable to hazards leading to near-crash incidents and crashes.

4–2.3.3. Cognitive overload

For the most part, we do drive safely with an average cognitive workload; during the majority of our trips nothing too great ever happens to us. Hence, over time we develop a 'false sense of security', and subsequently, we believe that we can safely engage in a variety of secondary activities while behind the wheel. For example, we all assume that we can adequately compensate for talking on our *Samsung Galaxy4*, texting on our *HTC One*, web-surfing on our *LG Optimus*, or scrolling through music playlists on our *iPhone5*. We assume that by lowering the speed, or putting some space in between our car and the one ahead, we have remedied any trouble and we are prepared for whatever might arise. Because we perceive that driving rarely requires our full attention, we also look out the window, listen to the radio, sing with our favourite tracks, discuss current events over the mobile phone, talk about our troubles with passengers, read texts, view pictures, and seek out navigational information – all at the same time as eating, drinking, smoking, grooming, and at times perhaps engaging in 'something more intimate'. Long ago, Shinar (2007) astutely commented: "as we gain driving skills we learn to pay less attention to the road and traffic and to share the driving demands with more and more non-driving tasks" (p. 556). Yet cognitive overload often occurs when a secondary task, typically not related to driving, makes exclusive use of one or several processing modules. When we engage in mental, perceptual, or motor operations – for other in-vehicle activities that are beyond the requisites of the primary task of driving – we weigh down and sometimes strain (or 'drain') our resources because of interaction effects between the systems. Hence, by adding other undertakings to the driving task, no matter how mundane, such as listening to music or singing, we do in fact increase driver workload (Patel et al., 2008; RoSPA, 2007; RSC, 2006; White et al., 2004). To highlight the problematic nature of cognitive (over)load, Jamson and Merat (2005) outline two well-accepted theories: Wickens' *Multiple Resource Model of Divided Attention* and Baddely's model of *Working Memory*. In Wickens' model there are three limited capacity resources: a Processing Resource (early perceptual/later central); a Modality Resource (auditory/visual), and a Response Resource (spatial/verbal). The model forwards the concept that optimal task performance occurs when conflicts between resources are kept minimal. Thus, within the context of a primary visual driving task, engagement in secondary tasks will produce less disruption when visual components are low (i.e., faster RTs and efficient vehicular performances). Alternatively, Baddely's model accounts for a Central Executive that selects and/or rejects incoming information in a Short-term Memory module, as well as oversees two subsequent slave systems for storage of Long-term Memory. One is the Phonological Loop that attends to linguistic information, while the other is the Visuospatial Sketchpad that attends to visual and spatial information. This model forwards the concept that optimal task performance occurs when only one of the above tasks is operational. Thus, in the event of concurrent dual tasks, when both tasks share the same working memory modules, performance in one or both tasks may be compromised (i.e., accuracy of vehicular performance deteriorates to the degree of total disruption).

When considering the above, we need to enter 'driving experience' as a crucial component. For example, Crundall and Underwood (2001) raise the point that most likely some drivers – especially young(er) novice drivers – already drive in a state of overload simply by trying to keep their car in the correct lane at the appropriate speed. Heck and Carlos (2008) feel that young(er) drivers are developmentally less able than adults to cope effectively with distractions during driving. Therefore, when supplementing driving with 'ordinary' endeavours (such as listening or singing to music), the previously overtaxed system may be easily extended with further perceptual demands that promote obstructed responses to hazardous events (Horrey, 2011). In this connection, Horrey and Wickens (2006) conducted a meta-analysis of 16 behavioural studies examining dual-task driving. They found that among young drivers, the ill-effects (i.e., 'costs') associated with driving with concurrent cell phone conversations were larger for reaction time tasks (such as braking) than for tracking tasks (such as steering). Moreover, Fagioli and Ferlazzo (2006) found that when drivers have to constantly relocate and reallocate attention between *personal space* (where conversation, reminiscing about a favourite tune, or singing song lyrics takes place) and *extra-personal space* (where driving is executed), there is a cost in target detection tasks. Lee (2007) points out that inexperienced drivers take an average 250 milliseconds longer to detect peripheral targets than do more experience drivers – due to cognitive overload resulting from the lack of automatized driving skills and spare attentional capacity. In addition, there is evidence that young(er) drivers tend to hold off reallocation of attention between dual tasks for up to 400 milliseconds, whereas more experienced drivers refresh in a much quicker manner (Crundall & Underwood, 2001; Underwood, Chapman, & Crundall, 2009).

In an interesting study, Underwood et al. (2009) found poor performances and noticeable declines in vehicular control among young novice drivers resulting from impaired peripheral scanning. The research team offer several conceivable explanations for this finding, and one in particular is highly pertinent to the present discussion. The researchers raise the possibility that driving-related stimuli may in fact trigger unnecessary action employing a sub-vocal phonological code among young(er) drivers in their attempt to interpret traffic signals and road signs. Accordingly, not only is this act redundant to realistically discriminate the printed graphic information, but it is also an additional all-consuming process that reduces global response times. Underwood et al. assert that as the driver gains more driving experience, they also develop the ability to 'override' such uses of their phonological resources. Yet, as far as novice drivers are concerned, while they are still obliged to undertake such operations, clearly such phonological encoding would ultimately cease when engaged in conversation or when listening to music and singing song lyrics (two secondary tasks that engage almost 100% of the age-group's driving activity).

Many studies (Engström et al., 2005; Harbluck et al., 2007; Recarte & Nunes, 2000, 2009) found that higher cognitive loads significantly affect visual aspects of driver performance, including: visual detection, visual discrimination, and visual response selection capacities (i.e., taking evasive action to avoid objects). Although there is evidence that points to cognitive-auditory demands as leading to cognitive narrowing (demonstrated by increasing drivers' spatial gaze to road ahead), this 'improvement' for scrutinizing the roadway also comes at a cost of reduced hazard perception and situational awareness. Therefore, while most drivers who engage their phonological reserves (i.e., cell phone and passenger conversations, or singing along with songs heard in the background) appear to be highly confident in their multi-tasking capabilities and certainly show no fear of taxing their own attentional proficiencies, the reality of the situation may simply be that such courage is no more than bullish arrogance and ignorance in recognising their own limitations.

4–2.4. Social Diversion

While the advancement of technology has supported the use of mobile communication devices for a number of secondary activities (such as phone conversations, sending and receiving e-mails/text messages, accessing the Internet), and these have been seen to cause distraction leading to diminished driving performance, there are also circumstances unrelated to communication devices that are just as detrimental for drivers – and mixing music with either of these seems to be highly potent for increased risks. In this connection, the effects of *passengers* on crash involvement has long been recognized, and while it may be perhaps difficult to accept, there is much evidence that demonstrates the impending danger that is directly related to carrying friends in the car (Regan, Young, Lee, & Gordon, 2009; Williams, 2003). Actually, there has been much deliberation between researchers as to what extent mobile cell phone use may differ from live conversation with passengers (Strayer & Johnston, 2001). On the one hand, mobile phone usage has a number of distracting effects. These are structural mechanical interference (from holding the phone or dialling a number), perceptual blindness of visual driving field (from looking at the phone display or keypad), auditory masking of car monitoring sounds (through listening to the conversation), and cognitive overload (from being immersed in detailed dialogue). On the other hand, McEvoy, Stevenson, and Woodward (2007) claim that passengers in the car also instigate a comparable level of obstruction. There is manual interference (by reaching for and handing something to a passenger), visual interference (by turning away from road to gaze at the passenger), auditory interference (by paying increasing aural attention to the passenger during conversation), and cognitive interference (by heeding more mental thoughts to the discourse than to the road ahead). Of course, the same can be said about in-car music. The driver is hampered by engagement in manual activity (such as reaching for a disk box and placing the CD in the music player), visual activity (such as turning away from the road to gaze at the player when choosing a track or scrolling through MP3 files), auditory activity (such as paying increasing aural attention to the songs heard in the background), and cognitive activity (such as heeding more mental thoughts to the names of the songs/artists in the music playlist than to the road ahead).

Certainly, the presence of more than one occupant inside a motor vehicle creates a social structure that can affect driving behaviour. The simple truth is that in the company of passengers, drivers may encounter more stimuli than they can handle – including conversation, laughter, motion, and commotion. All of these have the potency to encourage a driver to turn around, or even join in the revelry. Simply said, while the driver is driving, *there is a party goin' on* in the back seat. At the same time when we are doing our best to drive carefully down main street, a social scene is in play around us involving singing favourite tunes to tracks that hard-rock through the car interior, a feeding frenzy with bags of munchies or burgers with fries and cans of *Coke* or a beer, smoking cigarettes or roll-ups, a friendly discussion while surfing the net for info on the nearest club, and then a communiqué for stayin' in touch with the buds by texting them *where* and *when* to meet. Stutts et al. (2001b, 2003) initially identified driver distraction as the main reason for documented associations between passengers and crash risk. Williams (2003) demonstrated that teenaged drivers transporting teenaged passengers is a high exposure activity; more than half of deaths among 16–17 year-old drivers occur while they carry passengers < 20 years old. Police reports indicate driver inattention secondary to interactions with passengers as a contributing factor in 32% of crashes. For example, Williams found crash risk four times greater (i.e., an increase by 400%) with 3–4 teen passengers than when driving alone. Accordingly, among 16–19-year-old drivers, the probability for a crash is from .001% rising to .004%. Perhaps at first such figures seem negligible, but the actual crash rates per 10,000 trips by the number of passengers, is: $M_{\text{0-Passengers}} = 1.25$ crashes; $M_{\text{1-Passenger}} = 1.75$ crashes; $M_{\text{2-Passengers}} = 3$ crashes; $M_{\text{4-Passengers}} = 4.2$ crashes. Later studies, such as McEvoy, Stevenson, and Woodward (2007) found that irrespective of age group, drivers with passengers are almost 60% more likely to have a crash, while the likelihood of a crash is more than double in the presence of two or more passengers. Spence and Ho (2009) outright state that the risk of an accident increases as a function of the number of passengers in the car; carrying two or more passengers increases the risk of a crash by twice the amount compared to travelling alone. Finally, Westlake and Boyle (2012) found that young drivers who typically had friends as passengers in the car were 1.37 times more likely to be involved in crashes than when the same drivers did not have friends in their car.

Yet, some researchers (Williams, Ferguson, & McCartt, 2007) indicate that the risks of carrying passengers may not be solely an issue of *if or not* or *how many* passengers are in the car, but rather *who* is in the vehicle and *what kind of influence* do they have on the driver and driving situation. First, depending on the personalities present, there may be quite distinct practises of verbal and/or physical interactions that occur between the driver and the passengers, as well as between the passengers themselves. These interactions can be either positive or negative by nature. That is, some passengers can directly encourage drivers to behave safely, or alternatively, incite them to exhibit risky behaviours (Simons-Morton et al., 2011). Lee (2007) feels that passengers might also have a 'protective effect' especially if they are more experienced drivers; there is a notion that people do tend to drive more safely when older more experienced passengers are present. Moreover, some passengers might support driver vigilance by negating fatigue, or enhance navigational procedures; some passengers might offer cautions and warnings regarding imminent dangers, or rebuke the driver for precarious behaviour and unsafe driving (such as lacking control over the vehicle); some passengers might operate entertainment devices and communications systems (such as turning on the radio, adjusting the volume, switching between channels, searching through playlists, selecting preferred music), or dial numbers on mobile phone; and of course some passengers can even substitute as the designated driver if necessary. Along these lines, Williams et al. delineate a number of studies that reinforce the fact that the presence of passengers for drivers above 30 years old decreases crash risk. Then again, we might just surrender to the possibility of passengers providing a defensive shield as essentially invalidated when considering the extremely large incidence of young drivers under 20 (and their passengers within the same age group) that are involved in traffic accidents when passengers are present (Ho & Spence, 2008).

As far as young(er) drivers are concerned, not only do passengers distract them, but often 'inspire' them to take risks and drive faster. Passengers often incite young drivers to display a wide array of arrogant behaviours – all of which can hamper the driver's performance and vehicular control. And not the least, the norms that guide young drivers' behaviour are both those from inside *and* outside of the vehicle (Lee, 2007). The truth is that most previous research studies consistently find the presence of passengers in vehicles do increase the crash risk of young drivers less than 20 years of age (for example, see: Heck & Carlos, 2008; Underwood, Chapman, & Crundall, 2009; Williams et al., 2007; although see Simons-Morton et al., 2011). In this connection, Williams et al. revealed that the risk of having an accident while driving indeed increases by, and is a function of, the number of occupants in the vehicle (see Figure 4.6). As can be seen in the figure, the percentages of USA 16–17 year old driver fatalities from crashes in the year 2005 indicate a clear increase of single car crashes, as well as due to speeding and driver error; these are all associated with an increased number of

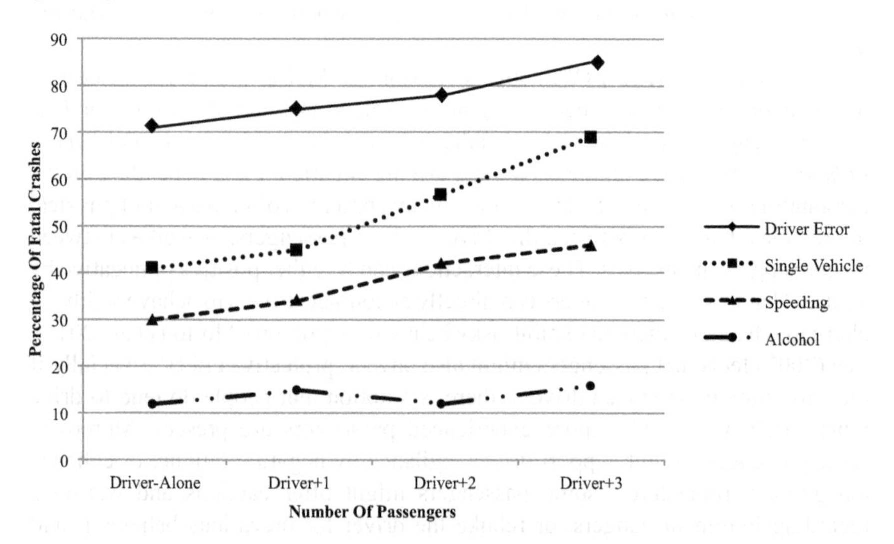

Figure 4.6 Young driver fatal crashes by number of passengers
Source: Adapted from Williams et al. (2007).

accompanying passengers in the vehicle. Nevertheless, it should be pointed out that there was no interactive association between the level of fatal crashes and the number of passengers crashes for driving under the influence.

Arnett (2002) reflected on the over-involvement of teenage drivers in near-crashes and crashes. Accordingly, the age group is not only plagued with fatalities, but also with incidents that result in physical injury and prolonged hospitalization. As a side comment, Arnett reiterates that this age group is also highly represented among events for which the consequences lead to property damage of grave financial outcomes. Tabulating crash data from the *Fatality Analysis Reporting System* (USA) for year 2000, Williams (2003) reported that crashes among 16–19-year-olds were 89/1,000 drivers with a fatality rate of 6/1,000 drivers, and that 40% of teenaged car passengers who died were driven by teenage drivers. In a later comprehensive review of the literature, Williams et al. (2007) point out that passenger deaths among 16–19-year-olds is higher than any other age group, with a fatality rate of 9.3/100,000 (compared to 2.9/100,000 for older ages). Further, Peek-Asa et al. (2010) examined the *Iowa Crash Data Base* (USA) finding over 82,000 crashes among drivers aged 16 to 18 between the years 2002–8; there was a crash rate of 88/1,000 with a fatality rate of 3/1,000. To summarize, it seems clear that there are overriding risks when teenage drivers carry passengers from the same age group. Moreover, there seem to be distinctions based on gender. That is, the presence of female passengers seems to have a minimal effect on crash risk, whereas the presence of male passengers generate the greatest threat. As a matter of fact, the presence of one male passenger has been seen to double the potential fatality rate for both male and female drivers, while the presence of two or more male passengers more than triples the chances of a disastrous outcome.

There is but one study that presents a different picture – Simons-Morton et al. (2011) implemented a *Naturalistic Teenage Driving Study* with forty-two 16-year-old newly certified drivers from Virginia (USA). In the study at least one parent per participant was also recruited. Prior to the onset, all 42 cars were outfitted with a data acquisition system, including kinematic (g-force) accelerometrics, multiple video recorders, and GPS tracking. As audio was not recorded, this study is unfortunately a missed opportunity for exploring in-cabin driver behaviours related to music. The study collected data over a 26-month period, covering more than 6,000 miles, reflecting an average 366 miles per driver per month. During the study there were 37 crashes and 242 near crashes – albeit just four resulted in a police report, and only one necessitated hospitalization. The results indicate: (1) combined rating scores of crashes/near-crashes/elevated g-force events (referred to as 'CNC') were three to four times higher for teen drivers than for their adult driver parents in the same vehicles; (2) CNC rate was lower in the presence of parents, higher when other teen passengers were present, and highest when teen friends described as 'risk-takers' were in the vehicle; but (3) both CNC and risky-driving rates were 20% higher when no passengers were present in the vehicle. This later result is indeed a controversial finding. The research team suggested that for the most part, teen passengers actually seem to discourage risky driving through directly and explicit verbal reinforcement (i.e., cautionary statements), or through indirect and implicit influence (i.e., non-verbal communication to the driver that precarious driving is frowned upon).

Although the above study indicates that the presence of passengers may essentially be positive in nature, those findings are quite to the opposite of most others. For example, in a survey examining the utility and contribution of passengers to driver distraction among 2,144 high school seniors from thirteen high schools in California (USA), Heck and Carlos (2008) revealed that 623 (38%) young drivers report to have been sidetracked at least once in the past by accompanying passengers. The most frequently experienced circumstances of passenger-related driver diversion, were: talking, yelling, arguing, being loud (45%); fooling around, messing around, horsing around, being stupid (22%); and playing music, dancing, changing CDs/radio stations (16%). Nevertheless, other circumstances reported, included: deliberate distractions (8%), and accidental distractions (3%). Further, more than half (59%) of the high school seniors in the sample reported that they themselves had once been a passenger in a car when a driver was distracted and driving hazardously. Indeed the car may be the most opportune location for social gathering, and most certainly music plays a vital role in setting the ambiance for such interactions. Therefore, the interaction between passengers and background music, which can blend into a highly combustible mixture, is extremely important to consider.

4–2.4.1. Mental intrusion: Conversation

There is evidence (Heck & Carlos, 2008; McEvoy et al., 2007) that drivers are just as much at risk when conversing with a passenger as they are with a caller on a mobile cell phone. In a series of studies commissioned by the American *AAA* Foundation, Stutts et al. (2003) demonstrated the distracting effects of passenger conversation. Employing video analysis of the in-cabin behaviour for drivers aged 18+, the researchers found *conversing* with a passenger is a most commonly identified distraction occurring in at least 15% of all driving. Atchley and Dressel (2004) point out that while some exchanges between the drivers and passengers may be short and simple (such as giving directions), and others may be relaxed brief spontaneous chitchat, many conversations do escalate to emotionally challenging and even heated debates. These can be most debilitating. However, unlike driver-caller conversations, as driver-passenger conversations feature face-to-face interactions the nature seems to be one of verbal 'pacing'. That is, unlike cell phone callers who are only 'virtually' present in the cabin and hence blind to the traffic environment, passengers who are physically present during the conversation usually speak according to the pace of the driving conditions; they slow down and even pause when the traffic interchange is more demanding or congested (Spence & Ho, 2009). Certainly, as a cell phone caller is both unaware of the driving circumstances that transpire while engaged in the conversation, they cannot regulate the pace of the communication – the cadence of the exchange from either perspective of language use or content output – in the best interests of the driver. A study conducted by Crundall et al. (2005) elegantly demonstrated this phenomenon. In the study, 20 participants (90% female) were recruited to drive a 20-mile circuit on real roads under normal driving conditions while engaged in a 'conversational game'. The drivers, who were on average 26 years old with approximately five years driving experience, all drove under three conversational conditions: (1) with a sighted-partner passenger in the car; (2) with a blindfolded-partner passenger in the car; and (3) with a remote-partner caller via a speakerphone. Accordingly, the results showed that in normal sighted-partner conversations both drivers and passengers tended to pace and/or reduce their conversation on urban roads; such adaptation (referred to as *conversational suppression*) reflects a higher level of traffic demand and driver control. Yet, as Crundall et al. point out, social pressures often seemed to have forced drivers to maintain an on-going conversational flow while speaking with a remote-partner even under more demanding traffic conditions.

Finally, in a video analyses study of 7,858 clips taken from naturalistic driving among young drivers, Goodwin et al. (2012) found that loud conversation and singing was five times more likely to occur when multiple teenage passengers were present in the vehicle. Specifically targeting the 2,716 clips with passengers (35%), loud conversation and singing were observed in 12% of clips of which 4% were associated with serious incidents. Further, in nighttime weekend driving with one or more passengers present, loud conversation or singing was observed in 20% of clips; this combination caused drivers to be approximately six times more likely to have a serious incident. Most specifically to the present discussion, a host of video clips significantly linked loud conversation and singing with events triggered by high g-force, including: acceleration or speeding (n = 107, 14.6%), deceleration or hard braking (n = 324, 20.7%), sharp left turns (n = 335, 15.6%), and sharp right turns (n = 281, 14.3%). Namely, the mixture of passengers + loud conversation + music background and singing was directly seen to result in vehicular performances consisting of highly aggressive driving (as demonstrated by mechanical aerodynamic physics).

4–2.4.2. Physical intrusion: 'Horsing' around

For better and for worse, the overriding feature of the environment that young drivers live in has everything to do with their passion and overall concern for their mates. Given that so many other settings of their life are those in which critical judgments about themselves and their peers surface – such as criticisms of how they look, how they act, what music they listen to, how popular or unpopular they are, and how they should do more to prepare for their future – the car then serves as a locale to meet and convene in small crowds of which the driver is a member. In essence, the car is the basis for teenage lives outside of school and home (Arnett, 2002). Accordingly, there is evidence that points to a positive relationship between the social dynamics of friendship groups with accidents. Ironically, the importance of accompanying friends is all too often found in crash rates among teen drivers, especially when there are two or more passengers in the car. For example, Heck and Carlos (2008) reported that roughly 8% of peer passengers intentionally distract the driver. In the study, teenage drivers reported they have experienced a passenger who 'punched me in the head', 'tickled me', 'hit me in the face', 'squirt me with a water gun', 'threw stuff at me', 'threw a cup of water in my face', and 'poked me'. Moreover, several teenage drivers claimed that when more than one passenger were in the car, they made deliberate attempts to operate the controllers of the vehicle while the car was moving; namely, the passenger 'tried to take control the car', 'pressed buttons on the control panel', 'grabbed the gear shift' (putting the car in neutral), 'tried to get the car to crash', 'messed around with the mirrors', 'pulled the emergency-brake', 'tried to take control of the steering wheel', 'readjusted the driver's seat', 'turned on the car lights', 'operated the entertainment system' (i.e., changed radio channels, switched between cassettes, swapped CDs, watched DVDs, made mobile phone calls), and 'repeatedly switched on/off hazard lights'. Other reported activities of passengers included; 'wrestling with others in the back seat' and 'throwing things out the window at other passersby'.

Goodwin et al. (2012) found that rough, rowdy, and unruly behaviour – referred to as *horseplay* – was nine times more likely to occur when drivers carried multiple passengers in the vehicle (especially at night). Unfortunately, horseplay also entails a host of passenger behaviours, including: 'talking to someone outside of the vehicle', 'dancing', 'physical contact with the driver of an affectionate nature' (such as hand-holding or kissing), and 'physical contact with the driver of an aggressive nature' (such as pushing, pulling, and poking). Most ironically (or perhaps not), music has been seen to intensify such behaviour – especially loud fast-paced music styles. Albeit, thus far no study has been carried out that clearly documents this interaction, or even the frequency of in-car music as an incitement for such behaviours. Yet, Goodwin et al.'s 2,716 video clips of young drivers with accompanying passengers, horseplay was observed in 6.3% of the clips, and 2.4% of these were associated with serious incidents. Some specific behaviour observed in these video clips to a much lesser extent, was dancing ($n = 40$, 1.5%), communicating with someone outside vehicle ($n = 30$, 1.1%), affectionate physical contact ($n = 16$, 0.6%), and aggressive physical contact ($n = 8$, 0.3%). However, nighttime weekend driving with one or more passengers increased horseplay (seen in 11.2% of the clips). Finally, horseplay was consistently associated with events triggered by high g-force events, including: acceleration or speeding ($n = 215$, 29.6%), deceleration or hard braking ($n = 354$, 22.6%), sharp left turns ($n = 420$, 19.5%), and sharp right turns ($n = 458$, 23.3%). As stated by Williams (2003): "The presence or action of passengers (plus other influences such as loud music) cannot only distract but can influence more risky (or less risky) driving" (p. 14).

4–2.4.3. Behavioural intrusion: Accidents, substances, sex, and entertainment

As a rule of thumb, drivers use the car as a setting where they can be together, away from the monitoring eyes of others. In the case of young drivers, the overriding place where they can be together independently of their parents *is* the car

(Arnett, 2002). Therefore, the main issue here is not so much *where* the car is going, but rather *who* is in the car, and what the driver and passengers *do* when they are together. Sometimes, *what* is done is none other than simply riding around – which in the late 1940s was referred to as 'cruising'. Over a decade ago, Williams (2003) found that recreational driving was considered especially high-risk driving because both alcohol and nighttime driving was usually involved. Nevertheless, part and parcel to this package *is* music listening. In the 50s drivers cruised with Rock N' Roll; in the 60s meditative Folk or Psychedelic musics were heard in the cabin; the 70s brought Heavy Metal and Hard Rock styles to the autosphere; in the 80s the interior became a Disco club; and then later from the 90s onwards, styles such as Rap, Hip-Hop, Dance, House, and Drum&Bass caused vehicles to 'boom' throughout neighbourhoods.

Heck and Carlos (2008) found that the majority of accidental passenger distractions reported by teenage drivers were actually behavioural deviations, but yet most teenage drivers perceived these behaviours as 'innocent horseplay'. The drivers reported passengers to have 'put my car into neutral' and 'spilled drinks in my car', as well as making other references to drug use and sexual behaviour. In this connection, Goodwin et al. (2012) highlight evidence that links having friends who tend to be risky with more aggressive driving. Specifically, friends who exhibited at-risk behaviours while in the car, smoked (cigarettes or marijuana), drank alcohol, and refused to wear a seatbelt. It is interesting to note that previous studies (Arnett, 1991a, 1991b; Gregersen & Berg, 1994) point to an inter-relationship between music and driving among young adults; these directly associate specific musics with negative lifestyles, reckless driving, and traffic accidents. For example, Heavy Metal and Hard Rock musics have consistently been associated with significantly higher rates of driving while intoxicated, accelerated driving speeds, and traffic accidents.

4–2.4.4. Emotional intrusion: Thrill-seeking and risk-taking

Many drivers, especially the young, enjoy the excitement of risk-taking and are attracted to thrill-seeking activities. Risk-taking is often induced by the presence of passengers (Williams et al., 2007), along with high-intensity background music emphasizing particularly low frequencies (i.e., HI/LF music). Specifically, young novice drivers who tend to travel while carrying groups of friends are less familiar in handling peer pressure, and as the car is a context that offers an opportunity for social acceptance, these drivers eagerly accommodate to the behaviours of peers. Hence, many young drivers actually find themselves easily influenced to act in ways that they would not otherwise act (Arnett, 2002; Williams, 2003). Moreover, as Williams et al. point out, poor self-regulation capabilities of younger drivers reduce their rational decision-making process to lower levels beyond what might have been the customary. Certainly, regulatory control has much to do with drivers' ability to adjust and maintain emotion and attention (Heck & Carlos, 2008).

In the context of driving, an important constituent of emotional management regards possible interactions between heightened stimulation and the driver's self-regulatory system – referred to as *sensation-seeking*. Sensation-seeking reflects the degree in which a person seeks out novel and intense stimulation. Arnett (2002) claims that sensation-seeking has steadily been found to be higher among young males drivers and is related to driving in risky ways. Accordingly, a sensation-seeking temperament may not only lead some young men drivers to accept more risks while driving, but perhaps intentionally seek it out, as well as approve of other's tendencies for the same and hold them in high esteem. Arnett depicts a poignant picture of the above:

> It is easy to imagine a group of high school friends driving around together on a weekend evening. They are in that state of elation that adolescent friends often generate when they are together. But they are a little bored, too, and looking for something to do. They try a few places and not much is happening, but they are in a car, they can make something happen. The driver decides to show off how fast he can drive, or how quickly he can pass the car in front of him; perhaps his friends urge him on, perhaps they even initiate it: 'Let's see how fast this baby can go!' Maybe they have all been drinking and/or using drugs. He begins driving recklessly, and a crash results. Or maybe it just happens inadvertently, maybe they are all talking and laughing, and the driver is trying to join in and gets distracted, does not see the red light, or does not see the car coming the other way, and again a crash results (p. ii18).

Simons-Morton et al. (2011) claim that with regards to passenger-generated disruptive behaviour, research has shown that such conduct is predominantly flaunted simply because passengers find intrusions and violations exciting if not humorous. Namely, passengers frequently encourage or create dangerous situations for the driver – simply because it is fun!

There is clear evidence that factors associated with risky driving, include a disposition for sensation-seeking, the inclination for risk-taking, actual behaviour of an audacious nature, and membership in subgroups with peers who enjoy looking for thrills. In a very early study, Gregersen and Berg (1994) surveyed 3,000 young adults (about 20 years old) randomly selected from the Swedish population register. After excluding those without a driver's licence, the remaining 1,774 young drivers were questioned with the intent of identifying high-risk drivers to which they might then classify specific social interactions and lifestyle preferences. Subsequently, a Principal Components Analyses pointed out ten factors (explaining 48% of the variance), including: sports; alcohol; motives for using the car (showing off, pleasure, and sensation-seeking); culture (music listening, playing instruments, and going to the theatre); romantic inclinations (literature and film); social responsiveness (politics, peace movements, and human rights activities); being out and about (friends, parties, going to dances or movies); staying at home (watching films and TV); clothes (fashion); and cars (as a preoccupation, hobby, ownership, engagement in cruising behaviour). Then, employing a Cluster Analysis methodology, the study found 15 specific subgroup profiles of which four were seen to correspond with high accident risk. The final results indicated that high-risk drivers are seldom active in sports, not very interested in social responsibilities, exhibit only average interests in culture and clothes, are less involved with sub-groups that spend time being out and about, often drink, always have extra motives when driving, are very interested in cars, drive a lot both during the day and night (for both as transport to defined locations, as well as for social cruising), and are very typically male.

It is interesting to note that many studies have indicated dangerous driving linked to the presence of male passengers (Peek-Asa et al., 2010; Williams, 2003; Williams et al., 2007). Accordingly, young male drivers drive at faster speeds in the presence of other young male passengers compared to when they are alone or with either a female or adult passenger. While it is typical that teen-aged drivers drive faster as well as allow for shorter headways than the general traffic, long ago Brodsky (2002) found that such driving patterns are usually accompanied with music selections characterized by fast-tempo and percussive rhythms of an aggressive nature. Unfortunately, and as depicted in Chapter 1, especially in American society, reaching manhood seems to be linked with having been certified as a licensed driver, having partial ownership over an automobile, and a clear demonstration (to others) that he has the attitude, aptitude, skills, and overall knack for risky and ultimately dangerous driving. Over a decade ago, Arnett (2002) offered observational insights – and perhaps these should have been seen as cautionary warnings – that our entire social system is accompanied by media images linking fast driving and automotive manoeuvres to masculinity. Clearly these images come with a soundtrack accompaniment of loud, aggressive, frenzied-tempo music pieces from music genres and styles including House, and Techno music. Arnett perceived the distinctiveness of manhood – which as far back as our collective memory can go – as having always been equated with certain rites of passage that for the most part include personal courage and a willingness to take risks in the face of danger. We should be clear here, that in today's day and age, such a social conception may just involve deadly forms of dangerous driving, for which in-car music provides the most potent ambiance to develop the necessary emotional personification of these practices.

Chapter 5
Ill-Effects of In-Car Music Listening

In the current chapter three specific *ill-effects* of music on driver behaviour and vehicular control are considered. The chapter asserts that *music* is a distraction because the *sounds themselves* prevent drivers from making the best use of attention needed to drive a car. A drivers' hearing can be obstructed, their thoughts can be directed elsewhere to memories and connotations of past experiences, and their motor activity can be hampered by an overwhelming rhythmic organization expressed through body movement of the vocal cords, hands, pelvis, and feet. In short, the chapter reflects on the circumstances by which engaging in music activity might be incongruent for the perceptual timing, motor control, and cognitive judgement of occurrences subsequent to the optic flow of events necessary to synchronize with traffic. Within the chapter, individual differences related to age, gender, and personality (specifically sensation-seeking) are explored. The ill-effects of music on driving are: *music-intensity evoked arousal* (i.e., overly rambunctious driving), *music-tempo generated distraction* (i.e., hampered longitudinal and lateral control), and *music-genre induced aggression* (i.e., a habitual driving style beset with risk-taking and unfriendliness).

5–1. Driving with Music

What happens to the vehicle when we drive with music has as much to do with who we are, as it does to the fact that the context of listening is in a moving vehicle within a dynamic road environment containing a complicated and convoluted flow of traffic. As outlined in Chapter 2, some of us are *sensorial listeners* who react primarily to the raw materials found in music, others are *perceptual listeners* who react to the fundamental relationships presented within the music being heard, and still others are *imaginal listeners* who react to the associations activated by the music as it is heard. Nonetheless, the truth is that most likely we alternate between all three listening styles or employ them in parallel depending on the music pieces heard. For example, the densely voiced cantata 'Heart and Mouth and Deed and Life' (BMV 147a, 1716) by Johann Sebastian Bach is quite different to listen to than the more transparent folk ballad 'Diamonds and Rust' (1975) by Joan Baez. Therefore, during more focussed listening to specific measures of either piece – such as when suddenly following a voice, text, or instrument – we might switch between the listening styles. Arguably, the demands of road traffic intensity might also mandate the use or limitation of listening styles (as the case may be depending on low-demand or high-demand driving conditions). While some of us are easily overwhelmed by the general organizational structure in the background music, and thus encode the various music components as if experiencing one intact block of sound (referred to previously in Chapter 2 as the cognitive-based perceptual listening style of *field-dependence*), others are capable of breaching the surface cues in the music with a more analytic stance and thus encode the mass of intricate details found among the various music components all-the-while following each unique music part of the whole piece (referred to previously in Chapter 2 as the cognitive-based perceptual listening style of *field-independence*). Yet, as pointed out in Chapter 4, driving requires the effective orchestration of several executive functions, selective attention, and integration of information from a host of visual inputs as well as from additional sensory inputs (including the auditory), and the coordination of numerous compound behaviours such as divided attention, working memory, prospective memory, behaviour selection, and behaviour inhibition. Therefore, within such a context – which most definitely is distinct from all other possible venues in which everyday people may listen to music – the motivations for choosing music, the effects of music on the listener, and the reactions of the listener, are quite unique.

Everyday drivers not only anticipate taking music along for the ride but also perceive the automobile as a sanctuary. Interview and diary studies have often reported a highly common testimony: "the car is the only place where I can listen [to music] loud enough without annoying other people" (Sloboda, 2005, p. 324). Further, transcriptions of in-cabin recordings from on-the-road studies frequently describe driver perceptions about the impact of music on driving performance. For example, employing a car designed by the Swedish Road and Transport Research Institute, Oblad (2000a, 2000c; Oblad & Forward, 2000b) monitored spontaneous verbal commentaries of drivers, as well as measured performance parameters including clutch, brake, and accelerator pedals. The study found that accelerated driving speeds occurred as a result of the music being played in the car; when a driver liked the music, the sound level could never be high enough, and high intensity always caused accelerated cruising speeds.

Drivers reproduce music in the vehicle cabin for many various functions; music is heard for enjoyment and diversion (to reduce boredom monotony from low-demand traffic), for revitalization and stimulation (to maintain arousal and energy leading to increased on-task attention), for enhancing the meaning (to provide an emotional soundtrack that accompanies the occasion), and for soothing relaxation (to decrease the tension and stress of high-demand traffic). Nonetheless, some drivers may refrain or cease listening to music in an effort to decrease the amount of information flow from a secondary source that is irrelevant to the driving task. Such archetypal processes have been described in a model developed by Oblad (2000a, 2000c; Oblad & Forward, 2000b; as seen in Figure 3.1). Through everyday personal experiences, ordinary drivers learn about selecting music that is typical for driving, and for the most part, drivers sincerely believe that they can align their choices in such a way as to choose the most appropriate music that helps them attain the goal of driving in a more controlled manner with increased driver safety. Yet, listening to music while driving does engender a host of production behaviours that are essentially time-linked overt interactive responses at lower levels of consciousness – all of which are based on structural patterns within the music that mirror the beat, rhythm, tempo, or melody line. For example, Pecher, Lemercier, and Cellier (2009) observed their participants whistling, tapping, and singing along with happy music while driving. According to a survey conducted by *AutoTrader*, 76% of drivers revealed that they turn-up the volume when their favourite driving song comes on the radio, as well as engage in other music-related behaviours such as lyric singing, melodic humming, percussive-grunting, hand-clapping, arm-thumping, finger-drumming, head-banging, pelvic-thrusting, butt-jiggling, and foot-tapping. In short, drivers enjoy partaking in the band's performance *as if* they are accompanists in their own right. The band positions they take on, include: Car-aoke Singers with imaginary microphone (60%), Steering-wheel Drummers (38%), Driver-seat Dancers (7%), and Air Guitarists (2%) (AutoTraderBlog, 2012). Among the other band positions that can potentially be taken on by drivers, are: Knee Keyboardists, Sternum Saxophonists, Vocal-chord (beat-boxing) Percussionists, and Accelerator-pedal Timpanists. Mulholland (2013) claims that anyone who has ever 'warbaled their lungs out' while driving knows what it's like – you hear your favourite song and you transform from an everyday driver to a rock star.

The character of music effects (i.e., ill-effects) that surface when driving with music is co-dependent on a four-part interaction involving the *driver*, the *driving circumstances*, the specific nature of the *music* heard, and level of *music involvement*. Essentially, by accounting for these, the degree of variance between the primary task involving driving, and the secondary task involving music engagement, can be seen. Like a network of components whose recurrent interface has causal impact on specific outcomes, music responses of drivers are the result of systematic linkage among constituents that Hargreaves (2012) refers to as the *Reciprocal Feedback Model* (as seen in Figure 2.1). The consequences of in-car music are a host of instinctive, physiological, and emotional responses – which involve biological, motor, and affective parameters – in addition to the mental processes of perception, cognition, and conscious awareness. Hence, besides the primary driving task (that involves online navigation, judging and maintaining speed, maintaining lane position, following distances, adjusting instruments, scanning for hazards and information, evaluating risks, anticipating future actions of others, and reacting to unexpected events), *driving with music* provokes a vast array of psychophysiological reactions (that involve biochemical processes, functional activation of neuroanatomical systems, cardiovascular activity, facial gestures, frisson behaviour, gastric motility, muscular contraction, pupillary activity, sensations of body casing, and skeletal movement), as well as an unremitting succession of mood responses (such as elicitations and sentiments of emotion, affect, feelings, and dispositions), and an excessive preoccupation of cognitive resources (for example associations, connotations, intra-subjective imagery, and language-like sub-vocal verbal messages). All of these have been described in Chapter 2, and as pointed out in Section 2–2, the effects of music seem to be dependent on the nature and complexity of characteristics within the music structure. Namely, music comprised of a slow tempo, quiet dynamics, unvaried rhythm, minor tonality, and low register (perceived as sad, dreamy, and sentimental) is seen as *sedative* by nature in that it sooths and relaxes the driver, while music comprised of a fast tempo, loud dynamics, lively and varied rhythm, major tonality, and high register (perceived as happy, graceful, and playful) is seen as *stimulative* by nature in that it arouses and energizes the driver. However, in both cases, it may be difficult for us to judge the extent of such complexities, and we may soon find ourselves over-sedated or over-stimulated – neither of which is a rather well adaptive state for driving an automobile.

5–1.1. In-Car Music Activity as a Risk Factor for Inattention and Distraction

Several studies (ACF, 2009; The Telegraph, 2009; Dibben & Williamson, 2007; Milne, 2009; Quicken-Insurance, 2000) confirm that the majority of drivers who committed traffic violations were playing music in the car at the time, and were listening to fast-paced music of loud intensity. For example, the Quicken Insurance Survey found that 10% of drivers reported to drive faster when any music is playing in the car, while 6% specifically admitted to driving at increased speeds if the music was fast-paced and reproduced in a fairly loud fashion. Further, Quicken reported that 95% of those who usually drive with music received a ticket in the past 12 months (1998–1999) compared to 87% of those who do not drive with music. Ironically, nearly three-out-of-ten drivers (28%) who had received a traffic ticket, preferred to drive with fast-loud music in the background. Further still, those charged with speeding violations had been listening to loud Rock, Dance, or House music styles. Finally, many drivers indicated that both Rap and Hip-Hop music had an adverse effect on them; one-in-five disclosed that these genres prompted aggressive conduct. In the most widely cited survey to date, Dibben and Williamson recruited a UK representative sample (N = 1,780), finding that 23% of drivers who had had an accident reported to have been playing music at the time. It should be pointed out that this incidence was not in the least confounded by gender or age. Nevertheless, as these researchers found that only a quarter of those who reported accidents recounted the presence of music in the car at the time, they concluded that there is "no direct link between the presence of music while driving and involvement in an accident" (p. 581). Yet, perhaps such a conclusion is misleading. For example, if we look at this level of incidence simply from an epistemological point of view, the frequency of accidents in which music was reported to be present at the time of an accident is roughly equal to the same incidence of *inattention and distraction* to begin with. Therefore, in-car music listening may be far more worrisome than as conceived by Dibben and Williamson. Perhaps, an explanation for the demise of their findings is that not all of the drivers in this UK sample were comparable. For example, when comparing between the three age-groups of drivers participating in the study (18–29; 30–50,

50+ years old), statistically significant differences surfaced for driver-perceptions as to whether or not music: enhances calm ($M_{18\text{–}29}$ = 61%; $M_{30\text{–}50}$ = 58%; M_{50+} = 67%; $x^2 = 9.8$, $p = .007$); supports concentration ($M_{18\text{–}29}$ = 23%; $M_{30\text{–}50}$ = 21%; M_{50+} = 27%; $x^2 = 7.95$, $p = .02$); and generates distraction ($M_{18\text{–}29}$ = 9%; $M_{30\text{–}50}$ = 7%; M_{50+} = 3%; $x^2 = 10.58$, $p = .005$). Accordingly, drivers aged 50+ more often experienced driving with music as calming and supporting concentration, while younger drivers more often experienced driving with music as enjoyable but distracting. Curiously, the study found verbal conversation is equally distracting (roughly 7%) for all drivers regardless of age.

5–1.1.1. Partaking in the performance

Dibben and Williamson (2007) found younger drivers much more likely to sing along to music while driving than drivers aged 50+. One might then assume that singing mediates (at least to some extent) the higher incidence of distraction as reported by these drivers. However, as Dibben and Williamson found singing to music was equally distributed among both drivers with and without a *no-claims bonus,*[1] they adopted a position stating that 'music is not a significant form of distraction'. To support their claim they cite results from experimental studies showing that engaging in a *conversation* while driving has been found to be detrimental to attention (for example, Atchley & Dressel, 2004; Consiglio et al., 2003; Strayer & Johnston, 2001; Strayer, Drews, & Johnston, 2003) while other verbal activities such as *word-shadowing* have not been found to hamper attention. Thus, we are to understand that Dibben and Williamson view word-shadowing activity as equivalent to *singing-aloud.* Having said that, no other research study in the literature has ever compared word-shadowing to singing, and therefore, perhaps such a conclusion is rather far fetched. Moreover, the reported findings about word-shadowing do indeed demonstrate that the activity hampers attention. For example, Spence and Read (2003) recruited eight drivers between 24 and 42 years of age (with a driver's licence for at least five years) to participate in two 1-hour sessions employing a Rover *GTI 216* fixed-base driving simulator (Leeds, UK). Each participant completed a speech-shadowing task on its own (i.e., single-task), as well as while they simulated driving (i.e., dual-task). The speech-shadowing task required the participants to repeat as many 2-syllable triplets as possible (for example: 'olive, notice, topic' or 'jewel, reason, under') immediately after they were heard. The study found that the participants echoed a significantly higher per cent of correct words when the activity was isolated in comparison to when they were driving at the same time ($M_{\text{SingleTask}}$ = 77%; M_{DualTask} = 56.2%; $F_{(1, 7)} = 129.0$, $p < .001$). In addition, a host of studies (reviewed by Ho & Spence, 2008) demonstrated that shadowing a conversation slowed drivers' responses to critical visual events as much as 370 milliseconds when compared to merely

1 'No-claims Bonus' is a status among insurance companies indicating that the agency has not processed a claim to pay property damages or for incurred bodily harm in name of the specified driver for at least a period covering the last three years.

listening to a conversation. It should also be noted that studies employing the generation of speech sounds (as opposed to the repetition of sounds occurring during word-shadowing), or those utilizing a grammatical reasoning task (i.e., akin to Baddeley's 1968 paradigm), explicitly found verbal activities to impair driving performance – and hamper drivers most when the content of the speech was particularly demanding (Brown et al., 1969; Strayer & Johnston, 2001). To reiterate, although Dibben and Williamson conclude that singing activity while driving is far less distracting than conversation, unfortunately such a supposition may have no validity at all – despite being one of the most highly cited deductions about driving with music among the traffic research community.

Dibben and Williamson (2007) also related to a more intuitive conception about conversation. They claim that drivers are required to sustain a necessary level of attention throughout, in order to preserve a sense of content consistency and maintain an acceptable level of social politeness. On the other hand, as singing to music played in the car does not require either of these, drivers have the flexibility to start and stop singing at will with no ill consequences. Actually, previous research on driver-related conversation (Crundall et al., 2005; Ho & Spence, 2008) has confirmed two different modes of interaction. The first is *local in-cabin conversation* in which both participants of the discourse have identical access to the same visual scene, and therefore conversation can be modified, paced, or suppressed altogether according to the ecological context and demands of the roadway (or subsequent driver-need for an increased focussed attention). The second is *remote cell phone conversation* in which one of the participants in the discourse has no access to the visual scene, hence has no understanding of the current situation, and therefore is unlikely to adapt to the demands of the roadway (or subsequent driver-need for an increased focussed attention). Several research teams (Drews, Pasupathi, & Strayer, 2008; Fagioli & Ferlazzo, 2006) corroborate that mobile phone conversations between the driver and someone outside of the car is more difficult than conversations among passengers in the car – simply because regulating the conversation (i.e., the responsorial dialogue involving turn taking and paced verbal exchange) based on traffic complexity cannot occur. Considering these two types of exchange, then as far as music activity while driving on the road is concerned, we might also raise the ultimate question: Is singing in the car comparable to local in-cabin conversation in which drivers can easily regulate/adapt their engagement to the interchanges of a dynamic traffic environment at will with no ill consequences? This is obviously the claim made by Dibben and Williamson. Or, is singing in the car comparable to a remote cell phone conversation in which drivers cannot regulate/adapt their engagement to the events that suddenly and unpredictably require the driver to be more attentive to the road?

Now, to be truthful, all music that is heard in the car was previously recorded. Hence, it would only be natural to look at the occurrence as a *reproduction*. In this case, Dibben and Williamson (2007) would be correct in assuming that both participants (the driver and the virtually present recording artists) have identical access to the same visual scene, and hence participation can be modified or suppressed according to the demands of the roadway – at will with no ill consequences. But, if we enter a bit of metaphoric latitude into the current discussion, then we might also comprehend the event *as if* emulating a *live performance* – albeit within a concert environment that is actually a mobile venue right smack in the middle of cross-town traffic. In this case, the recording artists performing their songs (presented over radio broadcast, CD, or MP3 tracks) have no knowledge about the current 'performance stage' (i.e., the vehicle cabin, road conditions, and traffic congestion) on which the accompanying 'background singers' (who figuratively speaking *is* the driver) seems to be avidly performing. Clearly, the continuous audio stream, and subsequently the music tempo itself, cannot be modified or paced according to road congestion, nor can the 'performers' acclimate the temporal flow of textual lyrics to increased attentional demands. Theoretically, perhaps drivers could suppress listening to music and discontinue their vocal accompaniment at will with no ill consequences simply by turning off the music itself. In fact, many researchers conjecture that in highly demanding driving conditions, drivers do decrease the volume of the background music playing in the vehicle cabin or turn off the music altogether in an effort to compensate for necessary attentional demands (Brown et al., 1969; Engström et al., 2005; Jamson & Merat, 2005; Shinar, 2007; Strayer & Drews, 2004; Wikman et al., 1998). But, with the exception of Brown et al. all other reports are simply assumptions based on hearsay as no further investigation since the late 1960s has been implemented. Yet, if hearsay is an accepted method of report, then, as far as most drivers are concerned, in-car musical performance engenders a total eclipse of attentiveness towards primary driving tasks in favour of the secondary task. So unlike the foregone conclusion of Dibben and Williamson, we might estimate that singing in the car is at least comparable to, and possibly more hazardous than, a remote cell phone conversation. Therefore, singing to music while driving should be viewed in a more serious manner. In this connection Hughes, Rudin-Brown, and Young (2013) implemented a simulated driving study whereby participants drove without hearing music, while listening to music, and also while singing along with the music background. They postulated that drivers would indeed perceive the workload to be higher when singing along with the music, and that driving performance (as measured by speed, speed variability, lane position, lane excursions, and response times) would be more impaired when singing aloud.

Hughes et al. (2013) recruited 21 participants between the ages of 18–50 (M = 35 years, SD = 13.75, 95% female) for a single 40-minute simulated driving session while performing a secondary task. The participants were drafted via several networks associated with Monash University in Clayton, Victoria (Australia). The driving simulator was equipped with a driver's seat, steering wheel, accelerator and brake pedals, handbrake, gearshift, and rear view mirror; three screens provided a 120° field of view, and the ambient sound heard in the cabin was at 52dB (which increased to 60dB when playing music). During the session, each participant drove 6.6 kilometres (4 miles) on a straight undivided urban roadway with two lanes in each direction; the trip was split into six 1.1 kilometre segments, to which four speed zones were employed (40 kph, 60 kph, 70 kph, and 80 kph). Each participant drove under three conditions: (1) driving without music background; (2) driving while listening to the music background; and (3) driving while singing to the music background. The two music pieces employed in the study were chosen because they were considered to be widely popular and highly familiar; each was of a three minute duration, and did not include an instrumental interlude (as the purpose of the task was to sing the lyrics). The songs selected were: Neil Diamond's 1966 hit 'I'm a Believer' (Smash Mouth), and John Lennon's 1971 poetic ballad 'Imagine' (John Lennon). Prior to the experiment, all participants received the song lyrics and a *YouTube* Internet address to find and practice singing the songs.[2] During the session, music was reproduced free-field via a laptop computer positioned roughly 1.5 meters from the driver. Each session began with two practice segments for familiarization of the driving simulator and to become acquainted with the *peripheral detection task* (PDT). The PDT involved drivers activating the right/ left turn signal indicator when a pedestrian (icon) was detected (on either right or left sidewalks as viewed on the peripheral margins of the visual screen). The participants drove three consecutive trials (variegated by conditions counterbalanced across the sample) while concurrently responding to PDT at randomized intervals. Immediately following each trial, the drivers completed a questionnaire concerning their perceived level of workload. The results of the study revealed four overriding effects of driving condition (see Table 5.1). First, two significant effects surfaced for Speed ($F_{(2, 40)} = 51.08, p < 0.001, \eta^2 = 0.72$; $F_{(2, 40)} = 4.03, p < 0.05, \eta^2 = 0.17$). The former indicates that participants accelerated more when driving without music than when listening to music (while the average speed when singing to music is somewhere in between). The latter indicates that the inconsistency of vehicle speed was significantly greater while singing than when there was no music in the vehicle. Second, two significant effects surfaced for Lane Position ($F_{(2, 40)} = 8.44, p < 0.01, \eta^2 = 0.30$; $F_{(1.5, 39.9)} = 3.75, p < 0.05, \eta^2 = 0.16$). The former indicates significantly less deviations when driving with music than when driving without music. The latter indicates that the least amount of lane excursions occurred when drivers only listened to music rather than when they sang the lyrics. Third, a significant effect surfaced for Workload ($F_{(2, 40)} = 4.04, p < 0.05, \eta^2 = 0.17$). This indicates that singing while driving was perceived as significantly more demanding than driving while listening-alone or driving without music. Finally, while only trends surfaced for PDT ($F_{(2, 40)} = 2.79, p = 0.074, \eta^2 = 0.12$) and these show that RTs were slower when drivers were engaged in music, pairwise comparisons indicated that the overall time between detecting a pedestrian on the sidewalk and responding with the required driver behaviour was significantly slower during music listening. Although mean RTs for singing and listening were the same, as the variances were much greater for singing, RTs did not attain levels of significance.

---

2 'Imagine': http://www.youtube.com/watch?v=yRhq-yO1KN8. 'I'm A Believer': http://www.youtube.com/watch?v=0mYBSayCsH0.

Table 5.1 Effects of driving with music, comparisons of in-cabin listening versus singing

	Driving Without Music	Driving Listening to Music	Driving Singing to Music
A Speed			
Mean (KPH)	60.5	55.0**	57.4*
Variability (MNSD-KPH)	4.00	4.30	4.60**
B Lane Position			
Variability (MNSD-LP)	0.33	0.28*	0.30*
Excursions (%)	22.0	14.5*	20.0
D Mental Work-Load			
Score (1–10)	4.20	4.40	4.90*
C Peripheral Detection			
RTs (MS)	1.05	1.15*	1.15

Contrasts Versus Baseline NoMusic: * = $p < 0.05$ ** = $p < 0.001$.

Source: Adapted from Hughes, Rudin-Brown, and Young (2013).

It should be noted that Hughes et al. (2013) concede their findings could be skewed by gender (95% female), and biased by drivers who usually sing when driving (95% of the time). Namely, perhaps male drivers would have driven differently when singing to music, and there is always a possibility that drivers who do not routinely sing while driving would have spawned different results. Yet, the research team concluded that driving with music was found to cause a higher demand (i.e., mental workload), and therefore they did slow down in an attempt to counterbalance increased difficulty. Hughes et al. state: "In this study, singing does not appear to be more distracting to drivers than is listening to music" (p. 792). Nonetheless, if that were the case, then how would they explain deficient attempts to adapt driving behaviour for increased difficulty (seen in increased average speed, speed variability, lane position variability, and excursion to the shoulder) – especially during trips when the same drivers sang the lyrics of songs? *And*, how generalizable is such a conclusion when, for the most part, the drivers in the study were female drivers who habitually sing aloud when driving. The research team surmised:

> [Drivers] may overestimate the extent that singing impacts their driving performance; therefore even if drivers are not aware of the extent to which music may be affecting their perception, performance, or control of the vehicle (Brodsky, 2002), they should be likely to stop singing if they perceive the driving environment to be more demanding (e.g., in bad weather)… [After all], the potential detrimental impacts of singing are only relevant in situations where a driver does not feel sufficiently impaired to warrant the cessation of singing (p. 791).

But yet, their final words also included a warning:

> Drivers are encouraged to exercise caution when singing while driving especially if they feel that it is impairing the primary task of driving (p. 792).

We might, then, wonder why Hughes et al. didn't do more to strike a nerve in the minds of drivers by offering a more declarative statement, such as: *Drivers underestimate the extent that engaging in a music performance, such as singing lyrics aloud, can impact on their ability to efficiently control an automobile while in traffic.*

5–1.1.2. Driving music

Law firms nowadays publish blog-like newsletters for their clientele whereby they offer a wealth of information as well as case examples. Many of the vignettes presented in such newsletters specifically target young novice drivers in an effort to raise safety awareness. More recently, there has been a renewed focus on *driving music* (see Chapter 3, Section 3–1.2.) and the subsequent driver behaviours that might arise from listening to music while driving. In one example, a firm from Seattle, Washington (USA), profiled Jake, a local high school junior, who revealed that he was "so wrapped up in a song that was playing loudly in his car that he drove straight through a stop sign" (Nicholson, 2011). Subsequently, Jake was pulled over by a police officer and ticketed. It would seem that drivers do need to become more aware that as they get 'drawn-in' by a song, they move from an extra-personal space involving driving tasks to a more personal space of active music listening (Fagioli & Ferlazzo, 2006; Ferlazzo, Faglioli, Di Nocera, & Sdoia, 2008). They also need to be aware that evidence (as presented in Chapter 4) points to the fact that in-car listening tasks consistently hamper performance by producing significant declines in driver accuracy caused by processing auditory stimuli, as well as entice drivers to engage in verbal and phonological activity involving language resources. At the very least, drivers need be aware that anatomical brain activation in key regions that support driving are downgraded through listening tasks (Just et al., 2008). If the above are not enough in themselves, we might suggest that drivers need be more sensitive to the verified consequences of aural diversion, including: poor attention and planning, difficulty generating and implementing strategies, inability to utilize feedback, and inflexibility of thinking (Heck & Carlos, 2008). Hence, if one were to raise the overriding question about the effects of listening or singing to music while driving, then most certainly the answer that seems to come to the surface is: *In-car music engagement contributes to increased risks of inattention and distraction.* Perhaps for this reason, modern platforms that can easily reach mass populations – like electronic bulletin boards and social media such as *Facebook* and *Twitter* – have intermittently raised concerns over driving with music. For example, Power (2009) strongly advocates that drivers first think about what they want to listen to before heading out on a Sunday drive. His missionary-like stance, as if from a cyber pastor to the communal masses of blog congregants, forwards the proclamation that drivers must be more aware of their music choices. Power's stance is that drivers should be more discerning if the music they wish to hear in the cabin is suitable for driving an automobile, and perhaps they should take more care in selecting tunes more appropriately. But unfortunately, such an allocution falls on deafened ears.

Especially as musical preferences are somewhat influenced by previous exposure, we might then wonder how many times (if at all) a passenger has experienced going for a drive with someone whereby the background music was specifically chosen for 'increased driver safety'? Most certainly, the answer will be: 'Never'. Hence, if new generations of younger drivers who ultimately imitate a wide-range of driving patterns and conduct of other more experienced drivers – including in-car music behaviour – then, they will no doubt continue to perpetuate listening to music selections that place them at increased risk. Although North and Hargreaves (2000) strongly believe that people "have clear notions of which type of music is appropriate to different listening circumstances in order to achieve [their] goal, and this implicates prototypicality in preference" (p. 65), several researchers (for example, Oblad & Forward, 2000b) clearly indicate that this is not the case for operating an automobile. Accordingly, Oblad & Forward observed that while drivers usually do select the music to be played in the car (whether they choose a radio station based on overall music genre, or actually select specific tracks), drivers are clearly not aware of what kind of music is most suitable for driving. So, while we might inquire how can drivers become aware of more appropriate alternatives for vehicular listening, certainly the overriding initial question must be if such alternatives have been developed to begin with. Namely: Is there valid and reliable prototypically 'safe' music that we can listen to while driving? In their book *Music and Mind in Everyday Life* authors Clarke, Dibben, and Pitts (2010) presuppose that such music is not possible to create, as "...individual differences, and the constantly changing demands of the driving situation, make it impossible to specify with any degree of exactness 'safe' driving music..." (p. 97). At this point, suffice it to say, that Brodsky and Kizner (2012) developed such a music background custom designed for in-car listening, which was subsequently validated under the auspices of the Israel National Road Safety Authority. *In-Car Music*, an 8-item 40-minute programme, was found to significantly improve driver performance and vehicular control towards increasing driver safety (Brodsky, 2013; Brodsky & Slor, 2013). Chapter 6 highlights this safety application in detail. Nevertheless, there is much controversy among researchers about the effects and *ill-effects* of in-car music as a source of evoked arousal, the generation of inattention and distraction, and the induction of aggressive driving behaviour.

5–2. Music-Intensity Evoked Arousal

The extent to which music pieces evoke arousal is somewhat related to the structural properties found in the music itself. According to North and Hargreaves (1999), valence, loudness, tempo, complexity, and novelty are the main components that have bearing on 'arousal potential'. Hence, a driver's preference for specific music is mediated by the degree of arousal it evokes in their brain. For many researchers, the latter concept is supported by Berlyne's 1970s theoretical framework referred to as the *Objective Psychology of Aesthetic Appreciation*. Berlyne claimed that the liking of music is somewhat determined by moderate amounts of arousal, because it produces the greatest level of pleasure. Therefore, music with moderate arousal-evoking qualities increases 'liking' more than music that promotes extreme levels of arousal. Yet, anecdotal evidence illustrates quite the opposite phenomenon; that is, there are an extremely high number of drivers who prefer to drive with highly arousing music. Perhaps, then, Berlyne's conceptual underpinning is not ecologically reliable for the context of driving. However, another psychologist, Konecni (1982), suggested that music preferences are subject to interaction effects between musically evoked arousal and cognitive demands (such as cognitive load) subsequent to tasks that are concurrently performed while listening to music. This later conceptual approach might be a more operational framework with which we can assess the effects of music on driver behaviours. Namely, Konecni's conceptual model accounts for a context-dependent linkage of music choices, whereby the qualities of both the music and the task interact. In this connection, North and Hargreaves point out the dynamic relationship between the nature of the music, the nature of the task, the actual task performance, and the eventual preference for music. Moreover, they conclude that "the efficiency with which tasks are performed simply increases with the arousing qualities of the concurrent music" (p. 286). In short, it would seem that as we perceive success in our driving performances (i.e., we sense our trips have been safe without feeling that we have been placed at risk), the arousing qualities of the background music heard at the time increase, lead to an upsurge of perceived enjoyment from the playlist, and subsequent proliferation of preference for the same selections the next time we enter the vehicle. But yet, if the task conditions change, for example as a result of fatigue, traffic congestion, or experiencing hazards, so too will our playlist. In the later case, if we perceive the new playlist has enhanced our efforts to adapt and maintain a high level of driver safety, then, this new inventory will thereafter become our preferred music for driving in the car. It is warranted here to distinguish between the 'represented emotional state' drivers perceive to be inherently expressed in the music pieces *versus* the 'induced emotions' that are actually felt when they drive with music. Accordingly, Pecher et al. (2009) point out that based on our everyday listening experiences, we all seem to 'know' what kind of music we need for a particular trip, but not necessarily do we actually 'feel' the emotion that the specific music supposedly induces. Truly, these two aspects of emotional engagement involve different physiological, behavioural, and psychological mechanisms, as well as both being co-dependent on perceived arousal and valence of the music emotion.

5–2.1. Emotional Arousal and Valence

As pointed out in Chapter 1, the automobile is not only a means of transportation that allows us to get from Point A to Point B, but is a highly exploited vehicle of pleasure. The pleasure we receive from driving an automobile has much to do with arousal and valence. *Arousal* is the objective degree of bio-physiological activation we experience during driving (from calmness and exhaustion through to tension and excitement), while *valence* is the subjective value characterizing how we perceive the driving experience (from sadness through to joy). Although most drivers are indeed aware that their driving is likely to be predisposed by substances (such as tobacco, alcohol, and social drugs), or by additional irrelevant secondary tasks and activities (such as eating, conversation, text messaging, and operating infotainment technologies), Reed and Diels (2011) have astutely pointed out that drivers are not very sensitive to the fact that their emotions (i.e., feelings, sentiments, passions, moods) have similar consequences on their driving performance as well as on their ability to control the vehicle.

In an early study, Iwamiya and Sugimoto (1996) presupposed that many people like to drive simply because it allows them to listen to music: "The combination of music and the chance to look at the scenery is a pleasant experience for most people" (p. 309). But why is this so? Iwamiya and Sugimoto attempted to clarify the effect of listening to music on driver perception through investigating the emotional impression of passing landscape scenery. In their study Iwamiya and Sugimoto recruited ten participants (aged 18–25, 60% male) to judge ten video scenes – nine of which were presented with music accompaniment. The videos were reproductions of landscapes (i.e., riverside, residential area, and freeway) recorded from the front window of the car, adjusted in such a way that when viewed by drivers in a science laboratory setting the size of the objects in the landscapes would appear *as if* 'passing-by' as they usually do during real-world driving. The videos also included a soundtrack of engine noise and other sounds typically heard inside the cabin. One of the sequences was presented without music (as a control), while all other videos were viewed with one of three types of music background variegated by genre styles and pace (i.e., number of beats per minute). The exemplars were:

	Title (Artists/Composer)	**Genre**	**bpm**
1	'The Only Name Is…' (Malta)	Light Jazz (Instrumental)	60
2	'Petite Suite "En Bateau"' (Claude Debussy)	Classical Piano Duet	64
3	'Stop The Train (My Girl Is Gone)' (Saint and Campbell)	Reggae (Vocals, English)	88
4	'Living In Danger' (Ace Of Base)	Pop (Vocals, English)	102
5	'Femme Fatal' (Satoshi Ikeda)	Soft Rock (Vocals, Japanese)	102
6	'Walk The Dinosaur' (The Goombas and George Clinton)	Funkadelic (Vocals, English)	116
7	'Superstar' (Pizzicato Five)	Hard Rock (Vocals, Japanese)	140
8	'Carmina Burana' (Carl Orff)	Classical Orchestra	144
9	'On My Own' (Annerly Gordon)	Dance (Vocals, English)	148

The study found that driver impressions of landscapes were perceived as more 'pleasant' with accompanying music than without music. Moreover, pleasantness was highest when music excerpts were more relaxed (items 1, 2, and 5), and that slow-paced music (≤ 102bpm) contributed to make the landscape feel pleasant. On the other hand, the study found that landscapes were perceived as more 'powerful' when active music was present (items 6, 8, and 9), and that fast-paced music

(> 102bpm) contributed to make the landscape feel powerful. Iwamiya and Sugimoto concluded that the impression of landscape from the windshield of a moving car is systematically changed by the music, and therefore more than anything else, *in-car music* should be considered an 'audio-visual media'. They suggest that the interactions found in the study indicate physiological and emotional effects whereby one modality (i.e., audition) influenced the impression of another modality (i.e., vision) in the same direction.

Therefore, we might view the context of driving a vehicle as one in which the reciprocal influence between emotion and attention is of great importance, especially as operating an automobile functionally entails a high level of attentional resources to control and manoeuvre the car. Several researchers (Dibben & Williamson, 2007; Konecni, 1982) indeed claim that the essence of music's influence on driving performance may be explained by the *mechanism* in which musically evoked energy levels compete with attentional resources necessary for the primary driving task. Namely, the more music can induce higher energy or activation, the more music can compete for the attentional resources of the driver. Moreover, it would seem that such a mechanism is particularly dangerous in highly demanding driving situations whereby attentional resources are limited to begin with (North & Hargreaves, 1999). In this connection, Mesken et al. (2007) investigated the determinants and consequences of driver emotions. The scientific literature indicates that traffic is indeed a milieu in which emotions surface, but yet, the degree to which drivers experience moods, feelings, and sentiments is not well known. Actually, many studies seem to relate driver emotions to personality characteristics. For example, it has been presumed that people who have trait tendencies to get 'hot under the collar' and 'infuriated', more easily experience an ample amount of annoyance (and even rage) while driving on the road. There is a notion that when angry, such individuals drive with accelerated speeds, behave in a more risky manner, and perceive they have experienced a substantially increased number of near-accidents. Likewise, other personality traits might prescribe which types of people more often experience 'anxiety' or 'joy' while driving an automobile. In an early study, Mesken found that when comparing between driver subgroups, those who experienced the intensification of anger during driving also seemed to commit a higher percentage of traffic violations. We must bear in mind that while there is apparent linkage between emotion and perceived risk, and although anxiety is associated with higher levels of perceived risk, unfortunately both 'happiness' and 'anger' seem to relate to lower levels of perceived risk. Hence, there is every possibility that when we are angry, we *misperceive* or *miscalculate* concrete tangible driving risks – when for the purposes of maintaining safety it would have been especially prudent had we been able to act in a fashion reflecting clear-headedness. This was exactly the raison d'être for the study implemented by Mesken et al.

Mesken et al. (2007) recruited 44 licensed car drivers (resembling the typical characteristics of Dutch drivers) for a single 50-minute drive in an instrumented *Renault 19*. During the drive the participants provided ratings of their emotional state every three minutes, as well as whenever an emotion was identified as directly caused by the traffic context. Participants were asked to note which emotion out of three emotions they felt (i.e., 'angry', 'nervous', or 'happy') as well as to indicate the strength of the felt emotion on a 5-point scale (1 = Slightly, 5 = Very). Moreover, heart rate (HR) of the participants was recorded while driving, as well as during a 3-minute rest period after driving. Speed was measured using a GPS system. Sitting in the passenger seat next to the driver was a qualified driving instructor who continuously suggested the route, and there was an experiment-monitor sitting in the back seat who logged all emotions and risk-ratings as stated verbally by the driver. The results of the study found an average of five emotions for each driver within the 50-minute drive; the most frequently reported emotion per trip was 'anxiety' (2.6 times per trip), followed by 'anger' (1.5), and then 'happiness' (1). Nevertheless, the study showed that the strength of emotions was not very robust ($M_{TotalEmotion} = 1.4$, $SD = 0.5$ [$M_{Anxiety} = 1.3$, $M_{Anger} = 1.4$, and $M_{Happiness} = 1.8$]). The study found that 'anger' usually resulted from another car user (i.e., person), whereas 'anxiety' surfaced from the driver's attempt to maintain safety within the traffic milieu (i.e., situation). When comparing between active driving and the rest periods, no differences of HRs surfaced for intervals when 'anger' or 'happiness' was felt, but rather significant HRs increases were associated to 'anxiety'. In addition, the study found that 'anxiety' was most often associated to higher scores of perceived risk. Finally, speed was significantly faster when drivers reported 'anger' ($M_{Anger} = 91$ kph [57 mph], $SD = 4.2$; $M_{NoAnger} = 87$ kph [54 mph], $SD = 3.9$), and the percentage of time that speed limits were exceeded was significantly greater when drivers reported 'anger' ($M_{Anger} = 16\%$, $SD = 13.5$; $M_{No\text{-}Anger} = 2.4\%$, $SD = 4.4$).

Subsequently, Pecher, Lamercier, and Cellier (2009) studied the effects of emotions on driver's attentional behaviour. In a pre-experiment development stage they recruited 30 undergraduates at the University of Toulouse II–Le Mirail (France) to evaluate music excerpts for emotional valence; this procedure brought forward a test-set of 18 1-minute excerpts, comprised of six songs, each for one of three valences (i.e., 'Happy', 'Sad', 'Neutral'). The researchers assumed that simulated driving while listening to 'Happy' music would result in a more 'assimilative' processing style characterized by faster speeds with a tendency to take risks, whereas listening to 'Sad' music would result in a more 'accommodative' processing style indicating slower speeds with no risk-taking. In the active study, Pecher et al. recruited 17 participants ($M = 25$ years old, 63% female, with four years driving experience) for one 60-minute session in a fixed base simulator consisting of complete *Renault 19* with automatic transmission. The participants drove on virtual highways with two lanes in each direction on a dry sunny day with good visibility. After a short practice run and a 5-minute break, they drove roughly 25 minutes maintaining a speed between 80–120 kph (50–75 mph), controlling for lane position, and slowing down at curves. Every seven minutes a new aural background was introduced (i.e., one of the three driving conditions based on emotional valence). Each condition began with a 1-minute baseline period of silence, followed by a 6-minute period of music (comprised of six 1-minute song excerpts). The results of the study point to statistically significant main effects of speed. Namely, speed decreased when driving with music ($M_{Emotion\text{-}AloneWithoutMusic}$ = 100 kph; $M_{Emotion+CongruentValenceMusic}$ = 95 kph). However, speed variation was not consistent across all three aural conditions ($M_{Happy\text{-}Alone}$ = 98 kph, $M_{Happy+HappyMusic}$ = 87 kph; $M_{Sad\text{-}Alone}$ = 97 kph, $M_{Sad+SadMusic}$ = 101 kph; $M_{Neutral\text{-}Alone}$ = 99 kph, $M_{Neutral+NeutralMusic}$ = 100 kph). In addition, the results show the percentage of swerving (i.e., poor lateral control) that increased for 'Happy' music, while the contrary was seen for both 'Sad' and 'Neutral' musics ($M_{Happy\text{-}Alone}$ = 8.1%, $M_{Happy+HappyMusic}$ = 9.7%; $M_{Sad\text{-}Alone}$ = 8.9%, $M_{Sad+SadMusic}$ = 8.5%; $M_{Neutral\text{-}Alone}$ = 8.9%, $M_{Neutral+NeutralMusic}$ = 7.5%). Pecher et al. reported all drivers to have claimed 'Happy' music to be the most 'disturbing' of the valences: 13 (76%) drivers reported to be more seriously distracted as they subsequently followed the melody by singing, whistling, or clapping along. One participant stated:

> It was hard for me to control what I did with this kind of music because I had to follow the rhythm and lyrics. I felt good and, to be honest, driving was no longer my priority (p. 1,257).

Yet, the opposite effect was reported for 'Sad' music, to which drivers were calm, focussed, and quiet. One participant stated:

> With sad songs, I was calmer and I couldn't help feeling sad. The rhythm and melody caught my attention and I couldn't help listening to them. When sad songs were played, I tended to think about what was wrong with me, my job or my family (p. 1,257).

Finally, among 16 (92%) drivers, the 'Neutral' music had little to no effects on driving behaviour. One participant stated:

> Some of the [Neutral] music excerpts... didn't affect me at all. By this, I mean, I don't know whether I felt happy or sad or both. I didn't really listen when it was being played. I just kept on driving. The music just didn't exist for me. It was useless (p. 1,257).

Pecher et al. concluded that music valence significantly affected driving speed and lateral control. However, while they acknowledged the fact that the drivers were young people, and hence most likely sensitive to the emotional qualities of the Pop music presented in the study (for example, the song 'Everybody Needs Somebody' from *The Blues Brothers*[3] soundtrack represented the 'Happy' emotional valance), they also raised the possibility that many participants might not have perceived the intended emotional qualia from selections evaluated as belonging to a 'Neutral' valence.

One can only wonder if the emotional qualities of music (reflected in the characteristic valance and levels of arousal) do in fact affect driving behaviours, and if such music-evoked responses are different than other driving circumstances that

3 *The Blues Brothers* (1980, Universal) is an American feature film comedy starring John Belushi and Dan Aykroyd. The film was developed as a takeoff on characters that initially appeared in the *NBC* late-night show 'Saturday Night Live'.

might also generate emotion among drivers during everyday driving activity. Over a decade ago Recarte and Nunes (2000) raised a somewhat related question: How similar are different kinds of everyday mental activities that pre-occupy drivers while driving? Namely, are there differences between listening to a soccer match, listening to music, telling a story, doing mental calculations, or trying to remember something – while we drive a car? Along these lines, ten years later Reed and Diels (2011) from the Traffic Research Laboratory in Workingham (UK) implemented an excellent research package comprised of four studies, whereby each study was designed to evaluate driver behaviour and vehicular performance in an emotional state induced by a different but specific in-cabin context reflecting one typical situation that drivers encounter everyday. The four contexts were: (1) driving while listening to arguing children *versus* listening to radio sports commentary of a horse race; (2) driving while feeling stressed *versus* feeling relaxed; (3) driving while listening to four favourite preferred songs or music styles *versus* listening to four non-preferred songs or music styles; and (4) driving with different combinations of male/female passengers (i.e., single male *versus* single female passenger seated in the front seat *versus* both male and female passengers seated in front and rear seats), who engaged in passenger activities such as conversing with the driver, adjusting the heater/fan settings, and texting/calling a friend. All four studies employed the same route and driving environment; the simulator employed a Honda *Civic* family hatchback with a five-speed manual gearbox mounted on an electric motion system (inducing heave, pitch, and roll movements) delivering the approximation of g-forces and vibrations that would usually be experienced when driving a real vehicle. The road environment was projected onto three forward screens (210° horizontal view) and a rear screen (60° view). The car stereo sound system provided simulated engine, road, and traffic sounds @75dBA. In total, 74 participants between the ages 25–45 (M = 30, SD = 5.23, 50% male) were recruited. Two central variables were used as measures of driver behaviour and vehicular control: (1) Standard Deviation of Headway (SDHWs) or the degree of variance when keeping a constant distance between the car being driven and the car in front; and (2) Reaction Time (RTs) or the degree of time necessary to respond to an unexpected event in comparison to baseline control driving (see Figure 5.1). As can be seen in Panel-A, auditory distractions hampered drivers' ability to keep a constant distance between cars (increased SDHW-Δ). SDHW-Δ was significantly larger when drivers heard children arguing in the car (and this data might be all the more conservative as drivers were only exposed to 'recordings' of altercations rather than experience an actual dispute). Further, auditory distractions induced slower responses to unexpected events (i.e., increased RT-Δ). For example, Panel-B indicates that when listening to either arguing children or commentary of a horse race, RT-Δ increased. Further, Panel-B shows that when drivers were stressed their RT-Δ were slow (but not as impaired as when relaxed). The study found that drivers in a stressed mood not only accelerated driving speeds ($M_{Relaxed}$ = 68mph; $M_{Stressed}$ = 72mph; t = 2.663, df = 17, p < .05), but were all the more deficient in keeping a steady distance from the vehicles in front than

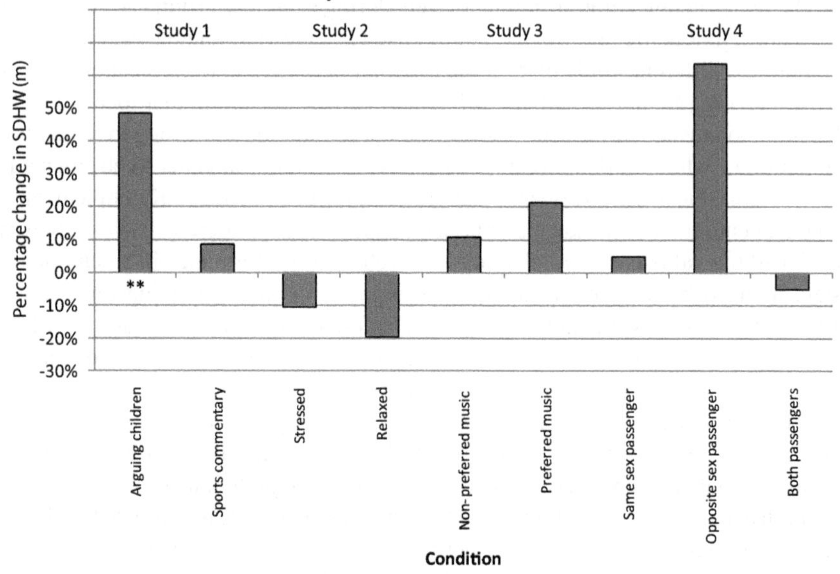

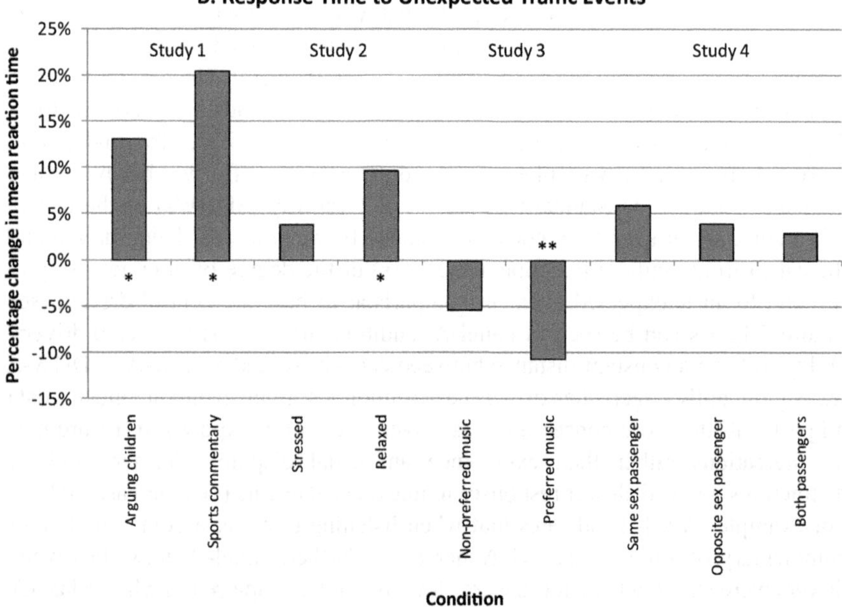

Figure 5.1 Effects of everyday driving conditions on driver behaviours
Source: Reed and Diels (2011). Reprinted by permission of the author.

drivers in a relaxed mood state (as seen in Panel-A). Moreover, the study found that listening to music improved RT-Δ to unexpected events – but mostly when driver-preferred favourite songs/styles were heard. While there was a tendency for participants to drive faster in both music conditions, these differences failed to reach statistical significance. In general, music listening caused a moderate effect on drivers' ability to maintain a constant headway. In addition, the results demonstrated both male and female drivers as equal concerning distractions when driving with a member of the opposite sex (albeit, female drivers found that driving with two passengers was more distracting). Finally, there was a consistent tendency for male drivers to drive faster than female drivers, and both male and female drivers tended to drive faster when accompanied by passengers of the opposite sex. Although the presence of passengers had very little effect on RT-Δ, their presence did typically affect SDHW-Δ when accompanied by a passenger of the opposite sex.

In the above-described study by Reed and Diels (2011) the participants drove to their preferred (*favourite*) music versus their reported non-preferred (*non-favourite*) music. Another study that employed the same music criteria was Wiesenthal, Hennessy, and Totten (2000, 2003). The research team claimed that a driver's 'favourite' music is apt to alleviate stress levels when driving in the car during highly congested driving in comparison to driving without music. Wiesenthal et al. (2000, 2003) recruited 40 participants from York University (Toronto, Canada) between 21–50 years old ($M = 26$, 50% female); all of the participants regularly travelled on Highway 401 (a major 14-lane artery for metropolitan Toronto). Each driver participated in a single 30-minute trip (on partly cloudy to sunny mid-week days, between Tuesday and Thursday) over a five-month period (October–February). Two segments along their regular route were identified for each driver; one was typically a section of flowing traffic with few bottlenecks, while the other was typically a highly congested jam-packed bumper-to-bumper section of the thoroughfare. In a counterbalanced fashion, the drivers were assigned to one of two subgroups variegated by segment order (i.e., low-to-high *versus* high-to-low congestion). Within each of the subgroups, every driver was also randomly allocated to one of two driving conditions – with or without in-car music accompaniment. The drivers in the Music condition were requested to select and listen to music cassette tapes or CDs of their 'favourite' music throughout the entire journey; the drivers in the NoMusic condition were requested to refrain from listening to any aural background including music and/or talk radio. Wiesenthal et al. claim that the music styles preferred by the drivers were mostly Pop, Top-40, and Country musics. However, given that the nature of the task was unaccompanied driving, we might wonder what controls were put in place to assure that these genres were in fact reliable descriptions of the music heard, or that the empirical driving conditions (i.e., engagement and/or abstinence from listening) were actually carried out. Continuously during the journey, at intervals of sixteen minutes, the drivers called the research team via a hands-free cell phone. Over the phone, each driver was surveyed on four principal issues; they verbally answered a number of questions about stress and aggressive behaviours, rated their level of arousal, self-assessed their perceived time urgency, and estimated their level of vehicular control. Wiesenthal et al. reported the study twice in a piecemeal fashion; in 2000 they assessed drivers' state *stress*, and in 2003 they evaluated drivers' *aggressive* behaviour. Taken together, the general gist of the study was that highway segments involving higher congestion caused significantly higher levels of stress ($M_{LowCongestion} = 33.9$, $SD = 2.69$; $M_{HighCongestion} = 43.3$, $SD = 8.13$), there was an effect of Condition *x* Stress ($M_{NoMusic} = 42.4$, $SD = 9.33$; $M_{Music} = 34.8$, $SD = 3.89$), and a significant three-way interaction between Congestion *x* Condition *x* Stress (see Figure 5.2). As can be seen in Panel-A, greater levels of stress were experienced by drivers who did not listen to music or talk radio during high-congestion traffic than as experienced by drivers listening to their favourite preferred music cassette tapes or CDs. The study found that during high-congested driving, significantly more frequent aggressive behaviours (such as honking and flashing lights out of frustration, swearing, and using obscene gestures towards other drivers and purposely tailgating) were reported; that is, there was an overall higher level of aggression during high-congestion traffic ($M_{LowCongestion} = 4.45$, $SD = 7.94$; $M_{HighCongestion} = 12.10$, $SD = 12.60$; $t = 5.9$, $df = 39$, $p < 0.05$). Although there were no meaningful differences of time urgency between highway segments (Low- *versus* High-congestion) or between the aural conditions (NoMusic *versus* Music), a significant interaction effect of Time Urgency *x* Condition *x* Aggression did surface. As can be seen in Panel-B, during high-congestion traffic, drivers who listened to music were significantly less aggressive towards others, but only when they perceived no urgency of time. But when drivers felt pressured to get somewhere in a hurry, listening to music made no difference at all. Hence, Wiesenthal et al. concluded that "music may provide a degree of distraction from potential irritating and frustrating traffic stimuli, but only when other salient personal stressors, such as time concerns, are minimal" (2003, p. 130). Unfortunately, a very different assumption (one that is an incorrect interpretation of the actual data) is the one most often cited in the literature: "music had an influence on mild driver aggression in high congestion but not low congestion" (p. 130). For example, Krahe and Bieneck (2012), who found that 'pleasant' music buffered the adverse effects of provocation in non-traffic-related tasks, deducted that their findings confirm other findings (such as Wiesenthal et al., 2003), and hence validate that pleasant music can countermeasure negative emotions that occur when encountering aggravated drivers or experiencing road-rage.

It should be pointed out that both of the above studies compared between conditions when drivers engaged in music listening compared to driving without music. We might question whether or not such contrasts yield reliable findings about *music emotion* within the driving context. Perhaps investigations employing methodologies exploring emotional (dis)similarities as measured between two aural backgrounds, rather than comparing music to naught, would be more strategically rigorous. In this connection, van der Zwaag et al. (2012)

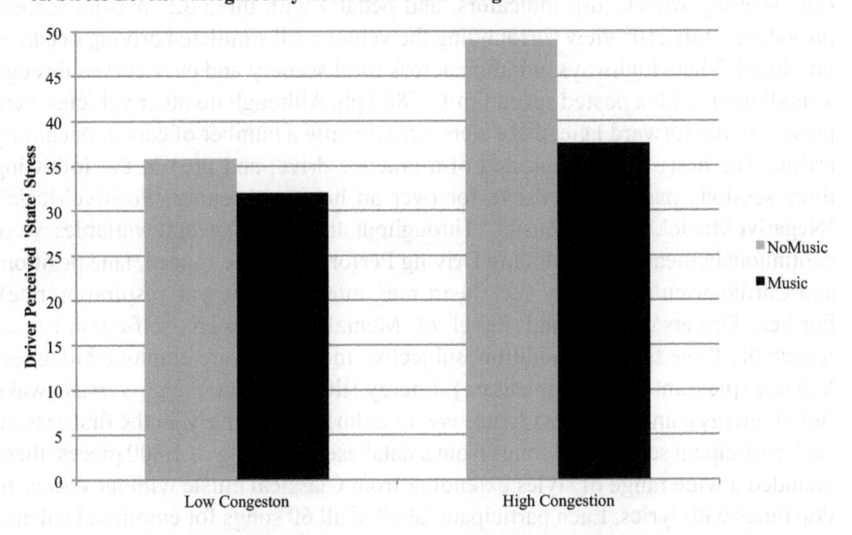

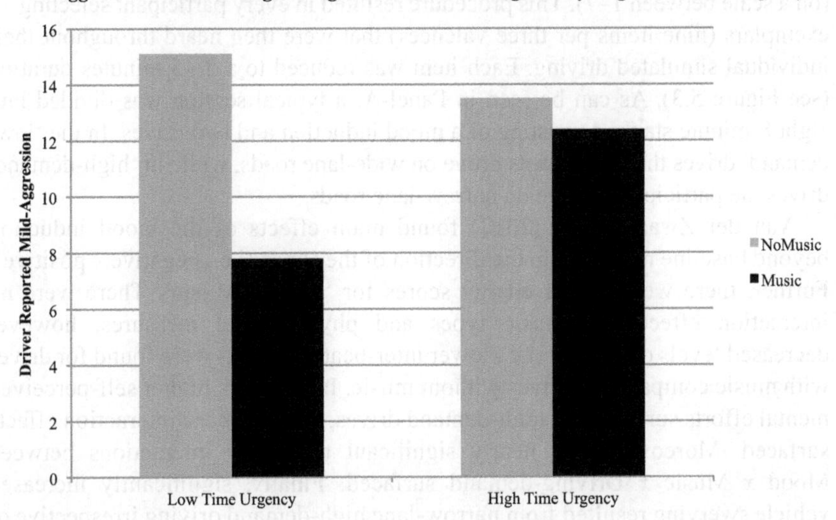

Figure 5.2 Listening to favourite music versus no music background
Source: Adapted from Wiesenthal et al. (2000, 2003).

investigated whether or not the presence of *positive* music versus *negative* music has a casual effect, and if music-evoked moods persist during both low- and high-demand drives. The research team recruited 19 participants for four sessions of simulated driving at the University of Groningen (The Netherlands); they were between 22–44 years old ($M = 28$, $SD = 5.2$, 84% male) with a licence for at least four years ($M = 9$, $SD = 5$). The study employed a fixed-base vehicle mock-up (i.e., steering wheel, turn indicators, and pedals) with three 32" plasma screens providing a full 210° view surrounding the vehicle. All simulated driving occurred on virtual 2-lane highways, winding across rural scenery and easy curves through a small town with a posted speed limit of 80 kph. Although no other vehicles were present in the forward lane, there were occasionally a number of cars in oncoming traffic. The first session consisted of a practice drive, and then in the following three sessions participants drove for over an hour with either 'PositiveMusic', 'NegativeMusic', or 'NoMusic'. Throughout the study several variables were continuously measured, including Driving Performance (i.e., speed, lane position) and Cardiovascular Activity (i.e., heart rate, inter-beat interval, respiration rate). Further, Drivers' Mood and Level of Mental Effort were self-administered repeatedly three times. In addition, subjective measures were employed to assess Valence (pleasant *versus* unpleasant), Energy (tired without energy versus awake full of energy), and Calmness (tense *versus* calm). Accordingly, in the first session each participant selected 60 songs from a database consisting of 1,800 pieces; these included a wide range of styles extending from Classical music without vocals, to Pop tunes with lyrics. Each participant labelled all 60 songs for emotional valence (i.e., positive *versus* negative character), and rated each one for level of energy (on a scale between 1–7). This procedure resulted in every participant selecting 18 exemplars (nine items per three valences) that were then heard throughout their individual simulated driving. Each item was reduced to a 2:45 minutes duration (see Figure 5.3). As can be seen in Panel-A, a typical session was divided into eight 8-minute stages consisting of a mood induction and two drives. In the 'low-demand' drives the participants drove on wide-lane roads, while in 'high-demand' drives the participants drove on narrow-lane roads.

Van der Zwaag et al. (2012) found main effects of the mood induction beyond baseline measures in the direction of the songs (i.e., negative – positive). Further, there were higher energy scores for 'PositiveMusic'. There were no interaction effects for music types and physiological measures, however decreased levels of arousal (i.e., lower inter-beat intervals) were found for drives with music compared to drives without music. In addition, higher self-perceived mental efforts surfaced for high-demand drives, but only near-interaction effects surfaced. Moreover, only nearly significant three-way interactions between Mood *x* Music *x* Driving-demand surfaced. Finally, significantly increased vehicle swerving resulted from narrow-lane high-demand driving irrespective of the music, and no specific effects of music type surfaced for driving performance measures (such as speed or lane position). While the research team concluded that music and driving demand do seem to influence mood, physiological state, and driving performance, they also conceded to the fact that based on their findings, such effects are probably much less robust than could be expected. The truth is that the bulk of effects reported by van der Zwaag et al. only surfaced from analyses comparing trips with music versus trips without music. Therefore, we might understand that few if any meaningful differences were found between the two music types (that were specifically selected because of supposed

(A) Structure Of Simulated Driving Experiment Employed By van der Zwagg et al. (2012)

	Pre-Experiment		Experiment					
	Stage 1	**Stage 2**	**Stage 3**	**Stage 4**	**Stage 5**	**Stage 6**	**Stage 7**	**Stage 8**
Procedure:	Briefing	Habituation	Mood Induction	Park	Drive 1	Park	Drive 2	Park
Activity:	Attach Sensors	1. Watch TV 2. Mood Scale 3. Baseline HR & RR	Listen With Music <u>or</u> Sit-Relax Without Music	Mood Scale	With/ Without Music On *Narrow* Lane Road	1. Mood Scale 2. Mental Effort Scale	With/ Without Music On *Wide* Lane Road	1. Mood Scale 2. Mental Effort Scale
Duration:	8 min	8 min	8 min	8 min	8 min	8 min	8 min	8 min
Time Line:	0:00-------8:00---------16:00--------24:00---------32:00---------40:00---------48:00---------56:00-----64:00							

(B) Structure Of Simulated Driving Experiment Employed By van der Zwagg et al. (2013a)

	Pre-Exp	Experiment							
		Trial 1				Trial 2			
	Stage 1	**Stage 2**	**Stage 3**	**Stage 4**	**Stage 5**	**Stage 6**	**Stage 7**	**Stage 8**	**Stage 9**
Procedure:	Briefing Habituation	Drive A	Park	Drive B	Park	Drive A	Park	Drive B	Park
Activity:	Attach Sensors 1. Watch TV 2. Mood Scale 3. Baseline SCL & EMG	Happy Music During High Demand Driving	1. Mood Scale 2. Mental Effort Scale	Happy Music With *Abrupt/ Gradual* Change To Calm Music	1. Mood Scale 2. Mental Effort Scale	Happy Music During High Demand Driving	1. Mood Scale 2. Mental Effort Scale	Happy Music With *Abrupt/ Gradual* Change To Calm Music	1. Mood Scale 2. Mental Effort Scale
Duration:	10 min	8 min	2 min	8 min	2 min	8 min	2 min	8 min	2 min
Time Line	0:00---------10:00--------------------20:00--------------------30:00--------------------40:00----------------60:00								

Figure 5.3 Session structure for simulated driving experiments
Source: Adapted from van der Zwagg et al. (2012, 2013a).

distinctive valences). Perhaps, then, the matter is related to poorly developed empirical stimuli. After all, van der Zwaag et al. themselves raise concerns about the music selections:

> [While] the song stimuli varied in valance and to a lesser extent in energy levels, [we] note that the selected songs were not the most contrasting songs… [It was] impossible to select songs that solely varied in mood valence having equal energy levels (p. 20).

Hence, it would seem that although some of the songs were certainly *happy* and others were unmistakably *sad* (i.e., distinct emotional qualities), these selections were far from being matched in their level of intensity (i.e., recognized emotional strength). To be fair, the method employed by van der Zwaag et al. did assure that the songs selected were at least familiar to each driver, and did induce the targeted mood states. Nonetheless, van der Zwaag et al. generated music stimuli that by their own account were far from what might be expected for a tightly controlled empirical setting. For example, they did not account for musical characteristics such as: 'pace' (fast/slow tempo), 'tonality' (major/minor key), or 'complexity' (dense/transparent textures, internal music activity). In addition, van der Zwaag et al. raise the point that the participants in their study were all experienced drivers, and thus question if "the additional load added by music listening while driving could have been [far too] minimal" (p. 20). It is rather ill-fated that both the music used in the study, as well as the participants recruited, weakened the empirical demonstration of music effects on drivers' attentional resources in high-demand driving. Hence, the research team were ultimately left with a rather narrow final conclusion: "Listening to music… did not require more than the resources still available next to those required for driving, so no real competition for resources resulted" (p. 20). Ironically, they twisted their final conclusion by over-emphasizing *null-effects* of music on driving behaviour and vehicular control – never considering methodological error as having tainted the outcome. Yet, on face value, even though no impairment of driving performance surfaced, when considering that the majority of exemplars used were far from reflecting intense levels of energy, then perhaps the significant finding of this study is: *moderately emotive music facilitates driving better than intensely emotive music.* Namely, moderately emotive music *improved* driver mood and induced a more relaxed state of mind and body. This result, though underestimated by van der Zwaag et al., is truly a highly important find.

Because a calm mood often leads to safer driving conditions, we might also envision that it would be highly beneficial if drivers could moderate their music selections in an effort to reproduce a more musically-evoked calm mood. If drivers were listening to the radio when experiencing unfriendliness or hostility, it might be best that they change the channel to a station that broadcasts more relaxing music. Alternatively, if drivers were listening to music CDs when they experience irritation and aggression, it might be wise for them to switch to CDs with songs of a more serene nature. This leads us to the question of whether or not the intended mood change would be most adaptive if music was altered *abruptly* or *gradually*. The issue at hand regards the effectiveness of an intended mood change based on theories that underline *mood regulation* versus *mood congruency*. In the former theory, our deliberate attempt to influence affect occurs when we feel a discrepancy between our current mood and a target mood – especially when we perceive the current mood as hampering our driving. In this case, an abrupt change in music type would be highly efficient. Yet, the latter theory, *mood congruency*, furthers a claim that it is most effective to change our mood over time during the entire drive, by first listening to music that is compatible with the initial mood state and then ever so slowly, selecting music that gently induces the target mood. In this case, a gradual change in music would be highly efficient. Given the two theories, van der Zwaag et al. (2013a) explored the resourcefulness of both driver actions. The research team recruited 28 Stanford University (USA) undergraduates to participate in one 60-minute simulated driving session whereby music-change strategies (abrupt *versus* gradual) were administered as a within-subjects factor. The participants were on average 22 years of age (*SD* = 1.3, 50% female). Driving was implemented on a fixed-base vehicle mock-up (with a functional steering wheel, indicators, and pedals) employing a 1.83 metre (6-foot) rear projection screen. The order of the two music-change strategies was counterbalanced across the sample, and simulated driving was designed as high-demand driving (involving a relatively broad vehicle on a relatively narrow-lane road), with maximum speeds of 89 kph (55 mph) on straight sections and speed limits set to 64 kph (40 mph) for curves.

Prior to the experiment, the participants selected 62 songs from a database of 1,800 pieces. Each driver rated every song for valence and energy level (in a similar fashion as was implemented in van der Zwaag et al., 2012, described above). However, and unlike their previous study, the participant-chosen pieces were subjected to further analyses based on a valence-arousal model. That is, employing an algorithm accounting for a host of musical features (including, tempo, mode, and complexity), for each participant the six songs classified as having the highest dimension components were selected for 'Happy' music drives (see Figure 5.3, Panel-B, Stages 2 and 6), while the six songs classified as having the lowest dimension components were selected for inciting a mood change towards relaxation (see Figure 5.3, Panel-B, Stages 4 and 8). Each song was edited down to a 2:45 minutes duration. As can be seen in Panel-B, there were three 8-minute drives during the single session. Throughout the study several variables were continuously measured including Driving Performance (i.e., speed, lane position, near-crashes, crashes), and Physiological Processes (dry-finger electrode skin conductance levels, EMG facial muscle tension). Further, drivers' Mood and Level of Mental Effort were self-administered repeatedly. In addition, several subjective measures were employed to assess Valence (pleasant *versus* unpleasant), Energy (tired without energy *versus* awake full of energy), and Calmness (tense *versus* calm). Initially, the analyses demonstrated that there were no meaningful differences of mood between the two *A*-drives of Trials 1–2 (i.e., driving in a steady state with 'Happy' music, Stages 2 and 6). Nor were there meaningful differences of mood between the empirically manipulated *B*-drives of Trials 1–2 (i.e., counterbalanced *abrupt* versus *gradual* music-change strategies, Stages 4 and 8). Although significantly higher levels of self-perceived mental effort were invested in the first *A*-drive ($M_{\text{Trial-1}} = 66.04$, $SE = 5.9$; $M_{\text{Trial-2}} = 56.71$, $SE = 4.9$), no significant differences of mental effort surfaced for the *B*-drives ($M_{\text{Abrupt}} = 59.95$, $SE = 4.7$; $M_{\text{Gradual}} = 64.51$, $SE = 6.0$); these findings may indicate that such differences were simply artefacts of practice. The study found no meaningful differences for physiological processes (dry-finger SCLs or EMG facial activity) between the two *A*-drives. Nonetheless, the study did find that driving performances for the *A*-drives were significantly better in Trial 2 (again perhaps an artefact of practice); less accidents ($M_{\text{Tiral-1}} = 4.69$, $SE = 0.57$; $M_{\text{Trial-2}} = 3.30$, $SE = 0.59$), less line crossings ($M_{\text{Trial-1}} = 39.46$, $SE = 3.46$; $M_{\text{Trial-2}} = 31.78$, $SE=3.26$), but higher speed reaction scores (as measured as metres per second: $M_{\text{Trial-1}} = 20.03$ m/s, $SE = 0.37$; $M_{\text{Tial-2}} = 20.81$ m/s, $SE = 0.45$). However, and most importantly, when comparing between the empirically controlled counterbalanced *B*-drives, the *abrupt* music-change strategy caused significantly fewer accidents ($M_{\text{Abrupt}} = -1.38$, $SE = 0.47$;

$M_{Gradual}$ = 0.35, *SE* = 0.52), fewer line crossings (M_{Abrupt} = -4.78, *SE* = 2.01; $M_{Gradual}$ = 0.97, *SE* = 2.5), and lower RTs (M_{Abrupt} = -0.04 m/s, *SE* = 0.23; $M_{Gradual}$ = 0.39 m/s, *SE* = 0.17). Van der Zwaag et al. concluded that when drivers seek to induce a more prevailing relaxed state in their effort to achieve increased driver safety, then an *abrupt* change in background music is far more efficient than changing music pieces gradually. Namely, abruptly changing music excerpts affects driver mood in the expected direction, decreases physiological energy levels, decreases driving violations, and hence increases vehicular performance.

5–2.1.1. Affective music player

The above section clearly indicates that music valence affects driver mood (which in turn is linked to efficiency/deficiency of driving performance), that the optimal music for drivers to listen to are pieces with a moderate level of emotional energy (as intense emotional qualities of either positive or negative valance causes unwanted maladaptive driver behaviours), and that an abrupt change in music is most resourceful in inducing an adaptive change of mood (which may be required by the circumstances of the road environment and traffic milieu). Hence, the remaining question seems to be: How do drivers know which music to choose? After all, with the immense music storage capabilities that are found in today's digital music players, most drivers can have thousands of tracks available to choose from. And while drivers can use any number of software applications to organize their song database, the choices still seem to be endless – not to mention the fact that wrong song choice might be quite maladaptive and lead to serious consequences. Certainly, music listeners can personalize their playlists by tagging a *subject* or *theme* (creating a playlist such as 'Songs for Driving'), or a music *genre* (creating a playlist such as 'Ballads and Folk Songs'), or a specific music *feature* (creating a playlist such as 'Quiet Slow-paced Music'), or even a particular *mood* (creating a playlist such as 'Calming Relaxing Music'). Further, today there are countless Internet and satellite radio stations that offer structures for filtering music selections, generating compendiums, and actually streaming the music based on themes, genres, or activity (i.e., playlists for going on a holiday, for having a party, for studying, or for exercising and workout).[4] It should be acknowledged that while most everyday people are not really aware how to select music based on their own affective responses, some of the above platforms can choose music for the listener artificially in an effort to assist in either decreasing stress or increasing arousal. In most cases, the above formats are based on recognizing the pace and intensity of items in the playlist. In addition, some 'new-age' music players have *bio-feedback* capabilities that monitor heart rate, and subsequently adjust the music programme by selecting pieces based on a target tempo (for example, to enhance a specific activity such as bicycling or running). Two popular products that reorganize playlists according to the user's physiology are *BodiBeat* (Yamaha) and *MotoACTV*

4 Among the music services that listeners can subscribe to, are: *8 Tracks*, *Pandora*, *Intelligent Music Player* [IMP], *Media Monkey*, and *Moodagent*.

(Motorola). There has been only one attempt to develop a software tool (App) as a mobile application music recommender system (for *Android*) targeting users in an automobile. Baltrunas et al. (2011) supposed that a context-aware system, which could consider traffic conditions and the mood of the driver when recommending music tracks to be played in the car, should be highly beneficial to users. They developed a web-based system to rate music recommendations in advance, accounting for musical preferences and musical tastes; the music tracks offered ten different genres as examples for user ratings: Classical, Country, Disco, Hip-Hop, Jazz, Rock, Blues, Reggae, Pop, and Metal music styles. Then, these exemplars (as well as additional music items) were modified for the traffic situation by employing contextual factors and conditions, including: Drive Time (morning, night, afternoon), Driving Style (relaxed, sport), Road Type (city, highway, serpentine), Landscape (coast line, country side, mountains/hills, urban), Sleepiness (awake, sleepy), Traffic Conditions (free road, many cars, traffic jam), Mood (active, happy, lazy, sad), and Weather (cloudy, snowing, sunny, rainy). In two studies reported by Baltrunas et al., which reflect a development phase (N = 59) and a validation phase (N = 66), the research team found that their mobile application was in fact context-aware and that user ratings (i.e., drivers' behavioural responses) were dependent on the traffic environment. Therefore, they conclude that their software application could offer functional personalized music recommendations to users in a car scenario. Nevertheless, the reported finding of this mobile application was solely founded on a web-based user-response interface (Internet task) rather than involving a secondary task while listening to background music (i.e., emotional responses to background music while actively involved in a behavioural task), or employing the music recommender system within the environment of driving (such as lab-based simulated driving or on-the-road naturalistic driving).

Indeed, most drivers do not have the perceptiveness to pre-assign songs for driving, nor can they envision a diverse number of playlists each to be used in a unique set of circumstances encountered on the road. Most specifically, drivers cannot imagine which items they should select to design a playlist that might be needed to stabilize their emotional state in the aftermath of an incident. In this connection, the development of one assistive device stands alone. For some time, researchers in the Department Of Brain, Body, and Behavior at *Philips Research* (Eindhoven, The Netherlands) have been exploring the *Affective Music Player*. *AMP* has capabilities of distinguishing certain physiological measures that dictate song selection in order to induce a specific mood of the user to enhance performance (Thrasher, van der Zwaag, Bianchi-Berhouze, & Westerink, 2011; van der Zwaag, Janssen, & Westerink, 2013b). Early prototypes employed physiological measures that fed data into an 'emotional closed loop system' in order to direct the mood of the user within a workplace environment. Initially, research teams compiled emotional profiles based on features of seated communication, body position, posture, and movement. For example, eye contact, arm openness, distance, leg openness, hand and foot relaxation, backwards/sideways lean of torso, and leg-head-shoulder orientation all seem to differentiate between mood states. These early attempts led to deep-rooted beliefs about the potential of monitoring bio-physiological processes that ensue from affective experiences. But all early platforms in which music could be selected based on measurement of physiology (such as HRs, SCLs, and foot tapping) were flawed; their level of validity and reliability were reported to have been average at best. In their literature review of *smart players* by Janssen, van den Broek, and Westerink (2009, 2012) the authors claim that for the majority there are no theoretical frameworks surrounding affective states and their possible relationships to physiological changes (i.e., low *construct* validity), as well as weak demonstrations linking between the actual physiological measures monitored and the associated affective states supposedly elected by the music player (i.e., low *content* validity). Further, most previous attempts had been developed with data from participants in well-controlled sterile laboratory settings rather than real-world music listening situations (i.e., low *ecological* validity). Therefore, at *Philips* the research teams were steadfast to put forth efforts in developing a renewed concept:

> Music selection is based on physiological changes and a physiological goal state. The physiological goal state can be inferred from psychological studies… [whereby] increasing SCL will increase arousal and increasing skin temperature (ST) will increase valence. This overcomes the problematic inference step from physiology to affect but still allows to regulate mood in an affective physiological loop (Janssen et al., 2009, p. 2). The AMP uses physiological responses as input to personalize music selection and direct mood toward a target state selected by the user. The AMP thereby uses an affective loop that uses the physiological changes to adjust the music selection (van der Zwaag et al., 2013b, p. 58).

The Philips research teams used algorithms to select music in a probabilistic manner, whereby every time a song was listened to, the system learned from the resulting physiological changes of the user as 'how' best to engage the piece in future episodes. For example, if relaxing music was required, *AMP* would select particular music that, based on past exposure and the modelling of previous responses, would lower skin conductance levels (as SCL is known to be highly correlated with arousal or excitement). Then, after a period of training, *AMP* would be capable of pinpointing several music exemplars through a calculation of probability that will increase/decrease a specific effect within an identified target range and choose the exact specific song with the highest chance of achieving the target goal. Janssen et al. tested the full operation in a small pilot study (N = 3) employing 27 music pieces for each participant; SCL and skin temperature (ST) was employed as the physiological responses to be measured. In the pilot the participants listened to music during regular deskwork in their everyday work environment over several days for a total of 23 hours per participant. In a training procedure, songs with the highest probabilities for either increasing or decreasing STs response were recommended. Then, the participants rated each song for *positive* versus *negative* valence, and these songs were subsequently used as a test set for increasing/decreasing moods. The following playlists were reported as a typical set of 'Positive' and 'Negative' exemplars:

Songs With Positive Valence	**Songs With Negative Valance**
1 'Feel Like Flying' (Racoon)	1 'Old Tears' (Ilse De Lange)
2 'Blue Rondo A La Turk' (Dave Brubeck Quartet)	2 'De Waarheid' (Marco Borsato)
3 'Don't Stop Me Now' (Queen)	3 'Blowin' in the Wind' (Bob Dylan)
4 'Simple Creed' (Live)	4 'Carry Me' (Chris De Burgh)
5 'It's My Life' (Bon Jovi)	5 'Colorblind' (Counting Crows)

After the training period, the three participants engaged in four music listening sessions. Each session employed an 18-song playlist consisting of six songs for each of the three valences (i.e., 'Positive', 'Negative', and 'Neutral'). During the sessions, after 'Neutral' songs were heard (to standardize each listener's mood), two listening conditions were implemented whereby in one there was an attempt to increase ST while in the other there was an attempt to decrease ST. The study found that the songs in the low ST direction were more significantly rated as 'Positive' than the songs in the high ST direction (as high ST is related to low valence). The researchers claimed that the pilot study demonstrated the operational model as efficient in selecting music that first *detects* valence, and then through subsequent song choice, *directs* the listener to a specific emotional state.

In a further expanded study, van der Zwaag et al. (2013b) recruited ten participants (*M* = 26.5 years of age, 50% male) for a series of 28 sessions. In this investigation there was a 9-session training phase, and a 19-session active experiment phase. Accordingly, this investigation was branded as the first ever attempt to direct mood by means of song selection based on evaluation of physiological responses in real-time at a real-world workplace setting. Overall, the study employed 36 music pieces for each participant, personally chosen from the participants' own music database of 200 songs.[5] Subsequently, 12 songs were selected for every participant for each of three mood categories: (1) *Happy* (music items of high-energy positive-valance), (2) *Neutral* (music items of intermediate valance and energy levels), and (3) *Sad* (music items of low-energy negative-valance). Analyses revealed that songs in each category were statistically significantly different in both valence and energy levels. All songs were reduced to a 4-minute duration. The Training Phase operation was self-initiated by the participants during their workday in their regular workspace, but with the limitation that there could be no more than one training session per day. During the training dry-finger SCLs were continually monitored throughout, and the participants employed circumaural headphones to listen to the music (in an effort not to distract other workers in the same office environment). Training sessions ran for a minimum of 2:22 hours each, and this 9-session phase was spread out over a 3-week period. The training phase resulted in demonstrating that even in a real-world workplace environment, *AMP* was able to select an average of nine songs within the low SCL range, twelve songs within the neutral SCL

5 Each song had previously been rated by each participant for *mood impression*, and then validated for 'Valence' (unpleasant *versus* pleasant) as well as for 'Energy Level' (low *versus* high).

range, and eight songs within the high SCL range – for each participant. Then, the 19-session active Experiment Phase ensued with eight participants. The aim of the experimental phase was to manipulate SCL in target directions (i.e., to increase, maintain, or decrease levels) using the music exemplars that had previously been selected in the training phase that had the most reliable impact. The experimental phase employed a 24-song playlist, in which there were eight songs each for one of three moods: 'Energetic', 'Neutral', and 'Calm'. These were the eight songs with the greatest probability to increase SCLs ($M_{Valence} = 4.75$, $SE = 0.24$; $M_{Energy} = 4.67$, $SE = 0.16$), eight songs with the highest probability to maintain SCLs ($M_{Valence} = 4.08$, $SE = 0.23$; $M_{Energy} = 3.98$, SE 0.14), and eight songs with the greatest probability to decrease SCLs ($M_{Valence} = 4.02$, $SE = 0.24$; $M_{Energy} = 3.97$, $SE = 0.27$). Each mood state was the target for a complete 48-minute session, implemented in an alternating order, spread-out over a 4-week period (roughly one session per day). During the experiment each participant self-administered a pre- and post-session Mood Profile questionnaire, and both dry-finger SCLs and STs were continuously measured throughout. In each session, *AMP* first adjusted the playlist in real-time based on the initial SCL with an aim to direct SCL to a neutral state (16 minutes); then songs were automatically selected for either increasing or decreasing SCL depending on the targeted mood state (32 minutes). Van der Zwaag et al. found that music did affect SCL in the predictable direction, and that such changes were apparent even within a work setting with an ever-present background of room noise. It should be pointed out that the study is perhaps the first (and only) investigation to demonstrate *duration of effect*. Namely, once the music-evoked affect was achieved, the difference between the two physiological SCL directions was measured for over 30 minutes. Finally, the study found that listeners' subjective mood ratings were positively correlated with SCLs, and that the effects on STs were congruent with SCLs. The research team concluded that while it had been known for some time that listeners' perception of how a song influences their mood is not always accurate – namely the *affect expressed* by a song is not necessarily the same as the *mood induced* by that song – *AMP* seems to be highly capable of closing such gaps. Astutely, they also point out that an affective music player could be extended to environments that have been shown to benefit from the influence of music on mood – such as *while driving a car*.

Considering the above, Brodsky herein envisions an automotive version of *AMP* as outlined by van der Zwaag et al. (2013a). That is, by extending *AMP* to account for deficient vehicle performance, as revealed through modelling on-line analysis of pre-determined events generated by an in-vehicle data recorder coupled to the CANBUS of the car, music could truly become the ultimate mediated intervention to lower risk cued by drivers' affective responses. Moreover, the physiological measures already validated by van der Zwaag et al. could be accessed directly from sensors embedded in the steering wheel; prototypes that have already been developed in collaboration with automotive manufacturers such as BMW and Toyota monitor a variety of vital signs such as ECG measuring driver stress (Quick, 2011; Stomp, 2011). Brodsky proposes the following sequence (see Figure 5.4).

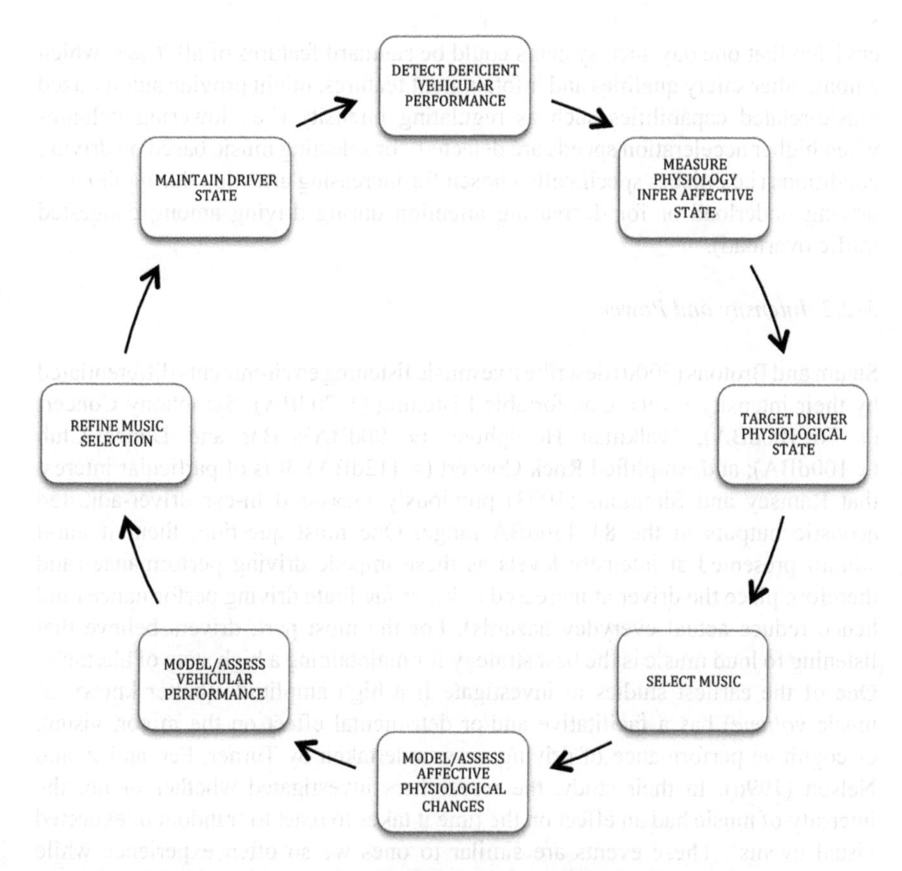

Figure 5.4 Brodsky's proposed closed loop system integrating vehicular performance and the Affective Music Player (AMP)

Source: Adapted from Janssen, van den Broek, and Westerink (2012).

Once driver deficiencies were detected, the triggered *Affective Music Player* could elicit data (i.e., physiology measures) to infer the driver's affective state, and then target a new physiological goal state. Based on Janssen, van den Broek, and Westerink (2012) the first music selection would be grounded to the inferred affective state and targeted physiological goal state. Thereafter, the system would model and assess the physiological change in an attempt to estimate if the music selected caused the targeted changes. If the physiological goal states targeted were not reached, then music would be re-selected in an effort to attain the intended objective. However, in the extended model suggested here, the system would also model and assess changes in vehicular performance by evaluating events directly produced by driver behaviour gathered through mechanical parameters elicited from the in-vehicle data recorder. Once these have been detected as meeting criteria of efficient safe driving, *AMP* could refine the music selection towards an effort to maintain the achieved affective and physiological status. We might envision that one day such systems could be standard features of all *iCars*, which among other safety qualities and infotainment features, might provide automatized music-related capabilities such as regulating intensity (i.e., lowering volumes when higher acceleration speeds are detected), or selecting music based on driving conditions (i.e., pieces specifically chosen for increasing attention during fatigued driving underload or for decreasing attention during driving among congested traffic overload).

5–2.2. Intensity and Power

Staum and Brotons (2000) describe five music listening environments differentiated by their intensity levels: Comfortable Listening (± 70dBA); Symphony Concert (± 76–100dBA); Walkman Headphone (± 90dBA); Bar and Dance Club (± 100dBA); and Amplified Rock Concert (> 112dBA). It is of particular interest that Ramsey and Simmons (1993) previously measured in-car driver-adjusted acoustic outputs in the 83–130dBA range. One must question, then, if aural stimuli presented at intensity levels as these impede driving performance (and therefore place the driver at increased risk), or facilitate driving performance (and hence reduce actual everyday hazards). For the most part, drivers believe that listening to loud music is the best strategy for maintaining a high state of alertness. One of the earliest studies to investigate if a high amplitude (better known as music *volume*) has a facilitative and/or detrimental effect on the motor, visual, or cognitive performance of driving, was undertaken by Turner, Fernandez, and Nelson (1996). In their study, the researchers investigated whether or not the intensity of music had an effect on the time it takes to react to 'random unexpected visual events'. These events are similar to ones we so often experience while driving on the road when suddenly, and without prior warning, the car in front of us comes to a full stop heralded by the onset of rear-end glaring red brake lights. Turner et al. recruited ninety 18–29-year-old undergraduates (50% male) from Wichita State University (USA) for a single 30-minute individual session. During the session, the participants sat at a table facing a computer monitor (at which they were to fixate their eyes on a cursor positioned in the middle of the screen), with a red light placed on the right of the monitor, and two foot-controlled levers labelled *start* and *stop* situated under the table on the floor (representing an 'accelerator' and 'brake' pedal). An experimenter sat behind the participants with a reaction-movement timer (connected to the red light, the start/stop pedals, and the computer), a cassette tape player, and a digital volume control unit (connected to the speaker output earphone-jack of the tape player). The participants heard music via two stereo speakers placed on the floor behind them. The music presented in the study were very short (8–120 second) excerpts of Pop music songs considered to be Top-40 Hits at the time of the study. The songs were: 'Still Rock-N-Roll To Me' (Billy Joel), 'Someone For Me' (Whitney Houston), and 'Lean On Me' (Club Nouveau). The participants were required to depress the accelerator pedal until a red light appeared, and then as quickly as possible depress the adjacent brake pedal. The latency between onset of red light and release of the accelerator pedal, as well as the movement time between the two pedals, were measured. In total there were 20 trials; five trials each for three conditions variegated by different levels of amplitude (music@60dBA, music@70dBA, music@80dBA), and a 'No-Music' condition (with room-noise@40dBA). The four conditions were implemented in random presentation orders. The procedure allowed for the measurement of task performances during a 5-minute period within each intensity condition. It should be pointed out, and in the very least raised as a critical fault of the procedure, that no details are presented as regards the actual stimuli employed. For example: Did every participant hear one song chopped up into fifteen 8–120 second slices? Or: Did the participants hear five 8–120 second slivers of all three pieces? Alternatively, we might ponder if the song portions were carved from a number of measures at the beginning of the instrumental introduction passage, or rather were these whittled down from vocal sections that appeared in either verse or chorus segments?

Turner et al. (1996) call attention to the popular notion that drivers (especially young drivers) who listen to loud music pay less attention to the road and therefore have decreased control over their vehicle, leading to increased risks, near-crashes, and crashes. One aspect of this viewpoint relates to a host of research efforts that have indeed demonstrated negative effects of noise and loud music above 95dBA. Namely, that high intensities have been seen to decrease cognitive-motor task performance as a result of boosted arousal (for a systematic review, see Dalton & Behm, 2007). As an example from a study outside of the traffic arena, Belojevic, Slepcevic, and Jakovljevic (2001) recruited 123 medical students from Belgrade University (Serbia) between 24–26 years of age (65% female), for a single session in which a set of mental arithmetic tasks were computed. The students were tested in small groups of five, under conditions in either a quiet room@42dBA environment or a loud room@88dBA environment. In the latter condition, recordings of traffic-noise were presented.[6] Belojevic et al.'s results indicated that all students in both conditions were just as precise (i.e., accuracy) in completing arithmetic tasks. But when examining the sample for individual differences based on personality traits, they found that those with higher levels of *extroversion* took far less time to complete the task (i.e., speed) under conditions of loud traffic noise than when the room was quiet, while students with higher levels of *introversion* demonstrated problems of concentration and fatigue under conditions of loud traffic noise compared to when the room was quiet.[7] According to Dobbs, Furnham, and McClelland (2011), both music

6 Survey data has found Traffic Noise to be 'highly annoying' (42%) while considered to be the 'most annoying' aural milieu (18%) compared to all other soundscapes, including work environments and social settings (Andringa & Lanser, n.d.). Further, road traffic noise has also been found to have a negative effect on health and well-being (for reviews, see: Andringa & Lanser, 2013; Ouis, 2001).

7 These two dimensions were measured by the Eysenck Personality Questionnaire (EPQ) developed by London psychologists H.J. Eysenck and S.B.G. Eysenck, published in 1975.

and noise have the potential to increase levels of arousal, and therefore introverts fall prey to conditions that cause them to be over-aroused beyond their adaptive level of functioning. Another study worth mentioning was conducted by Cassidy and MacDonald (2007) who found that high-arousal music produced distraction and reduced performance in comparison to either low-arousal music or noise. Cassidy and MacDonald also found that introverts were significantly poorer than extraverts in the presence of highly arousing loud music. Therefore, it would appear that more than anything else, while increased arousal can subsequently induce a circumstance of limited attentional capacity, other vital processes such as 'cue utilization' (i.e., selecting a range of cues that can be used to direct performance) is also hampered. Although Turner et al. point out that low(er) arousal allowed drivers to attend to both irrelevant and relevant cues, and increased arousal triggered a specific focus on the relevant cues allowing for drivers to ignore the more irrelevant ones, we must also note the possibility that when arousal becomes high and is maintained at higher levels for an extended period of time (such as when listening to music, especially music that is of the low-frequency high-volume *booming* genre), 'attentive selectivity' becomes so narrow that even relevant cues will be ignored. Hence, we might conceive that there is indeed an optimal level of arousal and music volume at which efficient driving occurs. In this respect, Ayres and Hughes (1986) long ago found differences between low-music@70dbA *versus* loud-music@107dbA. According to that study, visual acuity was significantly worse among college students during loud music; this finding suggests that the effects of loud music hampered *sensory processing* rather than cognition or *decision-making*. Yet, Ayres and Hughes also found visual acuity to be specifically impaired by loud-music@107dbA, but not by loud-noise@107dbA; they suggest that this finding demonstrates that more than anything else, the momentary 'peak' levels in music play a role in disrupting vestibulo-ocular control affecting uncontrollable eye movements – which have been subsequently demonstrated by Recarte and Nunes (2000) to reflect attentional states and apparent rate of thoughts. One report to confirm the above findings was Spinney (1997) who claimed that quiet music@55dBA is optimal for driving in comparison to either silence (i.e., in-cabin noise) or loud music@85dBA, and that driving with quiet background music is superior than driving without music.

To recap, the Turner et al. (1996) study was aimed at seeking an answer to the question whether or not music volume affects visual performance in a motor-reaction task (that is, the sudden onset of rear-end brake lights). Their findings indicated main effects of music amplitude for both RTs-ReactionTime (the measurement of lifting one's foot off an accelerator pedal), and for MTs-MovementTime (the total activation from onset of red light via lifting one's foot off the accelerator pedal to depressing the break pedal). Nevertheless, RTs and MTs were not found to increase as the volume of music intensified in a linear fashion, and therefore, decreasing the volume of the music does not predict shorter RTs or MTs. Hence, drivers do not automatically respond better to unexpected visual events when background music is heard at lower volumes. Rather, statistically significant shorter RTs and MTs surfaced for moderate amplitude music@70dBA

(M_{RT} = 0.33s, M_{MT} = 0.52s) compared to the other three stimuli: lower amplitude music@60dBA (M_{RT} = 0.38s, M_{MT} = 0.58s); higher amplitude music@80dBA (M_{RT} = 0.37s, M_{MT} = 0.57s); NoMusic (M_{RT} = 0.36s, M_{MT} = 0.55s); RTs ($F_{(3, 264)} = 26.73$, $p < .0001$); MTs ($F_{(3, 264)} = 19.26$, $p < .0001$). Finally, the study found that male drivers preferred louder music (M_{Men} = 71.8dBA, $SD = 4.87$; M_{Women} = 64.7dBA, $SD = 6.22$; $F_{(1, 88)} = 37.18$, $p < .0001$). The researchers offered several possible interpretations of the data: (1) Music played at a low amplitude level may cause the drivers to be in a position of straining to hear, and in such a circumstance they might require additional attentional resources; and (2) Music played at a high amplitude level might require more time to switch attention between aural and visual detection modes. In either case, they suppose that there are increased attentional efforts when background music includes words – which may be true, but as they themselves did not employ an instrumental piece in their empirical platform for comparative analyses, how can they make such a claim? Yet, they stated: "music with words, rather than without words, may be more detrimental to the perception of and motor response to an unexpected visual event" (p. 61). In what seems to be almost prophetic in nature (for 1996), Turner et al. foretold:

> Many automobile drivers maintain alertness by listening to music on the car radio or stereo. The results of our experiment suggest that a driver would react faster to unexpected visual events, such as the rear stop lights of a car in front, if the volume control was preset to his or her preferred level of music amplitude, rather than above or below that comfort level. The presetting of the volume control could involve a special device on the audio system of the car, to ensure that music amplitude would always be set at a constant ratio; the preset control would maintain the comfort level of the driver against other background noise present in the driving situation. Thus, the dynamic sound changes present in any driving situation, such as changes in speed, surrounding cars, or sound variations across songs or stations, could be controlled. The use of such a device might reduce the risk of accidents, by maintaining drivers' alertness while they listen to music (p. 61).

In a much later study Dalton, Behm, and Kibele (2007) investigated whether or not different music types, and music intensities affect performance during driving-related activities. They hypothesized that Hard Rock would hamper vehicular control more than Classical music, and that a low-intensity (i.e., soft volume) background would facilitate driving-related tasks whereas a high-intensity (i.e., loud volume) background would impair task performance. They rationalized their choice of music by claiming that Hard Rock music largely emphasizes heavy-bass low-frequency sounds, while Classical music features greater treble high-frequency sounds, but that 'industrial noise' (for comparison sake) is a composite of the whole aural spectrum. For the study, twelve people between 17–37 years old ($M = 25$, 50% female, with a driver's licence for at least four years) were recruited from The Memorial University of Newfoundland (Canada) for six 45-minute sessions. The sessions were variegated by three aural conditions (i.e., two distinct music playlists and industrial noise); each condition was also implemented at two levels of intensity (53dBA and 95dBA). The order of the six conditions was randomized for all participants, and every session took place within 48 hours of the previous session over a 2-week period (but at a similar time during the day to account for circadian rhythms). Dalton et al. were careful in presenting the selections from the playlists in a random order; they employed high quality music reproduction equipment (a Victor *VRX-2700* receiver amplifier coupled to Toshiba *HR-80* stereo headphones). The two playlists used in the study were:

'Hard Rock' Compilation

1 'Iron Man' (Black Sabbath)
2 'The Game' (Disturbed)
3 'Rise' (Hair of the Dog)
4 'Disintegrators' (Megadeth)
5 'Frantic' (Metallica)
6 'Holier Than Thou' (Metallica)
7 'Sad But True' (Metallica)
8 'The Shortest Straw' (Metallica)
9 'Dr. Feelgood' (Motley Crue)
10 'Kickstart My Heart' (Motley Crue)
11 'Blue Monday' (Orgy)
12 'Links 2 3 4' (Rammstein)
13 'Zwitter' (Rammstein)
14 'Dead Girl Superstar' (Rob Zombie)
15 'Dragula' (Rob Zombie)
16 'Scum Of The Earth' (Rob Zombie)
17 'The One' (Soil)
18 'Children Of The Grave' (White Zombie)

***Magic of the Pan Pipes*, Georghe Zamfir (Summit Records, 1996)**[8]

1 'Adagio' (Albinoni)
2 'Violin Romance' No.2 in F Major, Op.50 (Beethoven)
3 'Sicilienne', Trumpet Concerto in a (Telemann)
4 'Adagio', Concerto for Violin, Oboe, and Strings in D Minor, BWV 1060 (J.S. Bach)
5 'Ombra Mai Fù', Serse (Act 1) (Handel)
6 'Allegro', Trumpet Concerto in D (Loeillet)
7 'Andante', Trumpet Concerto in D (Loeillet)
8 'L'inverno' Largo, Concerto for Violin and Strings in F Minor, Op.8, No.4, R.297 (Vivaldi)
9 'Scènes D'enfants', Kinderszenen, Op.15 (Schumann)
10 'Ave Maria' (Schubert)
11 'O Mio Babbino Caro', Gianni Schicchi (Puccini)
12 'Addio Del Passato', La Traviata (Act 3) (Verdi)
13 'Hungarian Dance' No.1 in G Minor (Brahms)
14 'Donna Non Vidi Mai', Manon Lescaut (Act 1) (Puccini)
15 'Couleurs D'automne' (Zamfir)
16 'Le Chant D'evita' ('Don't Cry For Me Argentina') (Lloyd Webber)

Looking at the above playlist we must acknowledge (as do the researchers themselves) that music for Panpipe cannot, in the strictest sense, be considered Classical music (as is the case for music from composers like Johann Sebastian Bach or Ludwig van Beethoven). Yet, performances by Georghe Zamfir are in fact based mostly on orchestral arrangements of themes and melodies extracted from

8 *Magic Of The Pan Pipes* is an album of Gheorghe Zamfir (b. 1941). By the late 1960s, he had become a virtuoso panpipe player, and is without a doubt the world's best-known panpipe player. Zamfir's albums and CDs are strong sellers, perhaps because many consumers find the sound of the panpipes exotic with soothing sonorities. *Magic Of The Pan Pipes* was originally released in 1996 (Liberty Records), reissued in 1998–99 (Universal Music), and then again released in 2000–2001 (Phillips).

the classical music repertoire (from the early 1700s throughout the late 1890s). In addition, Zamfir's music is clearly associated with radio music platforms that broadcast classical-like background music referred with as *Light & Easy Classics*. During the six experimental sessions, the participants simulated automobile driving using a *PlayStation 2* (Sony) video-game console, a *GT Driving Force Pro Force Feedback Racing Wheel* (Logitech) with accelerator and brake pedals; the study employed the *Gran Turismo 4: The Real Driving Simulator* (Sony) computer software game (listed in Appendix C). Each participant drove the same vehicle on the same course. During the study, data from three driving-related variables were collected: (1) RTs that measured foot response to a light bulb, that when illuminated, required drivers to lift their right foot off the accelerator pedal; (2) MTs that measured the movement interval between the onset of the light bulb and the final depressing of the brake pedal; and (3) Driving Performance measurements such as completed lap-times, the number of crashes per lap, and the frequency by which the car swerved onto the road shoulder.

Foremost, the study implemented by Dalton et al. (2012) found main effects for intensity. RTs and MTs were significantly shorter during exposures of low-intensity: RTs ($M_{LowIntensity}$ = 0.282s, SD = 0.04; $M_{HighIntensity}$ = 0.324s, SD = 0.04); MTs ($M_{LowIntensity}$ = 512.1s, SD = 0.74; $M_{HighIntensity}$ = 554.5s, SD = 62.7). In addition, simulated lap-times were significantly shorter (i.e., quicker speeds) during exposures of low-intensity ($M_{LowIntensity}$ = 147.5s, SD = 11.4; $M_{HighIntensity}$ = 149.5s, SD = 12.4). Moreover, several significant two-way interactions surfaced. For example, RTs were shortest during exposures of low-volume Hard Rock music, and longest for high-volume Industrial Noise ($M_{Classical@53dBA}$ = 0.284s, $M_{Classical@95dBA}$ = 0.313s; $M_{HardRock@53dBA}$ = 0.278s, $M_{HardRock@95dBA}$ = 0.324s; $M_{IndustrialNoise@53dBA}$ = 0.284s, $M_{IndustrialNoise@95dBA}$ = 0.337s). This finding is in line with more common notions that subtle environmental sounds provide *proximal* situational awareness while rowdier sounds typically contribute to *distal* situational awareness (Andringa & Lanser, 2013). Then, lap-times were significantly shortest during exposures of low-volume Classical music, while longest for high-volume Classical music ($M_{Classical@53dBA}$ = 146.9s, $M_{Classical@95dBA}$ = 150.1s; $M_{HardRock@53dBA}$ = 147.9s, $M_{HardRock@95dBA}$ = 149.3s; $M_{IndustrialNoise@53dBA}$ = 147.6s, $M_{IndustrialNoise@95dBA}$ = 148.9s). Finally, a near-significant interaction (p = 0.56) was seen for the number of Crashes x Music per lap: more crashes seemed to have occurred when listening to Hard Rock music ($M_{Classical}$ = 1.36; $M_{HardRock}$ = 1.45; $M_{IndustrialNoise}$ = 1.28). Albeit, one can't help but notice from Dalton et al.'s data that an equally increased number of crashes also occurred during low-volume Classical music and high-volume Hard Rock music ($M_{Classical@53dBA}$ = 1.5, $M_{Classical@95dBA}$ = 1.25; $M_{HardRock@53dBA}$ = 1.4, $M_{HardRock@95dBA}$ = 1.5; $M_{IndustrialNoise@53dBA}$ = 1.1, $M_{IndustrialNoise@95dBA}$ = 1.46). Considering the above findings, Dalton et al. concluded that high-intensity backgrounds of any type impaired driver reaction times. With regards to increased lap-times (i.e., slower speed) of louder aural backgrounds, the researchers explain that high-intensity aural stimuli may require greater processing demands of the central nervous system, and hence deter attentional resources away from the primary driving task (i.e., distraction effects):

> Any form of loud sound whether it be irritating noise or preferred music will have a negative impact on simple vigilance task which could include activities such as the time it takes to apply the brakes or adjust the steering wheel while operating a moving vehicle (p. 163).

Yet, perhaps other explanations can account for the effects found above. For example, we might consider the possibility that drivers' emotional status subsequent to exposure of high-intensity sound caused the effects found by Dalton et al. (for example, see: Andringa & Lanser, 2013).

Taking a yet wider look at the findings presented by Dalton et al. (2012), one cannot but question the reliability of the data – as well as the interpretations they offer. Regarding the former reservation ('reliability of data'), one would not necessarily expect to see shorter lap times (i.e., quicker speeds) for quiet music compared to loud music, or for that matter an equal number of increased crashes (i.e., deficient driving) for both low-volume Classical and high-volume Hard Rock musics. Thus, while there is every possibility that the researchers attempted to be as diligent as possible when selecting the music stimuli they employed, they may have unfortunately chosen to contrast between two music genres wrongfully. Namely, within the context of 'exploring the effects of music intensity', it may simply be a useless exercise to contrast between exemplars of Hard-Rock Heavy-Metal distorted guitars with screaming vocals versus light and easy romantic classical themes played on the panpipe with lush string-orchestral accompaniment. These two genres are highly diverse in features, complexities, dynamics, use of lyrics, and affective parameters. Hence, employing two significantly conflicting music genres to explore *intensity* may simply be futile. Further, the fact that music tempo was either disregarded or totally overlooked, seriously breeches the cogency of music selection. Additionally, given the fact that 'level of intensity' (i.e., low@53dBA *versus* high@95dBA) was the critical parameter used to differentiate between experimental conditions, one would have expected a highly rigorous manner to measure intensity volumes. Yet, the protocol reported to have placed a decibel meter in-between ear-cups of a stereo headphone as a methodological procedure; it would appear that such a stratagem is simply not an acceptable method. As mentioned in Chapter 4, several studies have found that headphones hamper driving tasks that are dependent on auditory cues and can cause large interference effects with visual stimuli (Fagioli & Ferlazzo, 2006; Ferlazzo et al., 2008; Nelson & Nilsson, 1990). Hence, for the sake of empirical reliability as well as for ecologic validity, the study might have benefited more had the music been reproduced through free-field speakers. Concerning the second above mentioned reservation (i.e., 'reliability of interpretations'), both common sense and a review of the data presented in the report denote that Dalton et al. were careless in assembling their final conclusions. They state:

> Loud classical music was significantly more detrimental for RT compared to loud industrial noise. Due to the nature of classical music, the auditory stimulus is complex in design and may have greater arousal and higher processing demands compared to random noise (p. 163). The finding that loud classical music induced greater impairment than industrial noise is a unique finding (p. 164).

It is truly an unfortunate occurrence that such an inference is the final message from this study, especially given that the data indicate the exact opposite ($M_{Classical@95dBA}$ = 0.313s; $M_{IndustrialNoise@95dBA}$ = 0.337s). It is also saddening that the above false conclusion has found its way into several papers in the scientific literature.

5–2.2.1. Vigilance and mental effort

Much of the research going back to the 1970s about the effects of music intensity on cognitive and motor performance demonstrated that moderate-music@75dBA improves vigilance; for example, moderately-loud music increased correct detections and decreased detection latency. Among the tasks employed in these early studies were: arithmetic computations, college achievement tests, karate drills, reading comprehension, logical reasoning, visuo-spatial tasks, visual recall, and walking (for a review, see: Beh & Hurst, 1999). While the general effects were not always conclusive, at least as far as anecdotal evidence is concerned, people do believe that soft music facilitates cognitive performance (or at least does not cause undue harm), while loud music surely impairs performance. Further, as driving is seen as a complex task that includes a vigilance component, most people do expect that some aspects of driving will be expedited with background music when reproduced at low-to-moderate intensity volumes, while certainly the opposite is true about music reproduced at moderate-to-loud intensity volumes. Nonetheless, Beh and Hurst point out that, inarguably, many drivers operate a car while listening to high-intensity music (and it should be noted that their a keen observation was made in the late 1990s).

Beh and Hurst (1999) designed a study to investigate the effects of background music at different intensity volumes on driving-related performance tasks. In their study they recruited 45 undergraduates between 18–24 years old (M = 20, SD = 0.5, 64% female, with a driver's licence for minimum two years) who participated in a single session experiment inside a small soundproof room at the University of Sydney (Australia). Each participant was assigned to one of three groups variegated by aural background: NoMusic 'NM' (mechanical-hum@38dBA); low-intensity music 'LIM' (music@55dBA); or high-intensity music 'HIM' (music@85dBA). While the music presented in the study was described as a 'continuous series of 1990s Heavy Metal dance music', no titles or artists were documented. Hence, we might simply assume that there were tracks from groups such as AC/DC (Australia's foremost Hard Rock band) as well as other earlier imports that were still popular in Australia at the time of the study, such as: Black Sabbath, Iron Maiden, Judas Priest, Kiss, Metallica, Motley Crue, Slade, Slayer, Sweet, Twisted Sister, and Whitesnake. It should be pointed out that participants in both LIM and HIM conditions heard the same songs from the same high-quality cassette tape, reproduced by a Sony *XO-D50* music amplification system with two speakers situated on the right and left sides of the seated participant. During the session every participant in each group completed four tasks: (1) 'Stop-light RT Task', whereby participants were required to view a centrally located green circle on the computer monitor, and then respond as quickly as possible when the circle changed to red by pressing a foot pedal; (2) 'Vigilance Task', whereby participants were required to respond to a flashing red arrow that randomly appeared in one of five screen positions by depressing the foot pedal; (3) 'Low-Demand Single-Task Tracking', whereby participants were required to manipulate a joystick in an effort to keep a black crosshatch within a moving white circle as it moved to screen left or screen right; and (4) 'High-Demand Multi-Task Tracking', whereby participants were required to manipulate a joystick in an effort to keep a black crosshatch within a moving green circle as it moved to screen left or screen right, and simultaneously depress a foot pedal when a flashing red arrow randomly appeared in one of five screen quadrant positions.

In their study, Beh and Hurst (1999) found significant main effects for RTs. During both low-demand single-task and high-demand multi-task tracking, participants from both LIM and HIM music groups demonstrated quicker responses than participants in the NoMusic group; more than anything else, these findings demonstrated the beneficial effects of music on vigilance. Yet, on average the responses from participants from the LIM condition were significantly shorter than responses from participants from the HIM condition; these findings illustrate the beneficial effects of low-to-moderate intensity music on vigilance. When exclusively analysing RTs of the centrally located signal task ('Stop-light RT task'), similar results were obtained; participants from the NoMusic group took longer to respond than participants in either LIM or HIM music groups. Further still, the study found a two-way interaction between RTs *x* Intensity; that is, during low-demand single-task tracking, participants in the LIM group responded faster, but during high-demand multi-task tracking, participants in the HIM group responded faster. However, when exclusively analysing RTs from the peripherally located signal task ('Vigilance task') a completely different picture appeared. First, average RTs were shorter during low-demand single-task tracking than high-demand multi-task tracking. Second, RTs were not significantly different for participants in NoMusic, LIM, or HIM groups. Yet, while there were no meaningful differences between the two music groups during low-demand single-task tracking, statistically significant differences surfaced during high-demand multi-task tracking: during high-demand multi-task tracking activity, high-intensity (loud) music significantly slowed responses to peripheral signals. Beh and Hurst interpret the findings as supporting the notion that music does in fact have an effect on vigilance performance – but that music characteristics only partially explain how behaviour changes – given that *task characteristics* also appeared to account for a portion of the effect. Namely, that music of both low- and high-intensities do not necessarily facilitate or impair performances of a relatively naive undertaking when the task solely requires continuous visuo-motor coordination, but rather significant effects are seen when a more multifaceted operation requires both continuous motor involvement *and* intermittent motor responses to irregular signals. In this latter case – as is true for vehicular driving – there seems to be some advantage for listening to low-to-moderate intensity background music. It would seem that under low-demand conditions loud music does not necessarily interfere with performance, but then when conditions place a high-demand upon attentional resources, loud music facilitates performance on vigilance tasks for signals that are centrally located while impairing responses to peripherally located signals. Although Beh and Hurst do recognize that the ecological validity of their study is far from replicating real-life driving conditions, they conclude:

> The results of this study suggest that moderate-intensity music would be beneficial to driving performance in terms of shorter stopping times to critical signals in the driving environment. Although it could be argued that loud music would be of benefit to driving performance for signals located within central vision, the trade-off of an increase in response time to peripheral signals essentially nullifies any performance advantage under these conditions (p. 1,097).

Taking a rather broad perspective on the study, we might recognize that more than anything else, Beh and Hurst demonstrated that if normal driving means focussing all of our attentional resources on relevant driving information, then even the addition of secondary information from supposedly commonplace Heavy Metal songs can cause reduced visual processing. In an interview, Beh claimed: "If you play music at very high levels, you are going to miss things. There's no need to have your music thumping to get the benefits" (Henry & Lehrman, 1997).

The public has been made aware of many traffic-related studies in recent years, and experiments have 'trickled down' from the scientific literature through journalist reports in the electronic and printed media. For example, the BBC (2004) warned drivers against loud music: "listening to loud music while driving seriously hampers reaction times and causes accidents". Many of these communiqués briefly mention an unpublished study that was conducted at Memorial University in Newfoundland (Canada) by the research team Button, Behm, and Holms. That study examined ten participants who completed various tasks over several days while being subjected to noise levels at 53dBls and 95dBls. In both their original and later published report (Button, Behm, Holms, & Mackinnon, 2004), the researchers found that not only did it take up to 20% longer to perform physical and mental tasks during exposures of loud noise@95bBls, but that reaction times to tasks involving decision-making were also roughly 20% slower. Based on these findings, the British RAC Foundation motoring association heeded warnings that if motorists were delayed that long at the wheel they would no doubt suffer a fatal crash (Syal, 2004). If loud noise impacts reaction times, then the RAC feels that drivers who listen to loud music must be warned about adverse effects on their ability to react to unexpected events, such as sudden braking or a person running in front of the car (Shawcross, 2008). While the above statements may be true, it should be mentioned that Button et al. did not employ *music* in the aforementioned study, and that as pointed out above, noise and music do not necessarily affect drivers in a similar fashion. Moreover, in their warnings to drivers, the RAC proclaimed that some studies have shown that drivers are twice as likely to skip a red light while listening to loud music. Although drivers should be warned about the ill-effects of music, it is unfortunate that the research findings the RAC mentioned were carelessly misinterpreted from a study by Brodsky (2002), who actually never tested *intensity* but rather focussed on *tempo* (see study described below in section 5.3).

Nonetheless, several researchers claim that listening to a radio when played at a moderate volume does not in the least impair driving performance or driver vigilance (Beh & Hirst, 1999; Bellinger, Budde, Machida, Richardson, & Berg, 2009; Spence & Read, 2003; Strayer & Johnston, 2001; Turner et al., 1996; Wiesenthal et al., 2000). Such positions are based on the perception that listening to a radio does not require the driver to engage in increased mental effort or the focussing of attentional resources – as is required when drivers engage in an active discourse with passengers present in the vehicle or with callers over the phone. For example, results of an experiment by Consiglio et al. (2003) support the hypothesis that listening to music played on a radio would generate minimal interference, and thus would not affect braking performance. Their prediction was founded on the perception that being in the presence of an audible signal such as music does not necessitate the allocation of attention to the extent that engaging in conversation does. This study found that while RTs when listening to the radio (408 milliseconds) were just 4% longer than control conditions (392 milliseconds), such differences were not statistically significant. In all fairness, Consiglio et al. do concede that the attentional demand placed on a driver in real-world traffic would be quite different, and therefore their data might not be reliable. Also, they point out that the effects of radio on drivers would be highly contingent on the type of operation being performed (*tuning* versus *listening*), as well as what type of broadcast programming is heard (*talk radio* versus *music*). But yet, for the most part, the final take of their investigation is that listening to a radio while driving is quite positive in nature, and that relatively little performance decrement is experienced by drivers. In this connection, Unal and colleagues from the University of Groningen (The Netherlands) conducted a series of studies examining the influence of music on mental effort and driving performance. The researchers attempted to explain how drivers seem quite capable of controlling the vehicle despite the presence of a radio broadcast. Then again, when the study was completed, their press release announced something very different: "Listening to music while driving has very little effects on driving performance" (Groningen University, 2013; Science Daily, 2013). In a subsequent flood of Internet-related activity, the research team proclaimed: "Music doesn't hurt driving performance" (Mozes, 2013); "Listening to music isn't a distraction in the car" (Read, 2013); and "Listening to music has little effects on your driving – so crank it up!" (Styles, 2013).

Unal, Steg, and Epstude (2012) conceived of driving as a complex task, which if not carried out adequately could have serious safety consequences. Yet, they assert that while the presence of a secondary task often requires higher mental load and effort, drivers are usually capable of maintaining the primary task of car control. Specifically regarding the circumstances of driving with music, Unal et al. point out that listening to music is a habitual behaviour that accompanies driving, is most often perceived by drivers as helping them to pass the time, and is not perceived as a distracter that impairs driving performance. Hence, they explored music-generated effects on mental effort during driving, and raised the question if drivers are indeed able to cope with the increased mental load of music while maintaining an acceptable level of vehicle performance. The team recruited 69 psychology undergraduates between the ages of 18–31 (*M* = 21, SD = 1.96, 77% female, with an average driving experience of three years) from the University of Groningen (The Netherlands) to participate in a 45-minute single-session simulated driving task. Driving was carried out on intercity, rural, and residential roads, on a fixed-base simulator resembling a car cabin, with three screens offering a 180° view of the road. The study measured the effects of music on RTs to eleven unexpected traffic incidents, events, and hazards (such as a car approaching from left, a car emerging from right, merging with traffic, traffic pile-ups, traffic jams, and a parked car suddenly driving off from a parking space). In the study, data on driving parameters (such as deviations of speed, headway, and lateral position when following a lead car) were also collected. However, as the complexity of traffic flow has been seen to be a key factor that can increase the mental load of drivers (for example, high-density traffic is more challenging than low-density traffic due to congestion), the study implemented simulated driving in two qualitatively different scenarios (monotonous *versus* complex traffic flow). The sample was randomly assigned to one of the two conditions: driving with music *versus* driving without music. Drivers in the music condition created their own playlists by selecting songs via *Grooveshark.com.*[9] The researchers claim that such a strategy accounts for 'familiarity' in an effort to rule out any possibility that effects on mental efforts might be attributable to unfamiliarity. In a debriefing procedure (using a 5-point Likert response scale) the drivers reported that they enjoyed listening to the music (*M* = 4.53, *SD* = 0.56), that the music was similar to what they usually listen to while driving (*M* = 4.38, *SD* = 0.65), and that they did not find the music boring (*M* = 4.79, *SD* = 0.41). In the study the music was reproduced at fairly loud levels (*M* = 90dBA, range = 85–95dBA) to create a more demanding listening situation. Throughout the 45-minute session driver-perceived mental effort was repeatedly surveyed 13 times – every three minutes.

In their study, Unal et al. (2012) found a significant main effect of music for mental effort; but no two-way interaction effects surfaced for critical event types. Namely, drivers listening to music rated their perceived mental effort systematically higher regardless of any particular incident or traffic condition. However, as the performance data revealed that all drivers (in both music and no-music conditions) drove equally well in 9-out-of-11 scenarios (82%), Unal et al. concluded that music has no detrimental effects on driving performance. Having said that, looking at the two scenarios in which music did hamper driving performance (i.e., the 'car-following task' and the 'parked car suddenly driving off from a parking space' incident) would certainly offer further insights. First, RTs to accelerations and decelerations of a lead car were significantly shorter when listening to music

9 *Grooveshark* is an online music streaming service that provides a search engine, music streaming, and recommendations for music application. Users can stream and upload music to be played online, as well as add songs to playlists on personal digital players.

(M_{Music} = 3.44s, SD = 1.29; $M_{NoMusic}$ = 4.65s, SD 1.02; t = -2.82, df = 57, $p < 0.01$). Second, time-to-contact between the 'parked car driving off' and the 'expected collision' with the vehicle on the road was significantly longer when listening to music (M_{Music} = 1.12s, SD = 0.25; $M_{NoMusic}$ = 0.97s, SD = 0.30; t = 2.22, df = 67, $p < 0.05$). In recognition of these, Unal et al. did concede that listening to music at a high volume seems to add to a driver's mental load, and increases perceived mental efforts (during both monotonous and complex traffic environments). But yet, they reiterate their foregone conclusion that listening to music does not degrade vehicular control. Subsequently, their press releases, which can be found as citations in the scientific literature, declare: "Drivers who listen to background music perform just as well as drivers who do not listen to music". It seems quite unfortunate that other researchers do not find it warranted to look deeper in to such paramount conclusions, especially when they seem to be based on null-effects. For example, the primary independent variable – the *music* – raises many reservations and even doubts that might explain why no differences between drivers surfaced in 9-out-of-11 (82%) scenarios as reported by Unal et al. First, there is no mention of *what* music was selected by the drivers. We might assume that given the 35-minute drive, there were on average ten excerpts per person (as most songs are roughly of 3.5 minutes duration). Then again, theoretically participants could have chosen to hear only a few of their favourite pieces in a repeated loop presentation. In addition, there is no information as to the music genre of the participant-selected pieces (R&B, Pop, Rock, Jazz, Fusion, Hip-Hop, Rap, or Classical), or the music tempo (fast- or slow-paced songs), or the music valence ('positive' or 'negative' temperaments, 'happy' or 'sad' moods), or the music modality (major or minor keys), or even if the items were original songs or arrangements (well-known performing artists singing lyrics or instrumental covers). Further, there are no details of *how* the drivers heard the music: Was music reproduced via speakers or by headphones?[10] If speakers were employed, then we might ponder about the reproduction source itself: Was music reproduced by hi-fi audio equipment supplying the same quality sound to all participants with CDs prepared in advance? Or, was music heard through each participant's individually owned MP3 player or smartphone using custom-tailored playlists streamed from *Grooveshark.com*? Another possibility is that all music items were streamed from the Internet and reproduced on a laptop computer via on-board sound card through Microsoft's *Media-Player* and internal built-in speakers averaging about 2cm in diameter, although in this later case there is little doubt that the reported exposure volume (90dBA) could have been achieved. If speakers reproduced the music, then, we might also ask about exposure distance: Were speakers placed inside the mock-up car interior, or were they located elsewhere outside of the simulator cabin?

10 In light of the fact that Unal et al. reported that "a digital sound meter was used throughout the whole music condition to control for loudness" (p. 273) we might assume that the participants heard music from speakers. But then again, as described earlier, this was not the case for Dalton et al. (2012), who also reported to employ both headphones and a digital sound meter.

What about the audio quality? Was music reproduced from two hi-fi 2-way (5" and 12" diameter) full-frequency 20Hz–20kHz speakers, or from two laptop PC (2–3" diameter) satellite speakers? Finally, and most importantly, we must ask if additional ecologically relevant sounds such as engine revs and traffic noise were supplied to drivers; and if so, then was this same aural environment reproduced for participants in the NoMusic condition? Withstanding all of these, Unal et al. interpreted their findings as having demonstrated the facility by which drivers regulate mental effort towards maintaining the primary driving task despite the presence of distracting music in the car. Nonetheless, with the absence of details about the distractor, we can only wonder *how distracting* was the stimulus to begin with? Perhaps lack of differences were no more than artefacts of a ceiling effect.[11] The research team claimed that to some extent drivers regulated their behaviour, or their allocation of cognitive resources, as compensatory strategies to cope with the demands of driving. This proposition, then, led to a further exploration.

In a subsequent research package of two studies, Unal, Platteel, Steg, and Epstude (2013a) suggested that *blocking-out* auditory sources (by paying less attention to audio stimuli) might be a common strategy employed by drivers to handle increased task demands of listening to music. Especially if drivers perceive the subordinate task to pose a threat to their safety, there is every possibility that they will prioritize their attention to both primary tasks of driving (such as steering) as well as to secondary tasks of driving (such as using a GPS device), while 'ignoring' other driving-unrelated tasks (such as listening to the radio) altogether. Therefore, their main question of inquiry was: In high-demand driving conditions would drivers avoid paying attention to the radio in order to maintain the required level of driving performance, or would they still engage in active radio listening and attempt to cope with multiple task-demands? It is interesting to note that the same question was raised almost 50 years prior by Brown (1965) who recruited eight subjects aged 27–53 years (Md = 34, 88% male, with a driving licence between 1–33 years) to listen to radio broadcasts during six trips within two days; each day the participants drove the same 2.2-mile circuit in Cambridge (UK) three times. The circuit, which initiated in a residential area with light traffic, went through a more congested shopping area and then returned to a residential area. Driving was implemented under three aural conditions: (1) 'Quiet' – as a control for normal driving; (2) 'Speech Radio' – that consisted of one of six pre-recorded BBC broadcasts such as 'Round Britain Quiz'; and (3) 'Music Radio' – that consisted of a selection of pieces with characteristics of 'Ballroom Dancing'. The conditions were presented in different orders per day. It should be noted that three out of five participants (63%) had never owned or used a car radio prior to the study; hence Brown should have considered 'familiarity *versus* novelty' among the inclusion criterion. The overriding effort of the study was to

11 A *ceiling effect* in data-gathering occurs when variances in an independent variable are not measured because of constraints in data-gathering resulting from misconstrued stimuli or ineffective empirical methods.

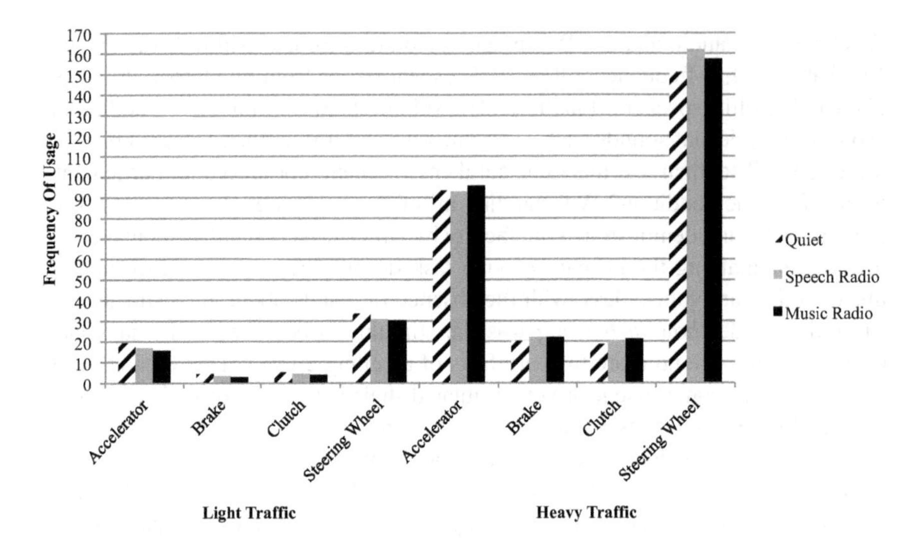

Figure 5.5 Frequencies of using car controls: Comparison of light versus heavy traffic by aural backgrounds

Source: Adapted from Brown (1965).

measure the usage of car controls: pedals (accelerator, brake, clutch) and steering wheel (movements). At the time of the study (in the 1960s) *control activity* was thought to be a highly valid and reliable measure of driver adeptness – ranging from efficient to deficient – and driver competence was viewed as adhering to a rather low usage of the vehicle controllers. The study found that the frequency by which drivers engaged car controls in heavy traffic was more than double than when they were in light traffic (see Figure 5.5). Moreover, while in low-demand traffic there was a tendency to reduce the overall number of control movements for trips accompanied by either Speech Radio or Music Radio; in high-demand traffic the data was completely opposite. Further, in a 15-month follow-up with six of the participants, the amount of attention given to Speech Radio when driving (the same circuit) was evaluated by comparing listening to speech radio while sitting in a quiet room. Brown found no effect of 'listening out of interest' versus active listening with the 'intent to remember' (original *versus* follow-up), nor were there meaningful differences of memory scores between listening conditions ($M_{\text{Driving+Listening}}$ = 54.3%; $M_{\text{ListeningAlone}}$ = 59.5%).

In their Study #1, Unal et al. (2013) recruited 15 psychology undergraduates (*M* = 21 years old, *SD* = 8.20, 73% female) from Groningen University (The Netherlands). As no simulated driving occurred in Study #1, the results also served as a 'non-driving baseline' measure for later comparisons with simulated driving in Study #2. For Study #1, seven 40-minute radio broadcasts were developed from recordings of actual Dutch-language radio stations; the mock transmissions were programmed as talk-radio interviews (16 minutes), with commercial fillers (14 minutes), and music items (15 minutes). It should be pointed out that all seven programmes were identical with the exception of the music items. Namely, each programme employed the same seven interviews covering a broad range of topics such as politics and food, as well as the same 41 commercials, but each provided 30 different music exemplars that had been specifically selected to represent one of seven target music genres. There were six sets of international hits from Rock, Electronic, Funk/Soul, Pop, Jazz/Blues, or Chill/Dance music styles, with an additional seventh playlist consisting of local Dutch-language songs. All music selections were chosen from Internet depositories featuring *Top-100* chart listings, and were then edited down to short (29–35 seconds) excerpts by choosing the most widely known fragment of each song. During Study #1, each participant watched a series of video-clips depicting a typical panoramic view of traffic on various roads accompanied by a mock radio programme with one of the seven soundtracks@75dBA; the participants chose the specific soundtrack based on their musical preference. The broadcasts were programmed as two alternating blocks, either beginning with music then leading to commercials and talk-radio or vice versa; the blocks were presented in a counterbalanced fashion across the sample. After the video-clip, each participant answered several questions regarding the broadcast. First, they recalled the content from the interviews, then they detected the advertised products from among a list of 99 brand names, and finally they identified the songs heard from among a list of 90 artist names (see Table 5.2). As can be seen in the table, the results indicated that 58% of talk-radio content was recollected, 40% of the name brands presented in the commercials were remembered, and 51% of the performing artists heard in the music were identified. The latter is quite low considering that the participants did claim that they were familiar enough to sing along with 84% of the tunes on the playlists.

Subsequently in Study #2, 46 psychology undergraduates ($M = 22$ years old, $SD = 2.44$, 54% female) from Groningen University (The Netherlands) were recruited for one 60-minute simulated driving session on urban and rural roadways in a fixed-base simulator resembling a car cabin with three screens offering a 180° view of the road. Driving was carried out on a single carriageway, with speed limits between 50–80 kph (30–50 mph). The simulated driving was implemented in two 20-minute segments variegated by complexity (i.e., low- or high-density traffic congestion) with traffic jams and critical incidents placed accordingly; segment

Table 5.2 Recall of radio programme broadcast variegated by driving tasks

	Study #1 (N=15) Radio-Alone	Study #2 (N=32) Radio + Driving	
		Low Demand	High Demand
Broadcast (Details Of Recall)	M% (SD)	M% (SD)	M% (SD)
Talk (Topic Of Discussion)	58 (16.8)	44 (21.3)	34 (14.0)
Commercial (Brand Name)	40 (17.5)	27 (16.3)	29 (13.9)
Music (Artist Name/Song Title)	51 (20.3)	32 (19.5)	36 (20.5)

Source: Adapted from Unal, Plateel, Steg, and Epstude (2013).

order was counterbalanced across the sample (i.e., low-to-high, high-to-low). Therefore, considering both allocation procedures (two traffic segment orders *x* two broadcast blocks), 32 (70%) participants were randomly assigned to drive with one of seven mock radio broadcasts playing in the background. As a result, there were roughly seven to eight participants in each cell sub-condition (i.e., music *versus* no-music driving, low-to-high *versus* high-to-low traffic density, music-before-talk *versus* talk-before-music radio broadcasts).

Unal et al. (2013) examined the extent to which drivers paid attention to the radio content while driving in traffic environments variegated by low- and high-complexity. Further, the research package evaluated differences of radio content recall for listening when not involved in a driving task compared to listening during simulated driving compared (i.e., Study #1 *versus* Study #2). Foremost, the research package revealed no influence whatsoever of radio listening on driving performance during critical incidents in either low- or high-complexity traffic (contradicting the findings of Brown, 1965). Unal et al. concluded that driving performance and vehicular control while listening to the radio is no different than when driving without listening to the radio – although it must be pointed out that no data to support this conclusion was ever presented in their published report. Second, a between studies comparison found that recall of content was significantly lower when driving (confirming Brown's findings, see Table 5.2). Accordingly, the researchers stated that drivers "prioritize the primary task of driving and pay less attention to the secondary task of listening to the radio" (p. 938). Finally, a within-subjects repeated-measures analyses between levels of traffic density demonstrated that while there were no meaningful differences between the presentation orders of the segments (low-to-high *versus* high-to-low congestion), main effects for traffic complexity surfaced for Talk Radio ($F_{(1, 30)} = 6.77, p < 0.05$, $\eta_P^2 = 0.18$). As can be seen in Table 5.2, the finding indicates that recall for topic of discussion was significantly lower (fell beneath levels of *chance* or random) when driving; albeit, it should be pointed out that also no meaningful differences surfaced for recall of product brand names, artist names, or song titles. Unal et al. claim their investigation is one that demonstrates *how* drivers maintain their driving performance while listening to the radio. Namely, that drivers effectively employ a strategy by which radio content is 'blocked-out' at will to reduce task-demands. Yet, if we take a step back to look at the research package from a wider vantage, it is possible that their results do not necessarily point to a 'mechanism' by which drivers cope with the interference induced by radio listening (as related to *attentional processes* such as blocking-out auditory distractors), but rather the study may have simply touched on aspects of 'mental effort' or 'cognitive load' (as related to *memory processes* involving faster decay in retention capabilities resulting from an increased information alongside decreased levels of storage capacity). Indeed, perhaps paying less attention to radio or music CDs in more complex traffic situations is a pertinent strategy for some drivers – and maybe some drivers capabilities to regulate attention are somewhat dependent on age, gender, driving experience, personality temperament, the number of passengers in the car, the music genre of exemplars, and even the pace (or tempo) of the music heard in the background. But, whatever the case may be, Unall et al.'s closing statement – that has been twirled and twisted to the press as a tag-line declaring that "*listening to music while driving has very little effects on driving performance*" (Groningen University, 2013; ScienceDaily, 2013; Unal et al., 2013c) – may be no more than a highly colourful but rather ill-fated dispatch.

5–3. Music-Tempo Generated Distraction

On face value, it might appear that the effects of music on drivers are solely attributable to affect or arousal (i.e., emotional potency of valence), and to volume intensity (i.e., power of aural and vibrotactile sensations). We do infer that all drivers turn down the radio volume in heavy traffic, as loud arousing music seems to require greater processing demands. Nevertheless, it is cogent to question the variance of music intensity on driving, especially since some music genres (such as Hip-Hop, Rap, Heavy Metal, and Hard Rock) are composed of highly dense musical characteristics that severely penetrate our soul and cause strong reactions including motor action, emotional involvement, and activity such as singing or tapping – and these occur at almost any level of reproduction volume. Therefore, how loud pieces of some music styles are played might not actually have any bearing whatsoever on drivers' responses. Along these lines, North and Hargreaves (1999) explored the effects of music complexity on driving performance. *Music complexity* accounts for a host of music features (like valence, intensity, pace), and reflects the fluidity of auditory combinations among music features that influence the responses of listeners. In their study, 96 psychology undergraduates between 18–25 years of age (M = 19, SD = 1.57, 50% male) were recruited for a single session to drive a race car with the *Indianapolis 500* motor-racing computer game[12] in a soundproof room at Leicester University (UK). During the session, each participant first underwent a practice trial (seven minutes), and then completed five laps around the racetrack. The participants were randomly assigned to one of four conditions: two types of music (low- *versus* high-arousal), and two levels of driving difficulty (low-demand *versus* high-demand). Two specially prepared 5-minute instrumental pop music pieces were employed; both pieces were in the key of C minor, with a 4/4 time percussion track, and futuristic sound timbre. By manipulating two acoustic features (i.e., intensity and speed), the music was capable of cueing two diverse psychophysiological properties: a low-arousal (LA) version was reproduced as a moderately-slow/soft piece (i.e., 80bpm@60dBA), while a high-arousal (HA) version was reproduced as a fast/moderately-loud piece (i.e., 140bpm@80dBA). Moreover, by requiring the participants to count backwards while engaged in driving, driving difficulty could be increased to high-demand (HD) driving, while low-demand

12 Indianapolis 500: The Simulation (Papyrus, 1989) is generally acknowledged as the original game to have kicked-off racing simulation on 16-bit PC hardware. Indianapolis 500 includes one track in which gamers could race the full 500 miles of the world famous 'Indy 500'.

(LD) driving simply occurred without any other simultaneous tasks. Hence, both LA and HA musics served as an accompaniment for both LD and HD driving. There were 24 participant-drivers in each of the four conditions. All participants wore earphones (Panasonic) to hear instructions and music, coupled to a *Linebacker L30* 30-watt amplifier (Laney). The participants controlled the race car via a specified fingering on the four keyboard arrow (❖) curser keys.

North and Hargreaves (1999) measured lap-times, subjective ratings of task difficulty, enjoyment of music, and perceived level of arousal. The study found a main effect for music type ($F_{(1, 92)} = 125.53$, $p < 0.001$) indicating that participants rated HA music as being more arousing than LA music. Further, effects of music type surfaced for Lap-Time ($M_{LA} = 83s$, $SD = 24.8$; $M_{HA} = 118s$, $SD = 3.54$; $F_{(1, 92)} = 11.32, p < 0.001$); this finding indicated that driving speeds were significantly faster in laps with the LA music background. Furthermore, effects of music type surfaced for Driving Condition ($M_{LD} = 90s$, $SD = 35.4$; $M_{HD} = 110s$, $SD = 14.1$; $F_{(1, 92)} = 38.45, p < 0.001$); this finding indicated that driving speeds were significantly faster in laps when participants drove without backwards counting interference. In addition, an interaction between Music *x* Driving Condition surfaced ($F_{(1, 92)} = 7.17$, $p < 0.001$); that is, the fastest driving speed (i.e., lowest lap-time) occurred for LA/LD driving conditions. Finally, it should be pointed out that while participants perceived LA/LD as significantly less difficult than the other three driving conditions, HA/HD was perceived as the least enjoyable condition. The research team concluded that as expected, more cognitive space was required to process the HA music, and while HA music was seen to lead to decreased performance the same was found for HD driving conditions (when backwards counting was added to driving). Namely, the quickest lap-times were measured for driving conditions that required the least cognitive processing (i.e., LA/LD), while the longest lap-times were measured for the conditions with the most cognitive processing (i.e., HA/HD). North and Hargreaves make a claim for the study as having demonstrated a broad concept relating to music responsiveness: "listening to music affects responses to a concurrent task, and carrying out a concurrent task affects responses to music" (p. 291). Such insights are obviously important for those investigating the effects of music in everyday contexts – and are especially critical for investigating music behaviours associated with driving an automobile. Nonetheless, while sound level (volume) and music pace (tempo) clearly mediated driving performance, and the results provide valuable information about how the alliance of these two parameters affected simulated speed (at least as far as a racing game is concerned), our ability to formulate some kind of analogy between the research findings and the expected effects of in-car music on driving performance would appear to be highly unlikely – even though the study is often cited by traffic scientists. There are two reasons for such criticism. Foremost, the controllers used to drive the vehicle were specified as finger-buttons (i.e., arrow curser keys ❖ on the keyboard). Certainly the research team did not even incorporate PC computer game peripherals approximating automobile controls such as a steering wheel, accelerator pedal, or braking pedal. Hence, while the current study may be an excellent demonstration of music-generated effects on a multi-sensorial computerized visual tracking task, the demonstration seems far-off from reaching what we would expect for ecologically valid simulations of driving. Second, it is most unfortunate that the study made no attempt to isolate each of the manipulated elements (i.e., intensity and tempo) as individual component factors prior to the exploration of accumulative effects; such data would have boosted the insight gained about the potency of distinct features in a more decompositional manner. Ironically, many who cite the study, do so as they believe it provides insight about the effects of 'music tempo' on driving, when in reality this is not what was implemented.

Several researchers (for example, Ho & Spence, 2008) refer to a study by Konz and McDougal (1968) as an early investigation highlighting the effects of musical pace on driving. The study is viewed as one that targets the effects of music tempo on the speed at which a car is driven. In the study, 24 undergraduates from the Department of Industrial Engineering at Kansas State University (USA) between 18–23 years of age (all male, with a driver's licence between two to nine years) were recruited for a single 60-minute driving session. The participants drove in a 1967 GM Pontiac *Tempest* 4-door sedan with a V-8 engine and automatic transmission. Each undergraduate drove a total of 46 miles (74 kilometres) on a 4-lane highway between the city-limits of Manhattan and Ogden (both in Kansas); they drove all four segments continuously in one session (to-and-fro twice, which is 4 x 11.5 miles per segment). The study employed a 1965 *Greenshields' Driveometer* (an early prototype of an in-vehicle data recorder) that calculated steering wheel position, brake applications, accelerator pedal applications, and speed (changes above 4 mph), individual lap-times, and total mileage. The music heard in the car was reproduced via a 4-track cartridge tape player (installed under the dashboard) using a single monaural car speaker. During the study the experimenter sat in the front passenger seat. First, the participants drove a trial practice run (one segment of the circuit), and then the other three segments were completed; each 15-minute segment was viewed as a unique driving condition variegated by aural background ('NoMusic', 'SlowMusic', and 'FastMusic') counterbalanced across the sample. The slow music was perceived as 'sweet' classical orchestra instrumentals, while the fast music was a 'lively' pop-oriented brass band. The playlists were:

	Slow Music Playlist: ***Our Winter Love***[13] **The Felix Slatkin Orchestra** **(Liberty, 1963)**			**Fast Music Playlist:** ***What Now My Love***[14] **Herb Alpert & The Tijuana Brass** **(A&M, 1966)**	
	Track Title	**bpm**		**Title Track**	**bpm**
1	'Our Winter Love'	72	1	'What Now My Love'	105
2	'I Left My Heart In San Francisco'	98	2	'Memories Of Madrid'	90
3	'Love Letters'	70	3	'Cantina Blue'	53
4	'Lollipops And Roses'	67	4	'Plucky'	84
5	'Fly Me To The Moon'	130	5	'Brasilia'	100
6	'Days Of Wine And Roses'	50	6	'If I Were A Rich Man'	99

Konz and McDougal (1968) found that both slow- and fast-paced music statistically significantly increased driving speed compared to driving without music ($M_{SlowMusic}$ = 45.9mph; $M_{FastMusic}$ = 46.4mph; $M_{NoMusic}$ = 45.6mph; $p < 0.05$); however, looking at the data from a 'clinical' perspective might suggest otherwise. Nonetheless, weaving was significantly greater among trips without music, and least seen for trips with slow-paced music. Moreover, the study examined drivers' usage of vehicle controls. By summating the recorded handling of fine- and gross-steering manoeuvres, as well as both accelerator and braking pedal use, the study found that slow-paced music improved driver competence, fast-paced music was detrimental to driving performance, and driving without music was somewhere in between ($M_{SlowMusic}$ = 663; $M_{FastMusic}$ = 680; $M_{NoMusic}$ = 675). Konz and McDougal concluded that although both music types resulted in faster speeds (compared to driving without music), the fast-paced music seemed to cause an increase in driver activity beyond the other two conditions. When taking a broad look at this study we cannot but question the independent experimenter-manipulated variable – the music. Although the researchers themselves did not view their study as an investigation of music tempo per se (but rather as an investigation of 'background music' on controller activity), they themselves clearly identified one driving condition as 'music-accompanied driving with slow music' – albeit the other condition was labelled as 'driving accompanied by Tijuana Brass music'. Yet, there is perhaps no other possibility but to view the study as related to *tempo*. Hence, the description in the text above refers to 'fast-paced music' (rather than 'Tijuana Brass Music'). Nonetheless, if we listen to the items employed in the study, we might notice that all pieces selected are rather similar melodic covers of well-known tunes, and that both music types found on the playlists often overlap with regards to music genre (i.e., Latin styles with multi-layered syncopated percussion accompaniment), solo instruments (i.e., trumpet and brass arrangements), and overall pace (i.e., slow-beat romantic-like ballads or a fast-beat energetic pepped-up dances). Moreover, given that the drivers heard all pieces as an intact playlist during the entire 15-minute 11.5-mile driving segment, we might note that analyses exploring variances between the average tempo of each playlist found no significant differences at all ($M_{SlowMusic}$ = 81bpm, SD = 28.5; $M_{FastMusic}$ = 89bpm,

13 *Our Winter Love* is an album by Felix Slatkin (1915–1963), arranger, conductor, and violinist active in Hollywood during the 40s, 50s and early 60s. Slatkin was the concertmaster of the 20th Century Fox Studio Orchestra, and then later conducted the Hollywood Bowl Symphony Orchestra. He was concertmaster and conductor for Frank Sinatra during his Capitol years of the 1950s. Our Winter Love was posthumously completed for Liberty Records after his sudden death in 1963.

14 *What Now My Love* is an album by Herbert 'Herb' Alpert (b. 1935), arranger, featured trumpet soloist, and the leader of the successful pop ensemble Herb Alpert & The Tijuana Brass. Alpert sold 72 million albums worldwide by 1976; there were five No. 1 albums, 14 Platinum albums, 15 Gold albums, and nine Grammy Awards. Alpert is a recording industry executive – the 'A' of A&M Records with partner Jerry Moss. What Now My Love was released in 1966 for A&M Records.

SD = 18.9; $t = .5853$, $df = 5$, $p = 0.58$). Therefore, in the best of circumstances, we might consider Konz and McDougal's study as an excellent demonstration of the more general effects of *driving with music* compared to *driving without music*. Having said that, the investigation offers little to our understanding of how *music tempo* affects driving performance.

As seen in Chapter 3, within the context of everyday environments background music variegated by tempo has been seen to modify human motor behaviour. That chapter outlined a host of research findings among supermarket shoppers (who moved around the store more quickly with fast-paced music), restaurant patrons (who ate their meals more quickly in the presence of fast-tempo music), and pub drinkers (who consumed drinks more quickly with a gradually increasing tempo of music). Traditionally, studies exploring the effects of music tempo have investigated both the speed and accuracy of human task performance. For example, fast music has been seen to increase the rate and precision of mathematical computations in stock-market environments (Mayfield & Moss, 1989), and accelerate self-paced line drawings (Nittono, Tsuda, Akai & Nakajima, 2000). Hence, what are we to expect of driving with fast-paced music?

Perhaps the most outstanding effect of music on driving performance is that which is related to *time perception*. When considering how perception of 'time' overlaps with the perception of 'velocity', and accounting for the fact that music itself is a temporal stimulus whereby sensory input of a temporal nature does interfere with other temporal perceptual impressions such as internal timing mechanisms, driving behaviours and vehicular control might be particularly at great risk. It should be pointed out that if music can influence how 'time' is perceived and estimated, then especially in a context where the visual field is a constantly changing stream (such as in the case when driving an automobile) the effects of music may be far more complex than we can fathom. For example, perceptual links between 'time', 'velocity', and 'distance' are essentially based on our early developmental experiences. Already in childhood we learned that faster moving objects travel a greater distance during a fixed time interval than slower moving objects. Namely, as children we often equated 'quicker' with 'longer duration'.

In general, humans are very poor at judging distance and speed. But, as Vanderbilt (2008) points out, this is unfortunately what driving is all about. Accordingly, we are often fooled into thinking that things are not as far way as they appear, and drivers perceive small cars to be farther away than they really are (because of mental images of bigger cars moving slower than small ones). Therefore, when both faster and slower moving objects start and stop together, faster moving objects are perceived as travelling for a longer period of time (Zakay, 1989). Actually, Kortum et al. (2008) bring forth a host of studies that collectively show ample evidence to indicate that the perception of time can be altered based on the environment to which the driver is exposed. Accordingly, how time perception and estimations are modified has to do with the size of the space in memory required for storing information about an interval. The more events that there are in a particular time interval, or the greater their complexity, the longer the interval seems. Therefore, the perception of time passage is directly related to the amount of information encoded during the period in question.

Specifically in regard to music, in the listeners mind *if more has happened then more time must have passed.* In this connection, North and Hargreaves (2009) claimed that 'louder' will lead to the (mis)perception that more has happened, and thus appear to the listener as if more time has passed. Similarly, pieces that are of a faster tempo might be perceived as having generated more information over a longer period of time. When music listening fosters a greater need from the driver to process more events (because the music might be overly loud, especially fast, or intensely complex), then, the music itself can distract from internal cognitive timers that match time passage. Hence, as our estimation of time can be hampered, the same might occur to our estimation of speed. For example, Cassidy and MacDonald (2010) implemented a study about the effects of music on time perception during a driving game. In the study 70 participants (M = 20.5 years old, SD = 2.1, 54% male) completed three racetrack laps with the *Xbox 360* at the *eMotion* lab of Glasgow Caledonian University (Scotland, UK). The study found that when driving with high-arousal fast-paced music@130bpm, lap completion times were the shortest (i.e., driven at the fastest speeds), and driver estimates for both lap-times, and lap-speeds were the most inaccurate. Vanderbilt (2008) alludes to the fact that we underestimate our speed when slowing down and overestimate our speed when speeding up – simply because of perceptual changes in optic flow. In fact, Underwood, Chapman, and Crundall (2009) stress that as levels of speed increase, with an environment consisting of more demanding traffic conditions, we are forced to adapt by increasing the sampling rate of visual information. Thus, as we drive and orient ourselves to a certain fixed point on the horizon or target (such as trees or light poles going by), ultimately it is the textural density (i.e., the swiftness of the optic flow) that prompts our perception of rate. Certainly, the finer the texture, the greater is our perception of our driving speed. Yet, in addition to optic flow, we are also aware that environmental sounds (such as the wind or road traction) offer feedback about acceleration. Hence, the louder the sound, the greater is our perception of our driving speed. Therefore, by the same logic, it should be understood that musical stimuli moving at higher levels of activity (i.e., at a faster velocity, pace, or tempo) are perceived as being longer in duration than musical stimuli moving at lower levels of activity. The point here is that our perception of time is co-determined by a host of variables – such as the stimuli that fill the time span, the context of that time span, and the person perceiving the time span. Consequently, time spans that are filled with content such as music are perceived very differently (for example 'longer' durations) than time spans filled with nothing; and time spans filled with fast-paced music are perceived as presenting 'more' music and hence are longer in duration than time spans filled with slow-paced music. To test such presumptions, North, Hargreaves, and Heath (1998) investigated the effects of slow (<80bpm) *versus* fast (>120bpm) Pop tunes on the perceived time spent in a fitness training centre. This study demonstrated that fast-paced music caused the participants to underestimate the time they spent in a gym – but unlike the participants in Cassidy and MacDonald's driving game, fast-paced Pop music caused them to offer much less inaccurate time estimates. North et al. claim their results confirm Zakay's (1989) conceptual model accounting for the effects of incongruent temporal information as distorting attention from internal cognitive timers. Moreover, they state that the degree of time inaccuracy is also directly attributable to the incongruence that may occur between the music heard and contextual nature of the listening situation. Therefore, there may in fact be quite a difference between perceptual obstruction as seen in a computer game compared to a fitness gym – and both of these may be highly different than when we drive a car. Namely, there is every possibility that music of a fast tempo will create specific temporal cognitive interference to driving performance. Yet thus far, only one study (Brodsky, 2002) is known to have investigated the role of music tempo on vehicular control (BBC, 2002; Campbell, 2002; Hamer, 2002).

5–3.1. The Effects of Music Tempo on Driver Performance and Vehicular Control

Brodsky (2002) implemented a research package of two studies that targeted the effects of music tempo on driving performance. The studies evaluated several dependent variables including Psychophysical Measures (cardiovascular activity), Driving Performance (speed), Vehicular Control (frequency of traffic violations and incidents), and Information Processing (perceived speed estimates). The first study was implemented among a sample of musician drivers, while the second study recruited drivers without formal music education or training.

5–3.1.1. Study #1: Drivers with formal music education and training

Twenty music education undergraduates (M = 33 years old, SD = 7.23, 70% female, with a driver's licence for at least five years) participated in the study; all had previous formal music training including private tuition on a musical instrument for over ten years. The musicians participated in a single 120-minute session of which 90 minutes was simulated driving. Accordingly, 65% of the participants reported they drive with music all of the time, and 87% claimed they usually listen to moderate-tempo music at medium-intensity sound levels. Simulated driving occurred in an everyday cruise mode, during daytime hours in bright sunshine weather, with moderate pedestrian activity, but without additional traffic. The trips were carried out in a virtual automatic transmission *New Beetle* (Volkswagen), with dashboard gauges, rear view mirror, life-like engine-motor revs, and vibrotactile feedback from tyre traction. The trips were along a 6-mile (ten kilometres) ring-road of Chicago (USA), without turns or exit ramps; the route consisted of a 3-lane boulevard expressway (3.5 miles) and an interstate highway (2.5 miles). Each participant completed eight laps of the ring-road for a total of 48 miles (77 kilometres). In each lap, driving was implemented with one of four aural conditions: (1) 'NoMus' – driving without music but hearing engine-noise; (2) 'MUS1' – driving with slow-tempo music ≤ 70bpm; (3) 'MUS2' – driving with medium-tempo music between 85–110bpm; and (4) 'MUS3' – driving with fast-tempo music ≥ 120bpm. While the engine noise itself was reproduced at roughly 35dBLs, by adding music the compound background was reproduced at an average 85dBLs. Each condition was repeated twice in a counterbalanced sequence order.

The music stimuli did not include vocal performances involving lyrics, or instrumental cover versions of well-known popular tunes. This is an important point as extra-musical associations and surface cues (such as those related to popularity, language, ethnic origin, gender, sensuality, and sexual preference) often contaminate music stimuli employed in empirical investigations. The music stimuli employed were a blend of several styles including Pop, Rock, Jazz, Blues, Funk, and Country. The exemplars presented a typical orchestration consisting of a rhythm section (such as electronic keyboard, electric guitar, bass guitar, and drums), with a leading solo instrument for the melody line (such as an electric guitar, electronic keyboard, woodwind or brass instrument). In addition, some pieces were augmented with a string section for mood painting. The music used in the study consisted of 12 items (four tracks for each of the three music tempos):

Condition/Tempo Rating	**Track Title (Artist Name)**	**BPM**
MUS1 – Slow-Tempo	1 'Stranger On The Shore' (Kenny G)	56
	2 'Being With You' (George Benson)	65
	3 'Like A Lover' (Earl Klugh)	63
	4 'Heart To Heart' (Larry Carlton)	66
MUS2 – Medium-Tempo	1 'That's Right' (George Benson)	94
	2 'Dinorah Dinorah' (George Benson)	112
	3 'Cashaca' (Spyro Gyra)	112
	4 'Cafe Amore' (Spyro Gyra)	100
MUS3 – Fast-Tempo	1 'House Trip' (DJ Jurgen/DJ Paul One VS Dave Scott)	132
	2 'Angels' (DJ Jurgen/Sequential One)	132
	3 'Kiss That Sound' (DJ Jurgen/Pulsedriver IV)	132
	4 'The Was It Isn't' (DJ Jurgen/Horney Bruce)	132

Simulated driving was implemented on a *Dynasty* (Mitzuba) 3D Multi-Media Notebook with 12.1" monitor, on-board Yamaha 16-bit digital stereo audio system, and two 7w full-range (20–20KHz) PC speakers (Mli). Driving was executed with the *Midtown Madness* (Microsoft) Chicago Edition software package,[15] and controlled by a *Side-Winder Force Feedback Steering Wheel* (Microsoft) coupled to accelerator and brake pedals.[16] Music was reproduced with an *UX-T150* (JVC)

15 *Midtown Madness* (Microsoft, 1999) is a racing game developed by Angel Studios. Unlike all other racing games published prior to 2000, *Midtown Madness* did not limit drivers to a race track but rather offered an open-map simulation of Chicago (ring-road and city centre).

16 This device is very similar to what has been used in other studies, for example the *MOMO Racing Force Feedback Wheel* (Logitech) employed by Ho and Spence (2007) and the *GT Driving Force Pro Force Feedback Racing Wheel* (Logitech) employed by Dalton, Behm, and Kibele (2007).

24w micro-component stereo with two detachable wooden-cabinet stereo speakers placed on the right/left floor at 45° upwards angles facing the driver (as if mounted on front-door panels). Finally, physiologic arousal was measured continuously, and stored every 15 seconds, with a wireless *Accurex-Plus*™ Heart-Rate Monitor (Polar Electro Oy) employing a chest-belt transmitter and wristwatch receiver.[17] Total elapsed time as well as simulated lap-time data was logged with stopwatch timer functions using split-recording features – all of which were synchronized with HR data recordings.

The results of Study #1 relate to three categorical areas: Physiological Data, Vehicular Acceleration, and Vehicular Control:

1. *Physiological Data.* Heart Rate (HR) and Heart Rate Fluctuation (HRF) were measured throughout; HRFs are calculated from the mean standard deviation of all HRs in each individual condition. The data indicated no main effects for either HR or for HRF. Namely, there were no significant physiological differences between the driving conditions, neither with music or without music, nor between the different music tempi themselves. However, HRFs tended to decrease with background music ($M_{NoMusic} = 3.16$, $SD = 0.86$; $M_{Music} = 2.92$, $SD = 0.99$).
2. *Vehicular Acceleration.* Average Lap Speed (MPH) was solely recorded while driving on the interstate highway segment. No main effects surfaced. That is, there were no meaningful differences of speed between the laps with music or without music, nor between the different music tempi themselves. However, MPH tended to increase with background music ($M_{NoMusic} = 65$mph, $SD = 12.2$; $M_{Music} = 68$mph, $SD = 14.5$).
3. *Vehicular Control.* Three variables were logged: (1) the frequency of accidents (ACs) including incidents, near-crashes, and crashes; (2) the frequency of lane wobble (LNs) including deviations from the lane and inappropriate excursions to co-joining lanes; and (3) the frequency of violated red-light signals (RLs) and traffic stop signs. ACs and RLs were solely recorded while driving on the 3-lane boulevard segment. No main effects were found for ACs; however a trend was seen whereby the frequency of collisions increased with music background ($M_{NoMusic} = .30$, $SD = .92$; $M_{Music} = .48$, $SD = .68$). Moreover, an upwards linear trend surfaced among the music conditions indicating that as the tempo accelerated so too did the number of ACs ($M_{MUS1} = .25$, $SD = .72$; $M_{MUS2} = .30$, $SD = .73$; $M_{MUS3} = .35$, $SD = .88$). Further, no main effects were found for LNs; however, a trend was seen whereby the frequency of lane excursions decreased with music ($M_{NoMusic} = .65$, $SD = 1.03$; $M_{Music} = .43$, $SD = .68$). Among the music

17 Changes in physiological parameters (such as HR) have been shown to predict driving performance impairment, and correlate to lateral position and speed changes (Brookhuis & de Waard, 1993; de Waard & Brookhuis, 1991). The precision and accuracy of the Finnish Polar portable wireless heart rate monitor has been found to correlate ($R = .98$) with various clinical apparatus including stationary ECGs and mobile Holter ECGs (Laukkanen & Virtanen, 1998).

conditions, the safest most controlled driving was seen for laps with medium-paced music ($M_{MUS1} = .65$, $SD = 1.03$; $M_{MUS2} = .50$, $SD = .76$; $M_{MUS3} = .70$, $SD = 1.08$). Nonetheless, a significant main effect of tempo was found for RLs ($F_{(3, 57)} = 3.04$, $MSe = .4651$, $p < 0.05$). Planned comparisons demonstrated significant differences between laps without music *versus* laps with music ($M_{NoMusic} = .25$, $SD = .55$; $M_{Music} = .67$, $SD = .65$; $F_{(1, 19)} = 7.26$, $MSe = .2330$, $p < 0.05$); this finding indicates that driving with music increased the number of violated red-light signals and stop signs by over 40%. While there were no linear effects of music tempo on RLs, it was observed that the least controlled driving performances were for laps with fast-tempo music ($M_{MUS1} = .55$, $SD = .89$; $M_{MUS2} = .55$, $SD = .69$; $M_{MUS3} = .90$, $SD = 1.02$). All in all, the above findings demonstrated that 20% of drivers defied red-lights on laps without music, 35% skipped red-lights with slow-tempo music, 45% infringed red-lights with medium-tempo music, and 55% of the drivers violated red-lights when driving with fast-tempo music.

To summarize, the main result of Study #1 is the effect of music tempo on the number of disregarded red-light signals. We might assume that higher levels of distraction caused these effects rather than increased physiological arousal – especially since no effects of HRs or HRFs were seen. However, it was surprising that null-effects surfaced to the extent that was observed – especially with regard to driving speed. Such lack of results requires further consideration. Brodsky felt that perhaps the readily available visual display of the dashboard digital speedometer restrained the participants from driving beyond a more modest speed ($M = 67$ mph), which no doubt is within reason for the declared 65 mph speed limits. Or perhaps, the music tempo itself might have a distinctive and unusual affect on drivers with formal music education and instrumental training. After all, neuroscience research long ago acknowledged a specific neuroanatomy and differential neuroprocessing ability among musicians (Bermudez & Zotorre, 2005; Bever & Chiarello, 1997; Marsui et al., 2013; Otte, Juengling, & Kassubek, 2001; Schmithorst & Wilke, 2002; Sergent, 1993). For example, differences of neurophysiological approaches, activation, expectances, and interference effects have been discussed (Berti et al., 2006; Besson & Faita, 1995; Gaab & Schlaug, 2003; Lopez et al., 2003). Most specifically, Micheyl, Carbonnel, and Collet (1995) demonstrated explicit perceptual processes for music among musicians in regard to their adaptive behaviour of everyday situations. For example, *loudness adaptation* – a functional sensitivity to intensity – decays to a much lesser extent over time in comparison to non-musicians. It is then entirely conceivable that formal musical training, daily practice, and instrument proficiency also leads to mechanisms by which psychomotor responses that are usually characteristic of momentary shifts in temporal pacing can be monitored and controlled more proficiently among musicians than non-musicians. Hence, Brodsky concluded that formal music training should be more carefully considered within empirical investigations involving music tempo and vehicular control. The above two considerations led to Study #2.

5–3.1.2. Study #2: Drivers without formal music education or training
Twenty-eight undergraduates (M = 25 years old, SD = 4.58; 65% female, with a driver's licence for at least five years) participated in the study; none had formal music training or private tuition on a musical instrument. The students participated in a single 120-minute session including 90 minutes of simulated driving. Of the participants, 89% reported they drive with music all of the time, and 73% claimed they usually listen to moderate-tempo music at medium-intensity sound levels. The music genre they most frequently drove with was a mix of 80s–90s Rock and local Hebrew-language Pop music. All the materials, equipment, and procedures were similar to Study #1, but with two exceptions: the dashboard speedometer was covered from the view, and simulated driving was implemented on a *DeskPro* (Compaq) desktop computer with 17" monitor, a Creative *SoundBlaster Live/Live!DriveII Platinum* soundcard, and two Yamaha 5w 2" *YST-MS28* broad-range speakers (40–20KHz) with an accompanying 15w 5" subwoofer.

The results of Study #2 relate to three categorical areas: Physiological Data, Vehicular Acceleration, and Vehicular Control:

1. *Physiological Data.* HRs and HRFs were measured. Although no main effects surfaced for HRs, effects were found for HRFs ($F_{(3, 81)}$ = 3.39, MSe = 0.7330, $p < 0.05$); planned comparisons demonstrated significantly increased HRFs for laps without music ($M_{NoMusic}$ = 3.43; SD = 1.30; M_{Music} = 2.86; SD = 0.71; t = 2.806, df = 27, $p < 0.01$). Yet, there were no differences between the three music tempi (M_{MUS1} = 2.92, M_{MUS2} = 2.73, M_{MUS3} = 2.92). When considering that decreases of heart rate variability are linked to cardiologic feebleness that can result from mental distraction and stress, these findings clearly show that even simulated driving with music gives way to somewhat adverse physiological states that hamper drivers.
2. *Vehicular Acceleration.* Lap Speed (KPH) and Perceived Speed Estimate (P-KPH) were recorded. Although no main effects were found for KPH across all four conditions, when re-analysing the data exclusively for laps with music, main effects surfaced ($F_{(2, 54)}$ = 3.61, MSe = 140.55, $p < 0.05$) (see Table 5.3). To reiterate, the speedometer was covered, and hence the drivers were not aware of their actual speed. As can be seen in Panel-A, as the music tempo increased so too did the speed of the vehicle. Further, significant main effects surfaced for P-KPH ($F_{(3, 54)}$ = 7.34, MSe = 49.194, $p < 0.001$). Also seen in Panel-A, as the music tempo increased, so too did the driver's perception and estimation of the speed. Namely, in every condition the drivers significantly underestimated their speed by an average of 45 kph, and speed estimates consistently increased with progressively accelerating music tempi. Finally, it should be noticed that speed estimates when driving without music were the most inaccurate of all the driving conditions. Similar results demonstrating underestimation of speed have been reported in many studies (for a review, see Groeger, 2000).

Table 5.3 The effects of music tempo on driving: Study #2

A. Acceleration

	KPH		P-KPH				
	M	SD	M	SD	t	df	p
NoMus	145	30.18	92	10.54	7.717	18	< .0001
MUS1	141	32.10	94	10.33	5.856	18	< .001
MUS2	143	26.97	95	10.65	7.319	18	< .0001
MUS3	147	30.98	102	12.19	6.186	18	< .0001

B. Violations

	ACs			LNs			RLs		
	M	SD	Range	M	SD	Range	M	SD	Range
NM	.07	0.26	0–1	2.43	3.36	0–14	.61	0.79	0–2
MUS1	.14	0.45	0–2	3.36	3.99	0–16	.72	0.90	0–3
MUS2	.14	0.36	0–1	4.68	4.04	0–14	.79	1.17	0–4
MUS3	.36	0.73	0–3	6.50	6.97	0–25	1.21	1.23	0–3

C. Speed Sub-Groups *x* Violated Red Light Signals

	Slower Drivers		Faster Drivers	
	M	SD	M	SD
NM	0.36	0.50	0.76	0.90
MUS1	0.55	0.69	0.82	1.02
MUS2	0.73	1.35	0.82	1.07
MUS3	0.45	0.69	1.71	1.26

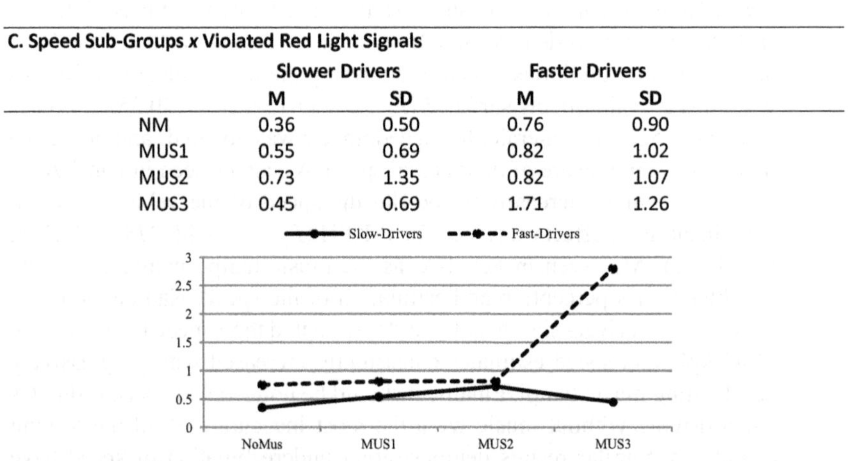

Source: Adapted from Brodsky (2002).

3. *Vehicular Control.* The frequency of three variables were logged: ACs, LNs, and RLs (see Table 5.3). While main effects for ACs were nearly significant ($F_{(3, 81)} = 2.21$, $MSe = 0.1940$, $p = 0.09$), as can be seen in Panel-B there were clear trends indicating that as the music tempo increased so too did the number of incidents. Further, main effects surfaced for LNs ($F_{(3, 81)} = 11.67$, $MSe = 7.4834$, $p < 0.0001$); this finding clearly shows that as the tempo of the music increased there was a decrease in vehicular control as represented through increased lane excursions. Finally, main effects RLs were extremely close to significant ($F_{(3, 81)} = 2.62$, $MSe = 0.7567$, $p = 0.056$); as can be seen in Panel-B there was a clear trend indicating that as music tempo increased so too did the number of disregarded red-light signals and stop signs. In a post-hoc analysis, differences among subgroups based on their overall driving speed were explored. As can be seen in Panel-C, 'faster' drivers were found to have violated more than double the number of red traffic lights and stop signs while driving with fast-paced music than 'slower' drivers listening to the same tracks ($F_{(1,26)} = 8.99$, $MSe = 1.627$, $p < 0.01$). Taken together, these above results illustrate the effects of music tempo on traffic violations:

Near-Crashes And Crashes

07%	Involved In	1.0 Incidents	Without Music
11%	Involved In	1.3 Incidents	With Slow Music
14%	Involved In	1.0 Incidents	With Medium Music
25%	Involved In	1.4 Incidents	With Fast Music

Deviations And Excursions From Lane

60%	Deviated In	4.0 Lanes	Without Music
72%	Deviated In	4.7 Lanes	With Slow Music
93%	Deviated In	5.0 Lanes	With Medium Music
93%	Deviated In	7.0 Lanes	With Fast Music

Disregarded Red-Light Signals And Stop Signs

39%	Disregarded	1.6 Signals	Without Music
47%	Disregarded	1.5 Signals	With Slow Music
39%	Disregarded	2.0 Signals	With Medium Music
61%	Disregarded	2.0 Signals	With Fast Music

The overriding picture of the study is that music tempo consistently affected driving. While music-evoked arousal was not confirmed in the traditional sense (by increased physiology), fast-paced music clearly induced accelerated driving speeds and more frequent traffic violations. Therefore, we might assume that as 'music intensity' has been conventionally seen to explain arousal effects, then perhaps 'music tempo' can be seen to account for distraction effects. Shinar (1978) long ago demonstrated that a high percentage of car accidents are due to attention and information-processing failures rather than to a lack of skill in driving performance responses. Therefore, Brodsky concluded that the study is substantially meaningful – if for no other reason – then because it demonstrates a distinction between *music intensity-evoked arousal* versus *music tempo-generated distraction*. Clearly the study highlights circumstances whereby drivers, who are not fully conscious of their actual speed, were rallied-on by fast-paced music to perform significantly more at-risk deficient driving behaviours.

The above findings seemingly touch on two concepts that may sit at the crossroads between psychoacoustic properties and human physiology. These are: 'rhythmic contagion' and 'rhythmic entrainment'. Since ancient times, writers of both Western and Eastern philosophies have acknowledged the effects of *rhythmic contagion* among music listeners (for example, see Plato, 380 BC/1930; Shiloah, 1978). Rhythmic contagion has been seen over the history of mankind, and relates to both the potential 'power' as well as the 'danger' of music exposure (Boxberger, 1962). In the context of driving, Connolly and Aberg (1993) illustrate the more common everyday occurrence by which we compare our own speed with that of other nearby drivers, and adjust our acceleration accordingly. Subsequently, contagion effects are seen as affecting driver behaviour (especially speed). Brodsky (2002) contended:

> If music stimuli which move at slower levels of perceived activity (i.e., slow tempo) cause drivers to experience 'time' and 'velocity' in terms of the subjective pace of the accompanying background music, then vital monitoring of internal cognitive timers may be hampered and distort perceptual information (for example, vehicular speed), and result in reckless driving-behavior or perhaps highway fatalities (p. 237).

The second concept, *music entrainment* (also known as the Huygens Phenomenon, ca. 1665), is the locking into phase of two previously out-of-step oscillators. Saperston (1995) defined music entrainment as the coupling of a biologically self-sustained endogenously generated rhythm with an external temporal music stimulus. Human behaviour is often seen as induced by external time-based phenomena that are sometimes found in nature, while for most part such stimuli are man-made. In the current study, one of the main effects for music tempo was the demonstration of heart rate fluctuations that increased for driving without music, but decreased for driving with music regardless of the quality of music tempo. This interesting effect may partially explain the vast number of violations occurring when driving with music in comparison to driving without music. That is, higher heart rate variability has been seen to indicate a healthier cardiovascular system and more agile response adaptability, while decreases in heart rate variability are linked to mental distraction or stress as well as contributing to unstable response aptitude. Therefore, listening to music while driving may, in general, functionally induce some form of rhythmic entrainment or synchrony constituting a decrease in variability that ultimately results in driver distractibility and subsequently impairs performance. Ironically, those drivers who characteristically drive faster demonstrated significantly more at-risk driving behaviours *especially* when driving with music of a faster tempo.

5–4. Music-Genre Induced Aggression

In a very early study, Corhan and Roberts-Gounard (1976) recruited 12 participants (50% male) for a single 60-minute experiment which required them to detect a visual signal (or change of a signal) seen on a static sine-wave while listening to one of two music backgrounds variegated by *genre*. Accordingly, half the participants were allocated to one type of music (described as relaxing instrumental music) while the other half were assigned to a second type of music (described as vigorous Rock music); both musics were reproduced at a 70dBA intensity volume. The signal-detection vigilance task was carried out during three 20-minute blocks. In each block-condition there was a different level of temporal presentation: a continuous signal, a fixed rate changing signal, and a randomized on-set signal. The analyses pointed to a main effect for Genre ($F_{(1,10)} = 10.29$, $p < 0.01$), indicating that signal detection was significantly better with Rock music than Easy Listening music. Namely, participants listening to Rock music demonstrated a heightened awareness that facilitated their responses to changes in the environment. A second main effect surfaced for Temporal Schedule ($F_{(2, 20)} = 164.36$, $p < 0.001$), indicating that more correct responses were made when signals were presented at random intervals ($M_{Continuous} = 15$, $SD = 2.81$; $M_{Fixed} = 17$, $SD = 2.04$; $M_{Random} = 22$, $SD = 2.75$). Yet, no two-way interactions between Music *x* Temporal Schedule surfaced. Corhan and Roberts-Gounard concluded that

> …vigilance is best when background stimulation is discontinuous and contains an element of uncertainty (as in the rock music which was more diversified, vigorous, and changeable than the easy-listening music) (p. 662).

From the above we might understand that there could be greater advantages to driving with Rock music than with Easy Listening music simply because driving in traffic is a volatile unpredictable environment, and in such circumstances Easy Listening might cause increased deficiency of performance. However, unfortunately the study is void of procedural information to consider the results as central. For example, if the data of responses (as reported above) are in fact average percentages of correct responses (i.e., $M_{\%}$), then even in the most advantageous of conditions, scores for Hits do not even bridge levels above chance (< 50%). Further, the report offers no details about the music heard by the participants: How was the music reproduced? What songs served as empirical stimuli? How many pieces were heard during the entire 60-minute 3-block within-subjects task? Considering the study was carried out in the mid-70s, the designated Rock music stimuli could have been anything from twenty 3-minute songs to three 20-minute extended instrumental solos. Moreover, the label 'Rock' could have been associated with as many as 30 derivative sub-styles (as presented in Table 2.1). Nevertheless, this study is often cited as a demonstration for music effects on vigilance.

We must understand that *music genre* has always been perceived as a highly complex feature that can prompt any number of effects on perception and human behaviour (both positive and negative). Ironically, music genre is perhaps the most elusive of all music features, and most research scientists who employ music within traffic-related empirical platforms are not musicologically endowed enough to reliably select exemplars as experimental stimuli – especially when their intention is that the *genre* itself constitutes a valid robust empirical condition. Therefore, although music is something experienced by the masses, traffic scientists must approach the concept of music 'style' and 'complexity' in a more diligent manner. One example will illustrate the point.

Matthews, Quinn and Mitchell (1998) studied the effect of Rock music and task-induced stress on simulated driving, specifically measuring driver mood and driver performance. Sixty-four drivers between 18–30 years old (52% female) were recruited for a single session to drive on a fixed-base *Aston* driving simulator at the University of Dundee (Scotland, UK). Stress was induced by drivers engaging in a high workload secondary task, prior to the main task in which they were required to detect pedestrians moving towards the roadway. The study was designed as a between-subjects comparison among four subgroups ($n = 16$ each); two of the groups listened to music through headphones (at intensity levels between 70dBs and 90dBs) while they drove. The music employed in the study consisted of two songs from the *Top Gear-Rock* (1994, Epic Records) music compilation: 'Brown-eyed Girl' (Van Morrison) and 'Two Princes' (The Spin Doctors). Matthews et al. found that music led to increased energy, enhanced happiness, maintained task interest, and lowered driver-perceived workload. Most importantly, they reported an increased frequency of pedestrian detection within a shorter time (decreased RTs) for drivers listening to music background. Therefore, the researchers concluded that listening to loud Rock music does not necessarily increase arousal nor distract drivers, but rather supports resourcefulness when increased attention is required. As such findings are highly contradictory, a closer look at the study is warranted. First, it must be pointed out that only two of the driver groups drove with headphones, and therefore the other two may have been biased by room noise or road feedback from the simulator. Hence, the findings may simply reveal increased perceptual attentiveness while donning headphones, and may have nothing at all to do with the effects of music background on driving performance. Further, no details are offered about music reproduction, and the researchers are highly illusive about their music choice. For example: Why were items chosen from the *Top Gear* CD? And, given that only two Rock songs were required, why were the two specific songs selected? Certainly, when employing music within driving experiments it would seem best to choose items that accompany the entire duration of driving. In this case, 'Brown-eyed Girl' (3:06) and 'Two Princes' (4:19) last long enough for the entire 2-kilometre Phase II drive (2km@30mph = 2.48 minutes); but yet, neither song seems to have been long enough in duration for the 4-kilometre Phase III drive (4km@30mph = 4.97 minutes). Moreover, if the mandate of the study was to highlight the effects of a collective 'genre' (such as the declared 'Rock' music style), then it would have been wiser to employ more than one single exemplar. Finally, when comparing between driving performances highlighting the dynamic interactive processes between affective states (stress *versus* non-stress) and loudness (70dBs *versus* 90dBs), which in fact represent the empirically-based manipulation of stimuli generating the specified experimental conditions, it certainly would be cogent to hold constant all other musical features such as stylistic complexity. For example, the song 'Brown-eyed Girl' performed by Van Morrison in a 'Folk-Rock' style is accompanied by a clear hollow-bodied guitar, a tapping tambourine, and backing vocal ensemble in chorus sections at a 76bpm slow-to-medium tempo,[18] while on the other hand, the song 'Two Princes' performed by The Spin Doctors in a 'Classic Rock' style is accompanied by distorted solid-bodied guitars, percussive snare drum rim-shots, and wailing guitar solos between the chorus sections at a 110bpm medium-to-fast tempo.[19]

As can be seen, conclusions about stylistic features of music (i.e., complexity) as they relate to genre, can often lead one astray. Even when suppositions about music selections are clearly defined, relevance as regarding the context of driving must be questioned. For example, the prominent research team North and Hargreaves (1999) allude to the fact that as the qualities of the music and the task-demands should interact, then when listeners are engaged in a complex task they will prefer un-arousing and undemanding music more than arousing demanding music, simply because it places less demands on the limited processing space. Therefore, if driving a car is seen as a complex multi-task action – requiring the effective orchestration of several executive functions, selective attention, integration of information from a host of visual and auditory inputs, as well as the coordination of numerous compound behaviours (such as divided attention, working memory, prospective memory, behaviour selection, and behaviour inhibition) – then we would certainly expect drivers to prefer un-arousing and undemanding music. But yet, everyday experience indicates that large numbers of drivers actually prefer highly complex arousing aggressive music styles while driving. To reiterate, music genre may be the most elusive of all music features for traffic science to come to grips with.

One UK website, *TalkAudio*, serves many *audiophiles* who spend thousands of British Pounds on amplifiers, speakers, and cable-leads. Already in 2004 they revealed a membership boasting of over 7,000 registered car stereo enthusiasts. When these figures were reported in the *Daily Telegraph* (Syal, 2004), it rekindled a public debate about loud aggressive music that is often blamed for contributing to fatal car accidents. Subsequently, a number of case-incidents surfaced in the press. For example, a coroner in Wigan (Lancashire, UK) claimed that Nicola – an 18-year-old female driver – had crashed and killed herself along with two companions in 1997 as the result of driving with 'intensely dynamic music'. The supporting editorials pleaded with parents to forewarn their youngsters about the possible dangers of playing aggressive hard-hitting music while driving a car. Although many people might have a rather limited awareness about the dangers of music loudness and music tempo on driving, most certainly there are huge beliefs vis-à-vis the ill-effects of 'Rock' music genres that have always been seen to induce a range of aggressive behaviours. From time-to-time personal stories are published that fuel the fire:

18 'Brown-eyed Girl': http://www.youtube.com/watch?v=kqXSBe-qMGo.

19 'Two Princes': http://www.youtube.com/watch?v=wsdy_rct6uo&list=RDwsdy_rct6uo.

> I was at a red light the other day and the car next to me was vibrating so much that even my vehicle was shaking from his 'musical' noise. I could hardly hear the sound of my engine despite the fact that the windows were closed on both vehicles. In addition to being irritating and annoying, is this not considered dangerous driving behaviour? – Kris from Windsor (in Will, 2012).

Syal reports that in 2004, Sony Corporation was accused of encouraging anti-social behaviour through its advertisements for car stereos. Accordingly, the Advertising Standards Authority[20] banned a Sony magazine marketing campaign for selling car-audio amplifiers that promoted the tag line: '*Maybe you like your music relaxing. Or maybe you like to shatter greenhouses and set off car alarms*'. So, while some drivers report they intentionally avoid certain kinds of music because they get 'carried away' by the loudness or emotional forces in the music (Bull, 2001), and others report to refrain from playing genres such as Disco because the tempos cause them to drive too quickly (Oblad, 2000c), for the most part drivers with booming car stereos dismiss such public warnings as little more than *scare tactics* of just a few people with political aspirations and a personal agenda.

The truth is that drivers who regularly play Dance music at loud volumes (i.e., ≥ 148dBA) from highly expensive car audio hi-fi equipment (costing up to £3,500 or $6,000) may be thrilled when others look at them while at the wheel. Syal (2004) reports that interview data repeatedly confirms these drivers to be delighted when reminiscing about having watched drivers in the next car turn around *as they wonder why their dashboards are vibrating.*[21] But, the question we must raise here, is: Does such an air of antagonistic belligerence convey *real* undeniable aggression and violence? For some time, studies have indicated that listening to Heavy Metal and Hard Rock music correlates to negative behaviours such as reckless driving and traffic accidents (Arnett 1991a, 1991b; Gregersen & Berg, 1994). If such observations are true, then, we might need to differentiate between 'music-generated distracted driving' and 'music-induced aggressive driving'.

5–4.1. Distracted versus Aggressive Driving

There is clear evidence that distracted driving and aggressive driving are dissimilar. During distracted driving our attention and/or judgement is impaired, while aggressive driving seems to imply that we have perhaps intentionally driven unsafely or inappropriately. Hanowski et al. (2006) reanalysed the data from the *100-Car Study* (Dingus et al., 2006; previously described in Chapters 3 and 4).

---

20 The Advertising Standards Authority (ASA) is the UK's independent regulator for advertising across all media (www.asa.org.uk).

21 We might assume that drivers do not really live for the thumping and rattling coming from their vehicle, but rather enthusiastically rally at the thought of competition. Williams (2008) reveals *dB Drag Racing* as a competition for serious car audio enthusiasts, who spend an excess of $4,000 outfitting their vehicle with amplifiers and subwoofers. The prize: to be recognized as the owner of the 'loudest car audio system'.

Table 5.4 The 100-Car Naturalistic Driving Study: Differences between incidents of distracted versus aggressive driving

	Distraction & Inattention		Aggressive Driving	
	N	%	N	%
A. A. Total Incidents (n=246)	46	19	37	15
B. At -fault Incidents (N=138)	31	23	34	25
1 Aborted lane change	1	3	2	6
2 Approaches traffic quickly	1	3	1	3
3 Conflict with oncoming traffic	1	3	2	6
4 Following too closely	2	7	–	–
5 Improper lane change	–	–	1	3
6 Improper passing	–	–	6	18
7 Improper stopping at an intersection	1	3	–	–
8 Lane change without sufficient gap	3	10	16	47
9 Late braking for stopping/stopped traffic	20	65	3	9
10 Late deviation of through vehicle	1	3	1	3
11 Merge without sufficient gap	–	–	1	3
12 Roadway entrance without clearance	–	–	–	–
13 School bus passing violation	1	3	1	3
Total	**31**	**100**	**34**	**100**

Source: Adapted from Hanowski, Olson, Hickman, and Dingus (2006).

The research team found a total 246 incidents, and among these 138 (56%) were caused by the drivers themselves. Distracted driving was found among 23% of incidents (14% for internal distraction, 5% for external distraction, and 4% for inattention), while aggressive driving was found for 25% of the incidents. It is interesting to note that the most frequent contributing factors for at-fault incidents were: Driving Techniques (70%), Distraction (23%), and Aggressive Driving (25%). In many cases these factors actually overlap (see Table 5.4). As seen in the table, the major factor for at-fault incidents resulting from distracted driving was 'deficient braking time when coming to a full stop', whereas the two overriding factors for at-fault incidents resulting from aggressive driving were 'failing to keep a sufficient gap when changing lanes' and 'willingness to pass another car just before a turn'.

To illustrate the above point, we can view the results of a UK survey implemented by Allianz Insurance (partially described in Chapters 3 and 4). While the full data has never been released to the public, recently Allianz Insurance (2014) has made the records available to Brodsky. The data point to a general trend in which 23% of drivers claim they are distracted by music playing in the car (see Table 5.5). As can be seen in the table, the columns reflect nine different music genres, which clearly indicate that the most prevalent music styles heard among British drivers, are: Pop/Chart (36%) and Rock/Indie (21%). These two categories account for over 50% of the 1,000 drivers polled about their in-car music listening behaviour. Furthermore, as can be seen in Panel-B, roughly 13% admitted to have had a near-crash due to being distracted by music, and 9% revealed that they had already been in a crash because of music distraction. It should be pointed out that the music genres appearing in the columns have been spatially placed in an ascending order (from left-to-right) by the frequency of incidents (i.e., near-crash incidents and crash accidents). The Allianz survey reveals that drivers listening to Jazz/Blues reported the highest (50%) frequency of incidents, followed by drivers listening to Country music (42%), followed by drivers listening to Hip-Hop/R&B (29%), and so on. As might be expected, a positive relationship can be seen among the Top-3 High-Risk Music Styles as far as the frequency of incidents and percentage of drivers reporting to be distracted by music. The table shows that as the frequency of incidents increase, so too does the overall level of distraction. Nevertheless, an unexpected converse relationship surfaced among the Bottom-3 Low-Risk Music Styles as far as the same relationship illustrating frequencies of incidents and distraction. Namely, as the incidents decrease, conversely distraction increases.

Top-3 High-Risk Music Styles			**Bottom-3 Low-Risk Music Styles**		
	Incidents %	**Distraction %**		**Incidents %**	**Distraction %**
1 Jazz/Blues	50	37	7 Soul/Reggae	13	19
2 Country	42	36	8 Rock/Indie	10	19
3 Hip-Hop	29	27	9 Classical	5	28

To summarize, the Allianz survey clearly denotes that drivers listening to specific music genres (i.e., Jazz, Blues, Country, Hip-Hop, R&B) may be more at risk for incidents and accidents due to music-generated distraction than other drivers. Finally, it should be noticed that an overall 60% of the sample across music styles become angry when confronted with other drivers' loud music. The drivers reporting the highest percentage of anger were those driving with Classical and Folk musics.

On the extreme outlay of aggressive driving is a phenomenon that came to the surface in the mid-1990s known as 'Road Rage' (Featherstone, 2004). Urry (1999) claims that whereas one aspect of automobility encourages drivers to be careful, considerate, and civilised (the *Volvo Syndrome*), another inspires drivers to enjoy speed, danger, and excitement (the *Top Gear Syndrome*). Yet, road rage seems to involve a general loss of control whereby angry drivers totally disregard the more acceptable forms of roadway courtesy in favour of reckless, aggressive manoeuvres. Thrift (2004) points out that while many drivers develop what they may regard as particularly shrewd ways of moving around, unfortunately others often perceive such tactics as a violation of basic moral and transportation codes. For example, some drivers interpret the way other drivers operate as aggressive, arrogant, disrespectful, ignorant, impatient, and irresponsible. Most specifically, some particular driving behaviours may be seen as disobeying road rules, endangering drivers, trying to control the road, or even as an invasion of personal space (Fraine et al., 2007). Accordingly, when some drivers see another doing something they deem as unacceptable or socially offensive – and this could involve as little as a discourteous rude gesture, shaking of the head, a facial expression, or simply playing music too

Table 5.5 Allianz Insurance music survey (N=1,000), driving behaviours (%) by music genre

		Jazz Blues N=38 %	Country N=78 %	Hip-Hop R&B N=85 %	Folk N=15 %	Dance N=43 %	Pop Chart N=349 %	Soul Reggae N=31 %	Rock Indie N=213 %	Classical N=86 %
	Preference of Genre among Full Sample	4	8	9	2	4	36	3	21	9
A.	Music IS a Distraction	37	36	27	7	16	14	19	19	28
	All The Time 100%	3	22	1	0	0	1	0	1	2
	Half Of the Time 50%	18	4	7	0	2	2	3	2	1
B	Near-Crash Incident Due to Music	29	19	21	13	9	7	10	8	3
	Crash Accident Because of Music	21	23	8	7	9	6	3	2	2
C	In-Car Music Aids Concentration	18	27	27	20	33	22	13	22	24
	In-Car Music Helps Pass the Time	42	47	53	73	53	63	52	64	52
	In-Car Singing Is Enjoyable	26	18	54	33	35	44	39	49	16
	In-Car Music Aids Relaxation	34	27	38	47	53	47	42	53	47
	In-Car Helps Block Out Road Noise	18	9	15	0	14	11	13	15	8
D	Annoyed by Other Drivers' Loud Music	74	62	50	80	47	61	61	62	84

Source: Provided by Allianz Insurance (2014).

loud – they not only become more *out-raged* but engage in an aggressive *road rage* pursuit. Ironically, as Featherstone brings out, road rage clearly demands a much greater level of driving skill; weaving through traffic at higher speeds, tailgating the car ahead, overtaking at great speeds, or coming to sudden halt against the car behind. Drivers justify such behaviour simply because they perceive they have a high level of anonymity on the road (Lutz & Fernandez, 2010; Shinar, 1998).

5–4.2. Angry Drivers and Music Background

Mesken et al. (2007) and Pecher et al. (2009) point out that for some time it has been known that experiencing anger while driving is highly common, and that such negative emotions lead to a more aggressive driving styles. As pointed out above, angry drivers deliberately jeopardize others with a surmountable level of hostile verbal and/or physical expression – as well as by using their automobile to harass other drivers on the road. In this connection, Wiesenthal et al. (2003) listed a host of aggressive driver behaviours, including: honking the horn, flashing the high beams, swearing/yelling at other drivers, and executing obscene gestures. But yet, this latter type of antagonistic behaviour needs to be distinguished from other at-risk behaviours and dangerous actions that drivers demonstrate – such as speeding, tailgating, overtaking, and manoeuvring without signalling. It would be warranted to explore the effects of background music on such animated conduct. For example, Pecher et al. documented that people drive according to the emotional valence of music heard in the cabin (see above Section 5–2.1). Accordingly, when driving with 'neutral' music, drivers were able to drive more efficiently. But, when 'happy' music was presented vehicular control deteriorated (demonstrated by lack of longitudinal and lateral control seen in slower speeds, increased excursions from the lane, and an increased tendency to stray onto the hard shoulder line). In the study, drivers claimed that 'happy' music affected their behaviour by inducing them to 'tap' on their wheel or 'whistle' along to the music. It should be pointed out that 'happy' musics are typically fast-tempo high-pitched songs with positive lyrics that attract attention and easily distract drivers. On the other hand, the same drivers reported that when 'sad' music was presented they became more withdrawn and their attentional focus that was progressively oriented towards the gravity of the lyrics and their own internal thoughts. The researchers concluded that it would be especially prudent to consider a range of emotions and their consequences on driver behaviour. In this connection, Pecher et al. envisioned that music-generated *positive* emotions such as 'joy' and 'happiness' would distract drivers causing increased risk-taking, decreased speed, and trajectory control; whereas 'excitement' would increase speed, but decrease RTs along with a higher frequency of errors. Alternatively, music-generated *negative* emotions such as 'sadness' would cause a more passive attitude involving a withdrawn attentional self-focus and longer RTs, whereas 'anger' and 'frustration' would lead to outraged aggressive behaviour including faster speeds, extreme use of the brake and accelerator, as well as verbal and physical viciousness.

Specifically targeting anger, several studies have proposed that listening to one's favourite music can reduce driver aggression. For example, Wiesenthal et al. (2000, 2003, see Section 5–2.1 above) compared listening to music versus silence under conditions of high traffic congestion and low time pressure. The study found that driver aggression was moderated and diminished by listening to preferred music radio or pre-recorded tapes. In another study, van der Zwaag and colleagues (Fairclough et al., 2014; van der Zwaag et al., 2011) explored the impact of music on affect during anger inducing drives. That study recruited 100 participants between 16–26 years of age ($M = 21$, $SD = 4.7$, 50% male) for a 45-minute simulated driving session. Prior to the session, music pieces were rated and selected as personalized playlists for the experiment. In the song selection procedure, each participant used a 7-point Likert scale to rate 80 songs for valence (classifying each item as 'NV' for negative-valence or 'PV' for positive-valence), as well as for level of energy (classifying each item as 'LE' for low-energy or 'HE' for high-energy). The ratings and a few song examples were:

Type	Valence Rating	Energy Rating	Feeling	Music Examples: Song Title (Artist)
PV/HE	6.7	6.6	Activating Joyous	'Rhythm Of The Night' (Corona) 'Breakfast At Tiffany's' (Deep Blue Something) 'Just Can't Get Enough' (Depeche Mode) 'Foundations' (Kate Nash)
PV/LE	5.1	3.5	Calming Relaxing	'What A Difference A Day Made' (Dinah Washington) 'Just My Imagination' (The Temptations) 'Jimmy' (Moriarty) 'Bound For Another Harvest Home' (Jay Unger and Molly Mason)
NV/HE	1.8	5.1	Activating Angry	'Everybody Clap Your Hands' (Euromasters) 'Wait And Bleed' (Slipknot) 'One Step Closer' (Linkin Park) 'Goddamm Electric' (Pantera)
NV/LE	1.8	1.3	Calming Sad	'The House Of Spirits' (Hans Zimmer) 'Silver Ships Of Andilar' (Townes Van Zandt) 'Now Let Thy Servant Depart' 'Vigiliile Nocturne', Op. 37, Vespers 5/15 (Sergei Rachmaninoff) 'Rias Kammerchor-Vergangen Ist Mir Glueck Und Hell', 'Seven Secular Choral Songs', Op. 62, No. 7 (Johannes Brahms)

Prior to the experiment each participant was assigned to one of five groups: four music conditions (PV/HE, PV/LE, NV/HE, NV/LE), and a NoMusic condition (for empirical control). Each participant heard the ten songs with the highest ratings in their specific driving condition. The music reproduction system employed two speakers on the right and left sides of the simulator (emulating ecologically valid mounted speakers located on the car door panels). During the session every participant completed a short trip under time pressure, in which they were also forced to deal with a number of complications – all the while being monitored for blood pressure and skin conductance levels (see Figure 5.6). As can be seen in the figure, there were five procedural stages. To induce an angry mood state, each driver was manipulated to perceive extreme time pressure. This was accomplished by three actions: (1) drivers were allocated only eight minutes (time pressure) to transport a child to an elementary school for a exam, with the understanding that lateness would result in a failing grade; (2) obstacles were strategically placed along the way, designed to prevent the drivers from completing the trip on time; and (3) monetary fines were enforced against exceeding the speed limit or causing a near-crash. While the data indicate that the levels of anger measured after the trip (Stage 4) were significantly higher than before the trip (Stage 3) for all groups, the highest level of post-trip anger was among NV/HE drivers and the lowest was among PV/HE drivers. Namely, although mood states generally became more negative during the trips themselves (as anticipated by the designed manipulation), positive-valence high-energy musics were able to counteract the subjective driving experience by distracting drivers from adverse events in the traffic environment (such as traffic jams). Thus, music seems to have been able to divert a range of negative thoughts and angry feelings from rising to an even higher intensity. It would also appear that negative-valence high-energy musics exploit the opposite, and do indeed allow for subjective feelings of anger to escalate.

In light of the above findings, we might question whether the impact of negative valence music comes from the sounds themselves, or are effects solely based on the semantic meaning of the content in the song's lyrics. In this connection, Fischer and Greitemeyer (2006) explored the direct impact of sexually aggressive song lyrics on aggression-related thoughts, emotions, and behaviour towards the opposite sex, by means of antagonistic hostile texts. The results indicated that vicious song lyrics caused participants to behave in a socially unacceptable manner against the opposite sex than did similar song styles with neutral texts. More specifically, harsh songs elicited circumstances by which male participants recalled more lyrics portraying negative attributes of women, and subsequently expressed feelings of vengeance against them. Ironically, women behaved in the same way towards men with aggression-related responses when exposed to men-hating songs. The literature is full of evidence that link exposure to music with antisocial content to a range of negative outcomes including aggressive behaviour. Hence, we might ask if music that seems to be aversive in nature, but does not have semantic content (i.e., instrumental music without lyrics), also triggers aggressive thoughts and behaviours. This was the main thrust of a twin study implemented by Krahe and Bieneck (2012). In Study-A, 111 undergraduates between 19–27 years of age (M = 23, SD = 4.3, 68% female) were recruited from the University of Potsdam (Germany) for a single 45-minute

	Stage 1	Stage 2	Stage 3		Stage 4		Stage 5
Procedure:	Briefing	Habituation	Mood Induction		Simulated Drive		Debriefing
Activity	Attach Sensors	1. Watch TV 2. Mood Scale 3. Anger Scale 4. Baseline BP & SCL	Listen With Music <u>or</u> Sit-Relax Without Music	Mood Scale Anger Scale	5-Group Study 4xMusic or NoMusic	Mood Scale Anger Scale	BP & SCL Remove Sensors
Duration:	8 min	8 min	6 min	2 min	12 min	2 min	8 min
Time Line:	0:00-----8:00----------16:00---------22:00---24:00-----36:00-----38:00--------45:00						

Figure 5.6 Session structure for simulated driving experiment
Source: Adapted from van der Zwaag et al. (2011) and Fairclough et al. (2014).

cognitive-psychology experiment. The participants were randomly assigned to one of three groups variegated by aural background: (1) 'PleasantMusic' (PM) consisting of Classical Music; (2) 'AversiveMusic' (AM) consisting of Hardcore/Techno music; and (3) 'NoMusic' (NM) for empirical control. The results indicated that AM caused the listeners to be more irritated; AM pieces were rated as more Unpleasant ($M_{PM} = 0.82$, $SD = 1.43$; $M_{AM} = 3.41$, $SD = 2.19$; $F_{(1, 75)} = 37.45$, $p < 0.01$), and less Uplifting ($M_{PM} = 3.87$, $SD = 1.49$; $M_{AM} = 2.00$, $SD = 1.81$; $F_{(1, 75)} = 24.32$, $p < 0.01$). All participants wore headphones throughout the session. They were required to role-play a marriage counsellor presented with a relationship conflict, and to recommend a solution in the form of a short written text. After completing the text (and handing the paper to a research assistant), each participant rated their own subject mood; the results indicate that positive moods were higher among those listening to PleasantMusic ($M_{PM} = 4.52$, $M_{AM} = 3.51$, $M_{NM} = 3.96$; $F_{(2, 108)} = 9.99$, $p < 0.001$). Then the texts, which allegedly had been evaluated by an associate, were returned with feedback; commentaries were either positive or negative by nature (counterbalanced across the sample). Subsequently, after reading the comments, the participants rated their level of anger – viewed as an emotional response to the feedback. The results reveal significantly increased levels of anger among participants listening to AversiveMusic. Finally, participants were instructed to evaluate a similar text, supposedly written by the same associate who allegedly evaluated their text – viewed as an opportunity to engage in hostile retaliatory behaviour. The results indicate that participants who listened to aversive music selections with associated angry moods demonstrated enhanced levels of aggressive behaviour. Therefore, the researchers concluded that music-dependent mood could indirectly modify a predisposition to engage in aggressive behaviour "even in the absence of aggressive semantic content by virtue of eliciting differential affective states" (p. 281). Nevertheless, this conclusion is somewhat puzzling when considering the music exemplars used in the study. To reiterate: PleasantMusic was described as instrumental Classical selections, while on the other hand, AversiveMusic was described as instrumental Hardcore and Techno pieces. The former consisted of selections from the 'Peer Gynt Suite' Op. 23 (1875, Edward Grieg), 'The Four Seasons' Violin Concertos (1723, Antonio Vivaldi), and the '9th Symphony, From the New World' (1893, Anton Dvorak). However, the researchers offer no details of the latter, except for the elusive description: "...range of different hardcore and techno pieces (e.g. 'King Deuce')" (p. 276). So, we may wonder what music did the participants listening to AversiveMusic actually hear? It is interesting to note that both a rapper and a heavy-metal band use the same stage name *King Deuce*, and in both cases the music they perform employs vocals promoting anti-social messages based on hostile lyrics. Given that the study was designed to target music without language cues, and the empirical stimuli used herein seems to be more than tainted by semantic content, is it any wonder that the researchers report AversiveMusic to have caused enhanced outrage? Yet, the demonstration of emotional states such as anger (and aggression) subsequent to hearing aversive music is still highly convincing.

In Study-B, Krahe and Bieneck (2012) recruited 142 undergraduates between 19–24 years of age ($M = 22$, $SD = 2.1$, 51% female). Study-B replicated Study-A in all aspects, with the exception that after receiving their evaluated texts the participants completed a word-recognition lexical-decision task. As in Study 1, AversiveMusic selections were rated as significantly more Unpleasant ($M_{PM} = 1.65$, $SD = 1.16$; $M_{AM} = 4.48$, $SD = 1.83$; $t = 9.20$, $df = 76$, $p < 0.01$), and less Uplifting ($M_{PM} = 4.35$, $SD = 1.46$; $M_{AM} = 2.76$, $SD = 1.61$; $t = 4.55$, $df = 76$, $p < 0.01$). After receiving feedback on their texts, the participants rated their level of anger; a main effect surfaced for Anger ($M_{PM} = 1.42$, $SE = 0.14$; $M_{AM} = 1.91$, $SE = 0.14$; $M_{NM} = 1.75$, $SE = 0.11$; $F_{(2, 135)} = 3.05$, $p < 0.05$, $\eta^2_P = .04$) indicating that participants listening to pleasant music reported significantly less anger. In addition, a main effect surfaced for Feedback ($M_{PositiveFeedback} = 1.39$, $SE = 0.10$; $M_{NegativeFeedback} = 2.00$, $SE = 0.11$; $F_{(1, 135)} = 15.39$, $p < 0.001$, $\eta^2_P = 0.10$); this indicated that participants reported more anger after provocation by negative feedback. Finally, a 3-way interaction for Music *x* Feedback *x* Anger surfaced ($F_{(2, 135)} = 3.38$, $p < 0.05$, $\eta^2_P = .05$); this finding indicated that when participants were provoked, those listening to AversiveMusic demonstrated significantly higher levels of anger than those listening to PleasantMusic. Finally, the word-recognition task required participants to detect meaningful German-language words from among 160 randomly presented strings of six letters. There were 40 antagonistic words ('weapon', 'knives'), 40 neutral words ('summer', 'meadow'), and 80 non-words. An analysis of the task found main effect for completion time RTs ($M_{PositiveFeedback} = 873.09$ milliseconds, $SE = 11.2$; $M_{NegativeFeedback} = 828.06$ ms, SE = 11.2; $F_{(1, 125)} = 8.36$, $p < 0.01$, $\eta^2_P = .06$); this finding indicated that all participants receiving negative feedback detected aggressive words faster. Moreover, a 3-way interaction for Music *x* Feedback *x* RTs surfaced ($F_{(2, 135)} = 5.43$, $p < 0.01$, $\eta^2_P = .08$); this finding indicated that provocation by negative feedback affected RTs differently depending on the type of music heard (PM *versus* AM). Although PleasantMusic did not cause meaningful differences in detection RTs whether provoked or not, both NoMusic control and AversiveMusic groups significantly decreased RTs when incited. To summarize, it would appear that aggressive responses do escalate from aversive music stimulation, referred to by Krahe and Bieneck as 'aggression-enhancing effects of music'.

5–4.3. Confrontational Music Styles

In a landmark series of studies, Arnett (1991a, 1991b) revealed that adolescents who preferred Heavy Metal music – with heavily distorted electric guitars, pounding rhythms, raucous raw screaming vocals, typically played at extremely loud volumes – reported a higher rate of reckless behaviour (specifically, driving behaviour), sexual promiscuity (unprotected causal sex), drug use (marijuana, cocaine), and anti-social conduct with minor criminal activity (including shoplifting, vandalism, damage to property). It should be pointed out that these results have been challenged by Roberts et al. (1998) who could not replicate the findings based on music preference alone, but rather only when they accounted for music preference and high negative emotion. Arnett also found that 'MetalHeads' and 'HeadBangers' were less satisfied with their family relationships, reported lower levels of self-esteem, and scored higher in Sensation-Seeking. Arnett noted that the song lyrics of this music genre promoted a variety of undesirable behaviours (including drug use, promiscuity, sadomasochism, Satan worship, murder, suicide), as well as supported an enthusiasm for listeners to reject and defy the norms and standards of adult society.[22] In one study, Arnett recruited 110 males between 16–19 years of age in Atlanta Georgia (USA); 50% (n = 55) identified themselves as preferring Heavy Metal music. In addition, 135 females between 15–19 years of age were recruited; 22% (n = 30) identified themselves as preferring Heavy Metal music. Among the total 85 who preferred Heavy Metal music, their favourite bands (in alphabetical order), were: Anthrax, Guns N' Roses, Iron Maiden, Megadeth, Metallica, Ozzy Osbourne, Rigor Mortis, Skid Row, and Slayer. The participants completed several questionnaires regarding the number of times they engaged in a variety of reckless behaviours over the past year. Of interest here are those items relating to driving behaviours (although we might also consider antisocial manifestations such as destruction of private and public property as highly relevant). The results indicated that reckless behaviours were far more pervasive for boys than for girls, and among the males who preferred Heavy Metal music, reckless behaviours were far more frequent. It should be noted that these differences were statistically significant for all driving categories (see Figure 5.7). As can be seen in the figure, the boys who preferred Heavy Metal music typically drove more frequently while drunk; a significantly higher proportion drove under the influence:

22 On the other hand, Arnett's (1991a) interview data clearly brought out that many Heavy Metal songs do employ content about emotions (such as distress after loss of a girlfriend), as well as social consciousness (for example, ecology, destruction of the environment, child abuse, accidental death from handling guns carelessly, threat of nuclear war, injustice towards ethnic groups).

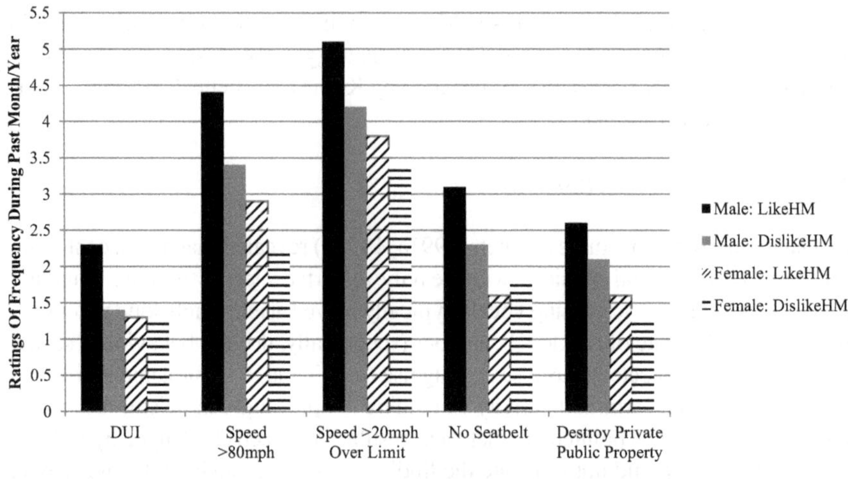

Figure 5.7 Frequencies of reckless behaviour and liking of Heavy Metal music

Source: Adapted from Arnett (1991b).

$M_{\text{LikeHM}} = 50\%$, $M_{\text{DislikeHM}} = 23\%$; $x^2 = 8.52, p < 0.01$. Further, these same males drove more frequently over 80 mph, and drove more frequently above 20 mph over the 65 mph Georgia State speed limit. Furtherstill, they drove more frequently without a seatbelt, and more frequently destroyed public and private property. Finally, also seen in the figure are girls who preferred Heavy Metal music that typically drove more frequently over 80 mph, and drove more frequently above 20 mph over the 65 mph Georgia State speed limit. Ironically, these same females drove less frequently without a seatbelt, but more frequently destroyed public and private property (with a significantly higher proportion from this sample destroying property). Arnett asserts:

> It would be a mistake to conclude that heavy metal music causes the problems [found] in this study. The music must be seen in the context of the entire fabric of socialization of American adolescents… Adolescents who drive recklessly, who have sex without contraception and with partners to whom they have no commitment, who use drugs, who destroy private or public property, do so among other reasons because they enjoy it… This is also true for heavy metal music. The high sensation of the music and the abandonment to sensuality both advocated and demonstrated by the performers has an intrinsic attraction to many adolescents, and there may be little in their socialization environment to prohibit them from following their impulses in this direction (pp. 591–2).

Nevertheless, the fact is that drivers who enjoy listening to music genres such as Heavy Metal, do demonstrate characteristic driving deficiencies and risks that cannot be overlooked or ignored – even if there are equitable sociological explanations for such behaviours. Arnett's (1991a) own interview data indicated that these same listeners do listen to more hard-core exemplars when 'ticked-off' – they claim that they listen to *full-blown thrashing metal pieces* when they are 'mad at the world'. As stated earlier in the chapter, for some time it has been known that experiencing anger while driving is highly common, and such negative emotions – especially when enhanced by aversive music – can lead to more aggressive conduct (including driver behaviour). It is of interest that Arnett reported approximately 54% of those preferring Heavy Metal music assert that listening to music 'relaxes them' and 'dissipates their anger'; albeit, some of these declared that their energy and negative emotions first increases and then dissipates after listening. However, we must consider that 46% of those preferring Heavy Metal music also assert that listening to music 'get's them going', 'revved-up', 'heightens their energy', 'makes them hyper', induces greater aggression by putting them 'in an especially good mood to go out vandalizing', and motivates them to 'want to go and beat-the-crap-out of someone'. One can only wonder, then, if perhaps we should be more aware about the effects of certain musical features and complexities (such as those components found in Heavy Metal music), as well as consider the fact that these same features might also be found in other musics styles?

It should be noted that the first study to report an association between music genres and driving performance among a wide range of age groups was the British research team Dibben and Williamson (2007). They recruited 1,780 participants for an online survey based on demographics representing drivers throughout the United Kingdom. The sample was evenly split by gender, and comparisons were made between three groups of drivers variegated by age (18–29 years [22%], 30–50 years [43%], and 50+ years [35%]). One overriding finding of the study was that most of the drivers saw music as an important influence on their levels of relaxation and concentration while driving on the road. However, an important facet of the study was that it placed music genre in a more central position than in any previous study. Most specifically, Dibben and Williamson analysed the frequencies of in-car music genres at the time of an accident compared to the *most listened-to genres while driving* for each of the age groups. The results revealed significant differences between the music styles playing at the time of the latest reported accident *versus* the expected norm for the same age group. For example, IndieRockPunk music, which Dibben and Williamson report to be the most listened to in-car music style among 18–29 year old British drivers, was heard significantly less by drivers who had been in an accident; those drivers reported listening to DanceHouse music significantly more than would have been expected. Further, drivers who had been in an accident from among the 30–50 year old driver group were listening to Classical, Chart-Pop, and Indie/Rock/Punk musics rather than genres reported to be most popular for this group of drivers. And finally, drivers over 50 years old who had been in an accident reported not to have been listening to Classical music – the most frequently listened to genre among older British drivers. Therefore, Dibben and Williamson suggested the possibility that *the most prevalent music genre* of a particular age group may actually serve as a pre-emptive deterrent against age-related distraction while driving.

In 2009, two consumer surveys were published attesting to ill-effects associated with music genre on driver behaviour. The UK's leading specialist sub-prime car dealer *ACF Finance* implemented the first survey. While no details are available about the sample or the methods themselves, the published results declare that 73% of drivers admit to having received a traffic fine for speeding, and that in 70% of the cases, the drivers were listening to 'pounding fast dance music' (ACF, 2009; Betts, 2009). In addition, *AutoTrader Magazine* announced findings from a reader's survey – the results of which were so sensational that a series of subsequent publications surfaced in the UK daily press. The main finding was that drivers who listen to Rap or Hip-Hop music styles are most at risk for car accidents and road-rage incidents (Noyes, 2009; *The Telegraph*, 2009). The details of the *AutoTrader* readers' poll are somewhat clearer than the survey by *ACF Finance*. Accordingly, *AutoTrader* questioned over 2,000 motorist-readers who voluntarily responded by e-mail about the music they prefer while driving. In general, the song 'The Real Slim Shady' (Eminem) was listed as the favourite driving track amongst 600 drivers, while another 500 drivers listed 'Dance Wiv Me' (Dizzee Rascal) as the tune they most often drive with. However, almost 50% admitted that songs by artists such as Eminem, Dizzee Rascal, and Jay-Z had an adverse effect on their mood and behaviour while driving; 20% (one-in-five) claimed that both Rap and Hip-Hop music styles make them highly aggressive behind the wheel, while only 6% reported to feel more relaxed when driving with those genres. It is interesting to note that about half of the sample felt Classical music relaxes them when driving, and hence perceived that such music styles offer drivers the highest safety experience. Most outstandingly, the *AutoTrader* survey found that for at least another 15% of motorists, driving with particular music genres induced aggression: Dance (6%), Classic Rock (6%), Pop (2%), and Classical (1%). Adding up all of the above, it seems that a total 65% of the respondents (i.e., 1,300 random auto enthusiasts who responded to a magazine survey) account for circumstances in which they apparently fall prey to the ill-effects of background music while driving – most specifically *music-genre induced aggression*. Paradoxically, at the time when these results were officially released, policymakers and road safety organizations rose to the occasion with commentary illustrating that for some time we have all witnessed drivers playing loud pounding music in their cars: '*you can hear the ground shake before they arrive*'. But then, in a seemingly apologetic fashion they point out that unfortunately British police crash investigation reports do not document the presence of music (styles and/or artists) that motorists had been listening at the time of an accident. Hence, for our purposes, the *AutoTrader Magazine* 2009 Reader Survey may simply be a ground-breaking pioneering account, if for no other reason than being the first wide investigative report about the kind of music drivers listen to while on the road. Reflecting on the findings, one accident attorney commented:

> Although I agree that music can play a significant role in distractions while driving, I personally am hesitant to classify all hip-hop listeners as bad drivers. I don't understand why they listen to that noise (I see little music in that genre) and why they have to listen to it so loud. The loudness of their hip-hop music at a traffic stop does increase my propensity for road rage – if not theirs (Noyes, 2009).

Then in 2011, the UK insurance company *Quotemehappy* examined the connection between driver preference for specific musical genres and their driving behaviour. To this end they subcontracted *Populus* (a UK polling firm) to run a questionnaire survey across a large (N = 2,050) sample who were recruited from the public without affiliation to a commercial organization (such as customers from an insurance company or a magazine subscribership). In general the findings indicated that British drivers who listen to Rock, Heavy Metal, or Hip-Hop were more likely to speed, tailgate, and be involved in accidents (Quotemehappy, 2011). Alternatively, drivers who listened to Classical or Pop music styles experienced less stress, and that drivers listening to Drum&Bass or Heavy Metal music styles were more likely to report aggressive driving behaviours than those listening to Classical music (Williamson, 2011). The data indicated that 73% of drivers listening to Rock genres, and 39% of drivers listening to Hip-Hop music, often swear and make rude gestures at other drivers (in comparison to the 32% of other drivers who listen to Pop or the 16% of drivers listening to Classical musics). Namely, the former were more than twice as likely to experience road rage. Most specifically, the *Quotemehappy* survey found that drivers listening to Heavy Metal reported to be the most aggressive; 75% admitted to 'speeding as a driving style', 62% revealed that they 'frequently loose their temper with other drivers on the road', and 11% disclosed they had recently experienced a near-crash accident because of music-generated distraction. Another interesting find, although ironically not reported in *Quotemehappy*'s PR materials but rather by music psychologist Williamson, was that drivers who listened to Jazz styles reported to have received significantly more speeding tickets than any other driver listening to any other music genre. This finding is similar to those of Allianz Insurance (described above in section 5–4.1). Williamson infers that drivers who prefer Jazz music styles may be those who listen more analytically (i.e., referred to in Chapter 2 as a *field-independent* perceptual style of listening), and become more engaged in the voicings and texture of the instruments; and as a result, they may loose concentration of the primary driving task, and eventually 'run-away' with the music. Additional findings of the survey were that drivers listening to Rock and Hip-Hop music had been more involved in accidents over the previous past three years period; 31% of those listening to Rock styles, and 20% of drivers listening to Hip-Hop, reported at least one road accident (compared to 13% of drivers listening to Pop music). In contrast, 42% of drivers listening to Classical music claimed they experienced driving as particularly relaxing, and were the least likely drivers to accelerate beyond the speed limits; only 19% of these claimed to tailgate, and just 34% admitted to becoming aggravated (i.e., 'lose their cool') when on the road.

On the other hand, the 2011 *Quotemehappy* survey also mentioned more positive effects of driving with music (Williamson, 2011). For example, 74% perceived in-car music to 'cheer them up' and 'make the journey more pleasant', while 47% of the respondents claimed that music helped them 'remain awake and alert'. Most specifically, 60% of drivers listening to Rhythm&Blues revealed that they rely on the music to 'relieve the monotony of driving'. Yet, as a music genre, Reggae is deceptive. For example, while 1,148 drivers (56%) reported that they often turn the music off altogether when driving in new surroundings – and perhaps this is an indication that background music does draw on the resources necessary to recruit temporal spatial navigation mapping abilities in unfamiliar terrain – Williamson reported that drivers listening to Reggae music never turned the music off for increased concentration. That is, although drivers who listen to Reggae music may perceive this genre as enhancing alertness, it is precisely these drivers who were significantly more involved in near-crashes. While Reggae music is mostly perceived as a predominantly laid-back relaxing background, perhaps these drivers do not recognize the fact they may simply be overloading their attentional resources as such musics also have as much potency to distract a driver as do any other type of vocal music. *Quotemehappy* released the following statement:

> We'd never base an insurance premium on someone's choice of music – but we are interested in driving behaviour. While in some driving situations, complete silence may be the most sensible option to maintain focus, music can ease the boredom of long journeys and keep people alert. We're not urging rockers to become pop fans but do hope that this research prompts motorists to think about how their choice of car music effects their driving – so that, ultimately, driving is as stress-free and safe as possible.

Nevertheless, a closer look at the *Quotemehappy* findings highlights two criteria, which as Williamson (2011) mentions, could be seen as confounding the data. First, the survey accounts for the fact that younger drivers are statistically more likely to drive with Drum&Bass or Heavy Metal music styles than older drivers, and there is evidence that links younger drivers to aggressive driving. Therefore, perhaps the survey is more of a demonstration about music effects on age. Second, we might recognize that both Drum&Bass and Heavy Metal musics are genres that embrace a performance style based on high technical prowess and an execution that is intensively loud and fast-paced. Certainly, boosted intensity volumes and fast music tempos have been found to divert attention from driving leading to increased risk and accidents (Brodsky, 2002). Therefore, perhaps the survey is more of a demonstration about the effects of musical temporal pace and/or sound reproduction. Yet, it does seem important to keep in mind that both of these variables (i.e., 'age' and 'complexity') are specifically *the* characteristic factors that link both Drum&Bass and Heavy Metal music genres to aggressive driving behaviour to begin with. Thus, while some critics may believe that music-induced driver aggression has little to do with the *music* itself but is rather solely related to the characteristic features of the *exposure*, and other researchers feel that if they could reproduce the sounds of the same tracks in a more quiet tranquil and slower tempo they would ultimately be able to neutralize the *ill-effects* of music aggression, these notions are quite imprudent and senseless. Certainly, the features of a particular music genre are exactly those qualities that weigh on a listener's ear, and these quite definitely include the more culturally accepted habitual forms of exposure (i.e., loudness) that are appropriately attuned for the contexts of listening (i.e., driving a car) for which specific musics are employed.

Finally, in the summer months of 2013, dramatic captions such as 'How Music Impacts Your Driving' splashed across the Internet, televised bulletins,[23] webcasts, and printed newspapers. Countless headlines boldly challenged readers to discover 'What Does Your Music Say About Your Driving'. This investigation was commissioned by Kanetix, a Canadian insurance agent for the online market, who subcontracted the online survey firm *Vision Critical* to recruit over 1,000 participants via a third party forum consisting of a large online sample group (CanadianUnderwriter, 2013; McGee, 2013; MetroNews, 2013; Mulholland, 2013; CNW-Newswire, 2013; Qureshi, 2013). The findings revealed that music preferences of motorists were somewhat linked to their driving behaviours. Accordingly, the genre of music that drivers listen to not only sheds light on their past driving record (retrospectively), but is an indication of their specific driving behaviours (prospectively). Kanetix released a massive PR campaign showering the public with figures describing profiles of drivers who listen to Classic Rock, R&B, House/Dance, and Country genres (Kanetix, 2013a). These consist of dozens of *infographics*, illustrating figures about which music genre causes drivers to commit more speeding violations, receive more tickets for driving under the influence of alcohol, engage in more at-fault accidents, and pay fines more often for dangerous driving (Kanetix, 2013b). Brodsky questioned the reliability of such claims, and found the original data set to be highly convoluted. Working in cooperation with Kanetix and *Vision Critical*, a new data set is herein reported below (referred to as the Kanetix-R [Revised] dataset). First, the original data was gleaned to include only respondents who had a driver's licence; the inclusion criterion of original survey base was simply 'Canadians Aged 18+'. Then, two irrelevant response categories were removed; the original data set included drivers who preferred listening to audio material other than music (i.e., talk radio and audiobooks). It must be pointed out that two lacunas regrettably remain embedded in the Kanetix-R data: the survey procedure allowed for respondents to indicate more than one music style when participants were in the role of either driver or as an accompanying passenger. Therefore: (1) there are actually more total responses (n = 2,026) based on categorical data of preferred music styles than respondents (n = 908); and (2) responses do not differentiate if reported incidents occurred when the respondent served as a driver or as an accompanying passenger. With the above reservations, the findings from the Kanetix-R data set are reported below. Foremost, the prevalence of preferred in-car music genres among Canadians drivers (in order of frequency) is:

23 http://www.youtube.com/watch?v=MRLl80zJOdE.

	Music Genre	%
1	Classic Rock	41
2	Pop/Top40	34
3	Country	27
4	Oldies	27
5	Alternative Rock	22
6	Classical/Instrumental/Easy Listening	14
7	Hip-Hop	14
8	R&B	12
9	House/Dance	10
10	Heavy Metal	10
11	Reggae	7
12	Folk	7

The Kanetix-R (2013) Music Survey was solely designed to target driving behaviour. Overall, 52% of the respondents had never been at fault for committing an accident, and 41% had never received a speeding ticket. But, when looking at those who committed driving violations while listening to their preferred music genres, a rather distinctive picture surfaced (see Table 5.6). Panel-A details the percentage of all At-Fault Accidents (from 'never' to repeated offenses up to four times); Panel-B details the percentage of all Speeding Tickets (from 'never' to repeated offenses beyond four tickets); and Panel-C details the percentage of all aggressive driving manoeuvres, including: Drunk Driving, Stunt Driving, Dangerous Driving, and Careless Driving. Especially when considering the inborn discrepancies within the data set, Brodsky felt it wiser to explore overriding trends than enter in a comparative analysis of variances between the categories (that most likely involve the same drivers in at least two to three categories). Therefore, both an *arithmetic mean* (of all three violation types) and *weighted mean* (comprised of Aggressive Driving manoeuvres doubly weighted than At-Fault Accidents or Speeding violations) were calculated. Table 5.6 presents the music genres spatially in columns (from left to right) as if in a ranked order by level of risk (see Panel-D). It should be noticed that in either case (Mean per cent, or Weighted per cent) the sequence of music genres by *Level of Risk* is almost identical. Namely, the Kanetix-R survey found that drivers behaving in the most dangerous manner listened to Heavy Metal, House/Dance, Reggae, or Hip-Hop genres, while those who drove in the safest manner listened to Folk, Oldies, or Pop/Top40 styles. It can be noticed that somewhere in between were drivers who listen to R&B, Classical/Instrumental, Classic Rock, Country, or Alternative Rock musics. To summarize, the Kanetix-R survey presents various trends of interest, which clearly pointed to aggressive driver behaviours as moderated (at least to some extent) by music genres.

There is one final question that must be raised at this point: Why don't some people who drive with specific types of music that lead to experiencing anger, road rage, near-crash incidents, or even crashes, consider changing their in-car music background? It could be proposed that at least one aspect of the answers is that those who drive while listening to music styles such as Rap, Hip-Hop, Reggae, Heavy Metal, and Drum&Bass are most likely young (male) drivers –

Table 5.6 Kanetix-R music survey (N=908), driving behaviour (%) by music genre

		Metal	House Dance	Reggae	Hip-Hop	R&B	Classical Instrumental	Classic Rock	Country	Alternative Rock	Pop Top-40	Oldies	Folk
		N=89	N=89	N=63	N=130	N=108	N=123	N=368	N=248	N=199	N=303	N=246	N=60
		%	%	%	%	%	%	%	%	%	%	%	%
A	At-Fault Accidents												
	Never	55	58	40	45	46	51	56	57	59	52	58	52
	1 Time	19	22	22	25	21	26	25	20	21	25	24	24
	2 Times	10	9	6	6	6	7	6	6	5	5	7	6
	3 Times	4	2	3	2	2	4	2	3	3	1	3	0
	4 Times	0	0	2	2	4	0	0	1	1	0	0	0
	Total % Accidents	33	33	31	33	29	37	33	29	29	31	34	30
B	Speeding Tickets												
	Never	31	31	30	36	28	42	37	39	38	42	42	53
	1–3 Tickets	55	48	46	43	46	42	45	43	44	44	44	35
	4+ Tickets	14	15	16	13	18	9	13	12	12	10	7	9
	Total % Tickets	69	63	62	56	64	51	58	55	56	54	51	44
C	Charged With...												
	DUI	6	3	5	3	5	3	5	6	2	1	4	3
	Careless Driving	6	5	5	6	3	3	2	3	2	2	2	0
	Stunt Driving	1	0	2	4	1	2	0	0	2	0	0	0
	Dangerous Driving	1	0	0	0	0	0	1	2	0	1	0	0
	Total % Charges	8	5	7	10	4	5	3	5	4	3	2	0
D	Accumulated Incidents												
	Mean % ([A+B+C]/3)	37	34	33	33	32	31	31	30	30	29	29	25
	Level of Risk:	5	4	4	4	4	3	3	3	3	2	2	1
		Very High	High	High	High	High	Moderate	Moderate	Moderate	Moderate	Low	Low	Very Low
	Weighted % ([A+B]/4)+(C/2)	30	27	27	27	25	25	24	24	23	23	22	14
	Level of Risk:	5	4	4	4	3	3	3	3	2	2	2	1
		Very High	High	High	High	Moderate	Moderate	Moderate	Moderate	Low	Low	Low	Very Low

Source: Provided by Kanetix (2013).

and unfortunately these drivers are exactly the group who are the least likely to listen to road safety advice. Nonetheless, another possibility may simply be that traffic safety agencies (and traffic research scientists) have not as yet embraced *music listening* as an important factor for serious concern, and therefore the public is only left with information from media communications and blogs, that report surveys implemented by groups and organizations with far less scientific rigor than required. Hence, every one involved with driving a car – from the young drivers themselves and their parents, to adult motorists, to driving instructors, and traffic law enforcement – have no other sources of hard evidence than the popular news media, which can be seen as less than systematic in their research attempts. In addition, neither fields of Music Psychology or Music Therapy have incorporated the effects of music on driving behaviour as a viable component within their professional agenda. Therefore, it would seem that theoretical and clinical investigations have not yet advanced our scientific knowledge base, nor have there been strides in developing alternative music backgrounds leading to clinical trials of proposed self-mediated driver interventions. Having said this, there is one exception for the above – a platform initiated by Brodsky and colleagues (Brodsky & Kizner, 2012; Brodsky & Slor, 2013) described in Chapter 6.

Chapter 6
Implications, Countermeasures, and Applications: Music Alternatives for Increased Driver Safety

This chapter will first look at efforts to promote safe driving through music choice. Wiser selection of music pieces has been advocated via popular news media, the electronic media, and social media. In addition, the chapter points to volume adjustment and control as a method to adapt to high-demand driving conditions. After considering both as a means for increasing driver safety, a list of applied recommendations are presented. Finally, the chapter describes a research attempt to account for the dynamic temporal flow that is required to achieve functional congruency between the aural conditions of driving and the critical perceptual/motor tasks necessary for safely operating a motorcar; this platform and music programme was developed by Brodsky (Brodsky & Kizner, 2012; Brodsky & Slor, 2013) as a background for driving a vehicle is presented.

6–1. Implications, Applications, and Countermeasures

For years drivers have described their automobiles as surrogate homes. Yet, these 'dwellings' are far more dangerous than the non-mobile houses that most of us live in. Although drivers might perceive the nature of the *journey* to be a rather unpredictable and volatile setting, for the most part they sense the *aural space* of the automobile cabin to be an intimate and safe environment – one that is ultimately controlled by the driver. Hence, from our earliest teenaged years that predate certification as a driver, we seem to have developed into motorcar operators with our mind set on the notion that we can control the journey precisely by controlling the inner environment of the automobile through sound (Bull, 2004). Yet, Lutz and Fernandez (2010) point out that there is more to the issue than simply control:

> When death by car is so common – when our chances of crashing are 1 in 5, and of then becoming permanently disabled are 1 in 83 – it is a wonder why we persist in believing that our cars are safe and that they won't kill us or our loved ones (p. 200).

One explanation is that we believe our cars are safe because of the pleasure we receive from them, and most certainly, among the crucial features that provide driver gratification is the car audio. We definitely have hedonistic desires that are satisfied when we crank up the volume to our most beloved album on a state-of-the-art hi-fi in-cabin surround-sound system, that subsequently washes over our body in an ethereal sonar bath of harmonic frequency spectra and vibrotactile sound waves.

Having said the above but returning to the reality of the issue, for many years drivers have had a basic understanding that they ultimately exhibit symptoms of *visual tunnelling* when engaging in secondary tasks that preoccupy their visual attention. Yet, more recently a host of studies have also demonstrated that drivers become less aware of their environment, and are in fact susceptible to missing much peripheral information, when engaged in cognitive tasks that are usually not considered to demand visual attention. For example, secondary activities that have been seen to affect drivers in a similar fashion are those involving mental processing of information delivered aurally – most specifically when engaged in mobile phone conversation. In this connection, Horrey (2009, 2011; Horrey, Lesch, & Garabet, 2009) have shown that drivers who are engaged in mobile phone tasks tend to fixate within a smaller region of space and neglect peripheral areas such as the side-view and rear-view mirrors. Moreover, both complex conversations, and those containing emotional content, have been seen to place the driver more at-risk mostly due to higher attentional demands that cause increased near-crash incidents (McKnight & McKnight, 1993; Yannis, 2012). The implication here is that reduced scanning often results in missing important road and traffic information, and this data only rise to awareness after a deferred interval. Hence, some cognitive tasks cause delays in reaction times as well as tracking (lane-keeping) performance (Horrey & Wickens, 2006). Furthermore, Victor et al. (Victor, Harbluk, & Engström, 2005; Victor, Engström, & Harbluck, 2009) found that auditory tasks cause drivers to look in a more concentrated manner at the road centre area (at the expense of glances to the road scene periphery), and decrease glances inside the vehicle (causing less inspection of the speedometer and operational gauges). The implication of all of the above is that eye movement behaviours vary and are contingent upon the task type. Most specifically, visual tasks tend to cause drivers to look away from the road, while auditory tasks tend to cause drivers to concentrate their gaze at the centre of the road.

Keeping these concepts in mind, and irrespective of the fact that music has not yet been embraced by traffic researchers therefore limiting the amount of data that can be brought forward, it should be clear that *comparable ill-effects* also emerge when we drive with music playing in the background. For example, a study by Pecher, Lemercier, and Cellier (2009) demonstrated that drivers who listened to music were gradually overwhelmed by emotions from the music heard in the cabin, reminisced about personal emotional events that were prompted by music associations and connotations, and subsequently drove with decreased vehicular control. Therefore, we might envision that one day, legislators will have no other alternative but to address the issue of driving with music in a more serious way. Perhaps, policymakers will demand hard data from driving simulations, on-road experiments, and naturalistic driving investigations all for the purpose of seeking resolutions and remedies. Or maybe politicians will further guidelines that prohibit in-car music listening altogether – just as regulations have already be put in place for curfews on late-night driving among novice drivers, restricting the number of passengers young drivers are allowed to carry, banning hand-held mobile phone use and text messaging, and prohibiting the use of alcohol and recreational drugs.

Nonetheless, unlike other restraints that have surfaced in the 'war against traffic accidents', it is clear that listening to music in the car will never be given up even if restricted or forbidden. Just as cars are here to stay, in-car listening will forever be an integral part of driving that integrates our motorcar experience (Brodsky, 2002; Brodsky & Kizner, 2012; Brodsky & Slor, 2013). No doubt, this premise will stay true regardless of uncompromising empirical findings that might demonstrate undeniable linkage between music and deficient driving, driver aggression, traffic violations, and even fatal accidents. Brodsky's inference accounts for a unique human behaviour that may be considered 'evolutionary' – in the sense that humans have adapted and developed a new means of relating to the environment through music – in such a fashion that sociologists could not have perceived or documented before the millennium. The first to make such a clear observation was Bull (2005) who pointed out that during the 1990s, when hardware for music reproduction were reduced in size to benefit movement, humans advanced a lifestyle in which they consistently and habitually accompanied themselves with music fixed to their ears through headphones coupled to mobile music devices. This, then, established the apparent human demand for music accompaniment as a constituent part of everyday life. Bull's interview data corroborates the fact that by the millennium music accompaniment had become a prerequisite for modern day human experience. Bull maintained that the essence of human experience itself seems to be recollected vis-à-vis the actual music selected for the soundtrack accompaniment of daily activity. That is, links between the internal emotive self as experienced inside the home, with an external image of self as experienced outside in the street, are formulated by joining the spaces in-between with music. Accordingly, it seems to be the use of mobile sound technologies (hereafter referred to as the *iPod*) that mirror how we inhabit the spaces in which we move. We have come to use music not only to simply fill spaces, but, through our personalized music playlists, we actively engage in music as a means to re-structure and manage the spaces we occupy. Music has become the platform for us to navigate and finely tune the physical and metaphysical dimensions of our experiences.

Users often mention feeling calm when they listen to an *iPod* as they walk about (Bull, 2005), and the street is often represented as a mere background having minimal significance. Perhaps, then, the most important utility of music accompaniment is to minimize the informational flow and feedback from the environment; this enables users to focus more clearly on their internal state of being. For example:

> I like to crank angry, loud music at night; the city seems so much more dark and brutal in the dark if I do that. Walking home, I sometimes listen to more soaring, passionate melodies and they make me see things differently (Sorten Muld[1] in particular, has a lot of ancient-sounding hymns which are sort of trance-like, and they make everything seem to be reduced to more elemental things, just metal, wind, clouds and sunlight). I listen to rhythmic and pulsating music sometimes, which makes me feel confident and secure... I don't have to be anything but 'following the beat', so to speak. Sometimes I listen to piano music, and because most of my piano music is kind of depressing/saddening (in a good way), it makes the world seem more fragile and on the verge of collapse. Delerium's music[2] always strikes me in this emotional, soul-searching way, and elevates even the smallest details to some greater significance; every movement of the people in the streets seems spiritual and sacred. Every once in a while I find songs that do this, and they really change everything. Massive Attack's 'Angel' is another one that does that – makes me swell up with this euphoric feeling! (Brian, in Bull, 2005, p. 349).

The truth is that many people do listen to their personal music players when driving a motor vehicle. Some even employ headphones, as they do when walking along the city streets, and even listen to the same playlists in the car as they do elsewhere. One can only wonder if such behaviour is adaptive for driving a car. It is apparent, that had Brian (above) selected the same tracks from Sorten Muld, Delerium, or Massive Attack to accompany himself when driving home, he and all others in the same crosstown traffic would have been at great risk!

6–1.1. Music Selection and Volume Control

Indeed, listening to music while driving a car has both potential benefits and costs. This was eloquently demonstrated by Oblad (2000c; Oblad & Forward, 2000b) who recruited five participants with the intention of investigating the impact of different kinds of musics on driving behaviour. In that study the participants each drove two 50-minute sessions across two consecutive days on a motor track outside Linköping (Sweden). The participants drove alone in an instrumented car while listening to music cassette tapes. The music items heard were chosen from a pre-study interview in which each participant described the particular music genres they liked most, as well as those they despised. For each of the two playlists, Oblad chose tunes considering a wide variation of characteristic stylistic features;

---

1 Sorten Muld (in English meaning Black Soil or Dark Ground) is a Danish folktronica band formed in 1995. Their music features sophisticated rhythms, floating sounds of typical folk instruments (such as bagpipe, flutes, and violins), within an electronica-based platform that supports dramatic lyrics.

2 Delerium is a Canadian musical group formed in 1987. Their musical style is wide ranging from dark ethereal ambient trance, through voiceless industrial soundscapes, to electronic pop music.

some selections were considered music of high arousal (described as fast, chasing, bouncing, vital, angry, exciting, and strong), while others were considered music of low arousal (described as slow, walking, smooth, stiff, tender, calm, and weak). The results indicated that changes in driving performances depended somewhat on the driver's judgement as to whether or not the music heard was according to their preferences, or if they experienced some form of 'unexpected surprise' in the music content. Oblad noted that the participants declared the same tune might have been suitable on the first day, but not necessarily the next. The study demonstrated that driving speed varied with the music pieces being played; the highest speeds were those when a driver's favourite music (genre) was being played. In general, the drivers felt the music influenced their rhythm of driving, but in retrospect, they were not always aware of their own reactions to the music. When recounting the experience after the driving sessions, the participants denoted only a limited perception of how much the music regulated their moods. Hence, Oblad concluded that when actively driving on the road, drivers act on impulses compelled by the music itself: "when a driver liked a tune, the sound level could never be too high, and they could never drive too fast". One of the female drivers mentioned that the ideal background music for drivers must be 'kind' to both the ear and the vehicle. For her, Country music and Pop tunes were safer to drive with, while music for Ballroom Dancing incited her to drive at accelerated speeds.

Regarding the selection of music for driving a car, Williamson (2011) surmises that two main factors need to be considered; these are the intrinsic qualities of the music itself, and the listener's personal experience as well as musical preferences. The former considers music containing lyrics to be potentially more distracting as verbal-based materials interfere more with driving tasks. Further, music containing an increased number of key changes, erratic rhythms, layered textures, and instruments/voices, are far more dangerous simply because complex music requires increased driver attention. Moreover, unfamiliar music is likely to take up more attention, as novel information requires mental processing that depletes cognitive resources. Finally, not liking music can have a negative influence on mood and state of arousal. Williamson highlights the fact that it is also important to consider the driving circumstances when choosing background music for driving. For example: Is the journey on a familiar route? Will the trip be driven under mental or physical stress? How complex is the road itself? How many other cars are on the road? What is the extent of road construction? Does driving occur in amicable weather conditions? Considering all of the above, it is suggested that drivers select music to accompany journeys in advance, as it is best not to rely on whatever comes across radio airwaves.

Finally, we must consider that research studies have found evidence demonstrating that when drivers engage in more emotionally and/or intellectually demanding phone conversations they are twice more at risk (for a review, see Ho & Spence, 2008). That being the case, then we might also consider what happens when drivers engage in particularly emotive listening and/or singing to specific music tracks of a more intimate and personal nature. In this connection, several groups have endorsed playlists contrasting *good* versus *bad* musics for driving, including: *Halfords* (NME, 2011), *The Royal Automobile Club* (RAC, 2004), *ACF Car Finance* (ACF, 2009) and *Confused.com* (Confused, 2013a, 2013b). As pointed out in Chapter 3, all of these are based on driver surveys, musicological analyses (i.e., ranking), and quasi-experimental methods.

6–1.2. Modifying Music Behaviours

Madsen (1987) perceived that when music becomes too demanding and hence distracting, it may be controlled by manipulating the volume level of the sound source. Obviously, as far as drivers are concerned, such strategies can only be interpreted as 'turning down the volume' or 'turning off the music'. Regarding the first proposal, measurements made by Ramsey and Simmons (1993) do indicate that in-cabin driver-adjusted audio output is as high as 83–130dBA. Although, few other studies have ever logged driver-regulated music intensities, one exception is Brodsky and Slor (2013) who described the average reproduction volume of music heard in the test vehicles as 85dBA – with maximal intensities measured well above 100dBA. One interesting find of Brodsky and Slor is that driver-adjusted volumes for preferred music CDs were significantly higher than for an alternative Easy Listening music background; that is, the same drivers significantly reduced the loudness intensity of the softer music to more comfortable levels ($M_{PreferredMusic}$ = 87.5dBs, SD = 4.24; $M_{AlternativeMusic}$ = 82.5dBA, SD = 3.92; $t = 7.01$, $df = 63$, $p < 0.0001$). Ironically, the participants committed fewer violations and drove in a safer style when listening to the alternative music (albeit the volumes differed only by ±5dBA). For a full description of this topic, see Section 6–2 below. Regarding the second proposal, many drivers above 35 years old, and especially those above 50 years of age, report more often driving in silence. Perhaps older drivers do not necessarily wish to drive with music, talk radio, or conversation (Dibben & Williamson, 2007). It would seem that many drivers do perceive the risk of masking auditory signals used to monitor the automobile as a highly dangerous circumstance. Such risks may be greater for older drivers who experience natural age-related losses of hearing acuity (Slawinski & McNeil, 2002).

When drivers are aware of informational processing limitations, then, they should attempt to compensate in some manner. For example, Shinar (2007) asserts that given the driver is aware of the demands of the traffic environment, when engaging in a secondary task (such as a cell phone conversation, navigation system, conversation with a passenger) they should adjust to the added demand by reducing the rate at which driving-related information must be processed. Namely, the driver should slow down as a strategy to decrease the rate of information flow, and subsequently increase the headway to the car ahead as a strategy to increase the time it takes to respond to sudden and unexpected events (such as the stopping or slowing down of the car ahead). Ho and Spence (2008) confirmed these tactics as countermeasures when drivers are otherwise preoccupied by a mobile phone. Similarly, we might ask about listening to music: Should drivers adapt to in-car music listening by deliberately applying actions (as a countermeasure) for diverted attention from driving? Dibben and Williamson (2007) suggest that many drivers indeed report turning down the volume of music in their car when performing a complex manoeuvre (such as reverse parallel parking). Conversely, drivers reported turning up the music when stuck in slow-moving traffic, or on long motorway journeys. Finally, it should be noted that several investigators advocate that drivers should adapt to increased road demand by lowering the intensity of background music (North & Hargreaves, 1999; Wiesenthal, Hennessy, & Toten, 2000), or by turning the music off altogether (Clarke, Dibben, & Pitts, 2010).

6–1.3. Applications and Recommendations

Considering the implications, applications, and countermeasures that have appeared among the social media and in sparse scientific reports, several insurance companies have collated listings as directives. Among these are UK-based *Direct Line Insurance* (Direct Line Insurance, n.d.), *Nationwide Insurance* (Nationwide, 2013), the *Royal Automobile Club* (RAC, 2009), and the *Royal Society for the Prevention of Accidents* (RoSPA, 2007). From all of these we might summarize the following suggested driver behaviours:

- Plan music programming in advance. Get everything ready before driving off. Select favourite CDs and radio stations in advance.
- Get familiar with the car audio controls (off/on, volume, timbre, channel selector). If need be, a passenger in the front seat can adjust audio controls.
- Make sure enough music is available to cover the whole journey so as not to engage in file-searching or 'hunting-down' songs through playlists or on CD covers. Load MP3 players with songs that are appropriate in length and suit the atmosphere for the journey.
- Downsize listening choices. Too many music styles are distracting. It is best to choose one genre, CD, radio station, or playlist of songs. Don't allow passenger requests for songs and playlists.
- Avoid reaching for the controls of a music player once underway. Only adjust audio equipment when safely stopped at the sideway or in a parking lot.
- Loudness can distract. Do not play music at a loud volume. Keep the volume at a reasonable level in order to hear car horns and emergency vehicles.
- Concentrate on driving. Don't sing to tunes or hold a car-aoke microphone. Don't drum on the steering wheel or play air guitar.
- Use technology sensibly. Invest in audio devices and entertainment systems that do not need manual tuning or changing CDs. Purchase equipment that features self-searching channels, disk-changers, USB sockets, *Aux* inputs, and steering wheel mounted volume controllers.
- Keep a simple diary-log of what songs are heard on every trip. This will bring attention to which pieces result in efficient driving performances, as well as which music styles and specific songs are more distracting.

- Buy CDs for the car more carefully. Don't listen to random music.
- Invest in music safety. Attend awareness-training workshops that assist in changing your personal music culture for driving an automobile.[3]

6–2. Developing '*In-Car Music*'

Given the recommendations above, it might seem as if the effects of music can easily be controlled, and that with a little more awareness and insight we should be able to countermeasure all ill-effects of in-car music. Yet, as presented in the previous chapters, there is evidence that drivers routinely select tracks that contain highly energetic fast-paced aggressive musics with accentuated beats, and these are reproduced at elevated volumes. It is rather unfortunate that drivers show little insight about how such music backgrounds influence perception, performance, and control of the vehicle. Truly, the crux of the matter is not the *use* of music per se, but rather the *misuse* of music by selecting exemplars that are highly inappropriate for driving a vehicle. We might inquire why drivers do not re-evaluate their music choices in an attempt to acclimate their CD collections and playlists specifically for driving activity? One answer is that unfortunately, even well-known respected seasoned music psychologists shun the possibility that a specific type of music could curb driver behaviour and/or possibly lead to more efficient safer driving. For example, Clarke, Dibben, and Pitts (2010) outrightly state that "individual differences, and the constantly changing demands of the driving situation, make it impossible to specify with any degree of exactness 'safe' driving music, much as insurance companies and music providers would find it profitable to do so…" (p. 97). Such cynicism aimed at research attempts to investigate the development of music programmes custom-tailored for driver safety is certainly destructive – to say the least.

It should be pointed out that all of the empirical efforts described in this book (with the exception of North & Hargreaves, 1999) have employed renowned familiar recognizable tunes from distinguished celebrated composers previously released on CD albums by famed groups and performing ensembles for prominent commercial record labels. That is, there has not ever been an attempt to compose, record, or mix music specifically for driving activity. Therefore, it would seem warranted that a more proactive approach be put in place to explore alternative musical backgrounds – and such efforts would most certainly be considered prime examples of *action research* as described by Robson (1993). In this connection, music complexity must be viewed as the crucial feature, whereby the notion 'the higher the complexity the greater the cost on attention and arousal' serves as the domineering precept. One such alternative music programme, presented in the

---

3 A new ongoing, as yet unpublished, study by Slor and Brodsky focusses on developing educational workshops to increase awareness towards changing attitudes of young drivers regarding how music affects their driving.

remainder of the chapter, was devised with features that were assumed from the outset to be optimal in the sense that it engages a minimum of driver resources. The development detailed below includes several field tests as pilot studies, and delineates a full clinical trial implemented under the auspices of the Israel National Road Safety Authority.

6–2.1. Music Featuring Optimal Levels of Perceptual Complexity

Consider, for a moment, that the car is a unique listening environment. Inevitably we must account for the dynamic temporal flow that is required for functional congruency between the aural conditions of driving and the critical perceptual-motor tasks necessary for safely operating a car. Certainly, *music complexity* is the crucial feature to be considered here. Music complexity is often portrayed as a bi-polar dimension. High music complexity is the consequence of a faster tempo, lower levels of structural repetition, increased syncopated rhythms, abundant melodic motives, lavish use of harmonic dissonance, dense instrumental layers, and great vocal meaningness. Conversely, low music complexity is the consequences of the opposite component features. Therefore, when bearing in mind levels of arousal, it would seem that musics containing features of a high complexity are stimulative in nature, while musics containing features of a low complexity are experienced as having sedative qualities. Long ago, Madsen (1987) found that the aspects of background music that seem to interfere with primary tasks involve incessant changes in volume, tempo, and instrumentation (including vocal text); these fluctuations demand momentary awareness, disrupting functions of primary tasks including decision making. Therefore, in an effort to provide an aural background with more moderate proportions of perceptual complexity and arousal, Brodsky envisioned a *music template* involving instrumental music with a stylistic character merging several music genres, composed and arranged with music elements and qualities that are well-balanced.

Initially, eight music pieces created by Israeli composer Micha Kizner were selected and re-mastered in a sound studio. Originally, professional recording artists and studio players recorded these items on 2" tape as accompaniment to songs for an admired vocal artist of yesteryear. But the released album was not commercially successful, and therefore the music remained effectively unknown to the public. After removing the vocals, all other tracks were reduced to a 2-track stereo digital format. The final 8-item 40-minute programme is a blend of easy listening, soft rock, light snappy up-beat smooth jazz, with a touch of ethnic world-music flavour. The music items attenuate medium-quality tone frequencies, instrumental ranges, arrangements, voicing textures, tempos, intensities, and rhythmic activity. Most outstandingly, each piece employs lush tonal harmonies sprinkled with sophisticated melodic fragments, but yet, without a specific melody line. Previous research by Brodsky (2002) found that these above characteristics provide a level of perceptual complexity that does not deter mental resources while driving. Hence, from the onset of development, it was assumed that such an aural architecture would furnish a driving environment that maintains alertness and positive mood without diverting cognitive assets. To reiterate: the experimental music background was unknown and hence had no previous memories for the driver to dwell on; there was no vocal content and hence no language-related mental processing ensued; there was no clear melody line to come to the surface, no 'tune' to sing along with, and hence no phonological or *tonological* processes were actively engaged. These characteristic features are indeed unique, and are rarely seen among background musics recommended for drivers.

6–2.1.1. Ecological face validity

The music programme was evaluated for *face-validity* via a listener's survey implemented at a university-wide gala social reception. Prior to the garden party, sets of cards and pencils were placed on 20 pedestal dining tables dispersed throughout the lawns. Given that the setting was an outside open-space environment, music was reproduced at roughly 90dBA (which is a volume loud enough for background music in an open-air celebration while not becoming too intrusive). The sound system was a DJ-quality CD player, 12-channel mixer, and four 2-way 100-watt powered speakers mounted on tripod stands. Out of approximately 60 guests, 22 participated in the survey (37% response rate); they were between 25 and 65 years of age ($M = 45$, $SD = 16$; 50% female). Using the survey cards, each participant judged the 'suitability' of the heard music as a background for five activities of everyday life: Home Chores, Learning, Office Work, Social Reception, and Driving In A Car. Suitability was rated on a 4-level Likert scale (1 = 'Not At All Suitable'; 4 = 'Highly Suitable'). The full music programme was heard twice. As is the case in garden parties, the music programme served as a background for a host of behaviours including talking, laughing, and eating.

The results indicate that the music was considered *highly suitable* for a Social Reception ($M = 3.48$, $SD = 0.75$), *moderately suitable* for both Driving In A Car ($M = 2.68$, $SD = 1.13$) and Home Chores ($M = 2.62$, $SD = 1.20$), but only *slightly suitable* for Learning ($M = 1.60$, $SD = 0.88$) and Office Work ($M = 1.\ 66$, $SD = 0.51$). There were no differences between genders or age subgroups. Although the listeners indicated the music background was more suitable for a reception than for driving ($t = 3.44$, $df = 20$, $p < 0.05$), they also envisioned the music more suitable for driving than for either learning ($t = 3.445$, $df = 19$, $p < 0.05$) or for office work ($t = 2.645$, $df = 20$, $p < 0.05$). These findings seem to point to *priming effects* or a response bias due to the circumstances in which the survey was carried out. However, respondents did perceive the experimental music as more energizing than *Muzak* background for offices or commercially available compilations to enhance cognitive learning. The study found that listeners of both genders across a wide range of ages envisioned the custom-tailored music background as suitable for driving.

6–2.1.2. On the road exploration

Two on-the-road field studies were implemented to explore how drivers responded to the experimental music background in a real-world setting. Field Study-A compared between driver-preferred music (PrefMus) *versus* the experimental music (ExpMus). Field Study-B evaluated the effects of repeated multiple exposure listening to ExpMus while driving in a car.

1. Field Study-A. Twenty-two undergraduates between 24–28 years of age ($M = 26$, $SD = 1.83$; 64% male), with a driver's licence for at least five years, participated in the study. None reported a driving history of traffic-related incidents (collisions, accidents, probation). Of the participants, 95% reported they listen to music *all of the time* when driving; 82% claimed they reproduce in-cabin music at an intensity volume described as *moderately-loud to very-loud*; 72% reported listening to music portrayed as *relatively-fast to extremely-fast*. In total, the participants drove 88 journeys, covering 5,280 kilometres (3,286 miles). The average trip was 45 minutes, across a distance of 60 kilometres (37 miles), at an average speed of 99 kph (61 mph). The study required each participant to drive four trips with another participant who served *as if* a passenger; when the 4-trip cycle was completed both participants switched roles for a second 4-trip cycle. All eight trips were completed within one month during dry weather between 6 a.m. and 12 a.m., using routes involving urban boulevards and highway intercity traffic. Short trips of less than 30 minutes were banned, as were two trips less than six hours apart. The participants drove their own automobile. All vehicles came fitted with a CD-player and a standard 2-pair set of four stereo speakers. It should be noted that each passenger also functioned as a field-monitor to ensure that driving conditions were executed, and countersigned trip-diaries as a confirmation of authenticity. In two trips drivers listened to PrefMus, and in two trips they heard ExpMus background. The order of trips was randomly assigned from one of six possible sequences: *abab*, *baba*, *aabb*, *bbaa*, *abba*, *baab*. Roughly half of the music brought to the study was local (Hebrew-language) Pop/Rock (26%), Hip-Hop (8%), Ethnic/Jewish-Soul (8%), and Reggae (6%) music styles; whereas the other half were international (English-language) Pop/Rock (34%), Reggae/World (9%), Classical (5%), and Soundtracks (4%). The drivers completed a diary-like journey-log for each trip. The data registered in the logbook contained trip information (date, time, duration, distance, estimated speed), a music playlist (performing artist, CD, track title), ratings of driver perceptions (such as: feeling at-ease, being in control of the car, awareness and enjoyment of music, level of distraction), and a profile of mood states.[4]

4 *Profile of Mood States* (McNair, Lorr, & Droppelman, 1971), 32-item brief research version. Positive Affect (PA) is the combined mean total score of 'Friendliness' and 'Vigour' subscales; Negative Affect (NA) is the combined mean score of 'Tension' and 'Fatigue' subscales.

The results indicate no differences between the music types for travel parameters (see Table 6.1). However, significant differences of driver mood surfaced. Positive affect (PA) was significantly higher than negative affect (NA) for both music types (PrefMus: $F_{(1, 21)} = 104.38$, $MSe = 0.2069$, $p < 0.0001$, $\eta_p^2 = 0.83$; ExpMus: $F_{(1, 21)} = 17.52$, $MSe = 0.3066$, $p < 0.001$, $\eta_p^2 = 0.46$), and mood states were more positive and less negative when driving with PrefMus (PA: $F_{(1, 21)} = 14.37$, $MSe = 0.1191$, $p < 0.001$, $\eta_p^2 = 0.41$); NA: $F_{(1, 21)} = 10.17$, $MSe = 0.0998$, $p < 0.01$, $\eta_p^2 = 0.33$). Further, differences for driver perceptions surfaced. Namely, when listening to PrefMus drivers felt more at-ease ($F_{(1,21)} = 78.51$, $MSe = 0.2110$, $p < 0.0001$, $\eta_p^2 = 0.79$), enjoyed the music more ($F_{(1, 21)} = 42.43$, $MSe = 0.5833$, $p < 0.0001$, $\eta_p^2 = 0.67$), and rated their awareness as significantly higher ($F_{(1, 21)} = 4.40$, $MSe = 0.2903$, $p < 0.05$, $\eta_p^2 = 0.17$).

To summarize, Field Study-A provided compelling *prima facie evidence* in a first effort to demonstrate criterion related validity between the drivers' own favourite driving tracks and the experimental music background. That is, while the drivers were clearly more passionate about the music they brought from home, no other meaningful differences between the two music types surfaced. Such findings corroborate previous studies (Hargreaves, 1987–88; North & Hargreaves, 1995; Parncutt & Marin, 2006; Schubert, 2007) documenting that the more familiar a listener is with music, the more distinct and intense will be

Table 6.1 Field study-A: Outcome by music type

	Preferred Music		Experimental Music	
	M	*(SD)*	*M*	*(SD)*
Time (min)	48	(14.4)	45	(14.3)
Distance (km)	65	(37.1)	57	(27.7)
Speed (kph)	99	(12.3)	99	(11.0)
Perceived Control over Car	3.93	(0.18)	3.73	(0.48)
Music-Generated Distraction	1.75	(0.67)	1.91	(0.73)
Music Effects on Feeling At-Ease	3.68	(0.42)	2.45	(0.65)
Music Effects on Driving	1.83	(0.71)	1.80	(0.68)
Enjoyment of Music	3.66	(0.63)	2.16	(0.81)
Awareness of Music	3.41	(0.65)	3.07	(0.62)
Attention to Music Elements	2.82	(0.81)	2.66	(0.73)
Positive Affect	3.00	(0.41)	2.61	(0.42)
Friendly	3.07	(0.44)	2.81	(0.42)
Vigour	2.93	(0.43)	2.48	(0.42)
Negative Affect	1.60	(0.37)	1.91	(0.48)
Tension	1.55	(0.37)	1.76	(0.48)
Fatigue	1.61	(0.43)	1.96	(0.56)

Source: Adapted from Brodsky and Kizner (2012).

their experience, subjective emotion, and level of liking for the music. Nonetheless, it is paramount to recognize the fact that drivers' levels of awareness to the aural environment were significantly higher when driving with PrefMus. Such a finding implies increased conscious attentiveness to the music that consequently denotes the allocation of greater cognitive resources.

2. Field Study-B. Thirty-one undergraduates between 24–28 years of age ($M = 26$, $SD = 2.07$; 65% female), with a valid licence for at least five years, participated in the study. None reported a driving history of traffic-related incidents. Of the participants, 94% reported they listen to music *all of the time* when driving; 88% claimed they reproduce in-cabin music at an intensity volume described as *moderately-loud to very-loud*; 93% reported listening to music portrayed as *relatively-fast*. The overriding question of Field Study-B was *familiarity*. Namely, would repeated listening experiences of ExpMus maintain the drivers' positive mood state (as seen in Field Study-A), or as often happens when the same repeated event occurs frequently, would drivers become more irritable and fatigued with a perceived decrease in controlling the vehicle? In total, the participants drove 310 journeys, covering 17,799 kilometres (11,060 miles). The average trip was 53 minutes, across a distance of 55 kilometres (34 miles), at a speed of 92 kph (57 mph). The participants were required to drive ten trips without accompanying passengers while listening to ExpMus. No other aural stimuli (such as driver preferred musics, lectures, audio books, or talk radio) were permitted. All ten trips were completed within one month during dry weather. As driving was expected to be *naturalistic* as occurs in everyday ordinary life, trips were allowed for all drive times (morning, afternoon, night) and road types (residential, boulevard, intercity highway). Short trips of less than 30 minutes were banned, as were two trips less than six hours apart. The participants drove their own automobile; all came fitted with a CD-player and a standard two-pair set of four stereo speakers. A journey-log similar to the one used in Field Study-A was allocated for each driver (with the exception that no playlist data was necessary). Although the participants were directed to drive alone, accompanying passengers were reported for 66 trips (21%); as a result of this unexpected methodological violation, the presence of passengers was entered as an independent variable for this subgroup.

The results indicate that the participants expressed only *moderate* levels of enjoyment ($M = 2.28$, $SD = 0.61$) and positive mood when driving with ExpMus; albeit PA remained higher than NA throughout ($M_{PA} = 2.68$, $SD = 0.55$; $M_{NA} = 1.63$, $SD = 0.32$ 56; $t = 7.731$, $df = 30$, $p < 0.0001$). Driver enjoyment of the music was associated to positive affect ($r = .50$, $p < 0.05$). Further, the participants rated their awareness of the music playing in the background as *moderate* ($M = 2.85$, $SD = 0.54$), and felt an overall *high* level of caution ($M = 3.69$, $SD = 0.31$). While there were no effects of repeated exposure for trip parameters or driver perceptions, effects for music awareness and positive mood surfaced ($F_{(9, 270)} = 4.4134$, $MSe = 0.4633$, $p < 0.0001$, $\eta^2_p = 0.13$; $F_{(9, 270)} = 2.7823$, $MSe = 0.1940$, $p < 0.01$, $\eta^2_p = 0.08$) (see Figures 6.1 and 6.2). As can be seen in the figures, across the ten trips drivers felt less aware of the music, but yet rated PA in a slightly escalating fashion. It is of interest to note that no significant differences of drive-time or road-type surfaced (see Tables 6.2 and 6.3).

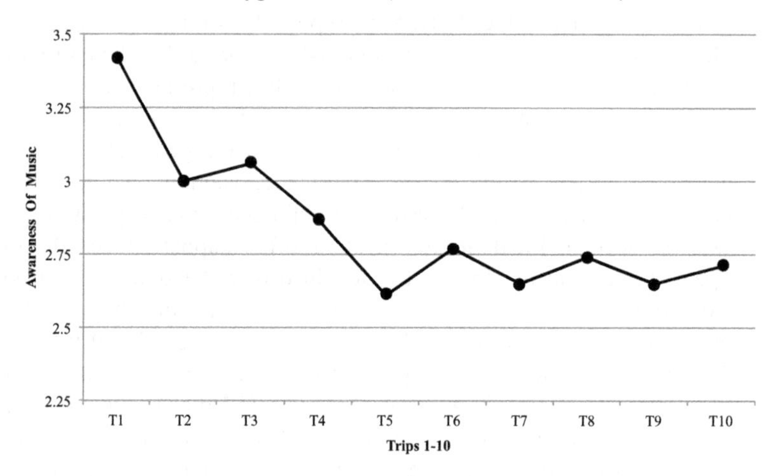

Figure 6.1 Field study-B: Repeated exposure of experimental music on 'awareness of music'

Source: Adapted from Brodsky and Kizner (2012).

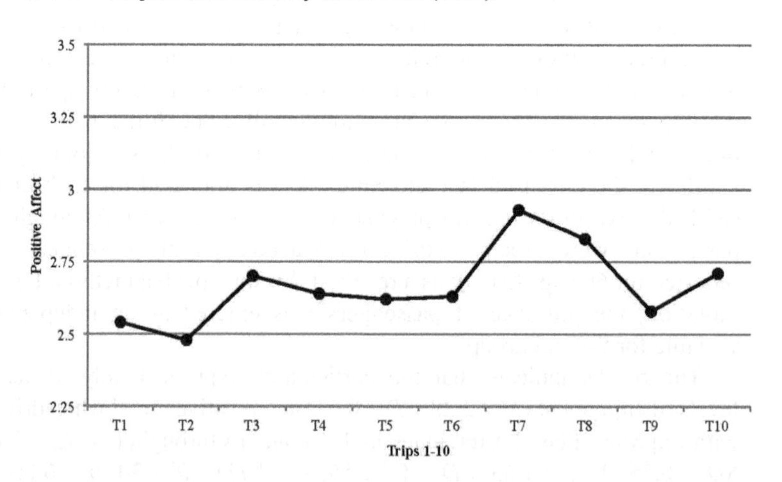

Figure 6.2 Field study-B: Repeated exposure of experimental music on 'positive affect'

Source: Adapted from Brodsky and Kizner (2012).

Table 6.2 Field Study-B: Outcome by drive-time

	Morning 06:00–12:59		Noon 13:00–18:59		Night 19:0–02:00	
	M	*(SD)*	*M*	*(SD)*	*M*	*(SD)*
Time (min)	53	(35.4)	52	(24.4)	56	(25.1)
Distance (km)	51	(33.5)	54	(31.5)	60	(28.7)
Speed (kph)	90	(15.0)	93	(14.5)	95	(12.0)
Perceived Caution	3.76	(0.37)	3.62	(0.44)	3.72	(0.37)
Enjoyment of Music	2.42	(0.70)	2.21	(0.78)	2.19	(0.70)
Awareness of Music	3.03	(0.56)	2.82	(0.65)	2.83	(0.76)
Positive Affect	2.66	(0.61)	2.67	(0.55)	2.66	(0.60)
Friendly	2.76	(0.64)	2.75	(0.61)	2.76	(0.70)
Vigour	2.57	(0.61)	2.59	(0.54)	2.57	(0.56)
Negative Affect	1.61	(0.42)	1.64	(0.35)	1.70	(0.39)
Tension	1.60	(0.44)	1.54	(0.34)	1.61	(0.45)
Fatigue	1.62	(0.45)	1.75	(0.45)	1.78	(0.44)
NUMBER OF CASES	28		31		30	

Source: Adapted from Brodsky and Kizner (2012).

Table 6.3 Field Study-B: Outcome by road type

	Local Neighbourhood Residential Driving		Boulevard Intra-City Driving		High-Way Inter-City Driving	
	M	*(SD)*	*M*	*(SD)*	*M*	*(SD)*
Time (min)	31	(14.1)	35	(10.3)	60	(30.6)
Distance (km)	22	(19.1)	22	(8.24)	66	(35.4)
Speed (kph)	58	(22.3)	67	(15.9)	100	(8.89)
Perceived Caution	3.67	(0.47)	3.67	(0.54)	3.69	(0.32)
Enjoyment of Music	2.60	(0.83)	2.30	(0.74)	2.23	(0.61)
Awareness of Music	2.67	(1.13)	2.84	(0.88)	2.79	(0.60)
Positive Affect	2.84	(0.57)	2.71	(0.54)	2.68	(0.58)
Friendly	2.98	(0.61)	2.74	(0.58)	2.78	(0.66)
Vigour	2.70	(0.64)	2.68	(0.54)	2.59	(0.55)
Negative Affect	1.45	(0.26)	1.46	(0.28)	1.64	(0.33)
Tension	1.42	(0.31)	1.45	(0.28)	1.56	(0.34)
Fatigue	1.47	(0.23)	1.46	(0.34)	1.73	(0.42)
NUMBER OF CASES	8		19		31	

Source: Adapted from Brodsky and Kizner (2012).

Nevertheless, among the twelve participants who violated the protocol and carried passengers, a comparison between trips with accompanying passengers *versus* driving-alone indicated two valuable issues. First, there were significant increases for speed estimations ($F_{(1, 11)} = 6.5390$, $MSe = 73.238$, $p < 0.05$, $\eta_p^2 = 0.37$), and increases in positive affect ($F_{(1, 11)} = 11.887$, $MSe = 0.0150$, $p < 0.01$, $\eta_p^2 = 0.52$) (see Table 6.4). Taken together, these findings indicate that drivers

Table 6.4 Field Study-B: Outcome by presence of passengers

	Driving Alone				Driving With Passengers	
	M	*(SD)*	*M*	*(SD)*	*M*	*(SD)*
Time (min)	49	(19.6)	46	(28.8)	63	(48.0)
Distance (km)	51	(25.0)	39	(33.0)	62	(52.0)
Speed (kph)	91	(12.5)	81	(12.7)	89	(16.1)
Perceived Caution	3.62	(0.40)	3.66	(0.39)	3.75	(0.23)
Enjoyment of Music	2.31	(0.62)	2.54	(0.68)	2.33	(0.78)
Awareness of Music	2.97	(0.57)	3.10	(0.54)	2.59	(0.72)
Positive Affect	2.63	(0.55)	2.64	(0.52)	2.87	(0.56)
Friendly	2.72	(0.60)	2.78	(0.52)	2.98	(0.60)
Vigour	2.55	(0.52)	2.67	(0.56)	2.76	(0.54)
Negative Affect	1.63	(0.31)	1.55	(0.24)	1.62	(0.30)
Tension	1.57	(0.32)	1.58	(0.30)	1.58	(0.39)
Fatigue	1.69	(0.38)	1.52	(0.24)	1.65	(0.32)
NUMBER OF CASES	31		12		12	

Source: Adapted from Brodsky and Kizner (2012).

perceived the trips with passengers to have been all the more friendly and energetic than when driving alone, and that they drove at faster cruising speeds. Second, as can be seen in Table 6.4, the drivers were significantly less aware of the music playing in the background with passengers ($F_{(1, 11)} = 8.4717$, $MSe = 0.1830$, $p < 0.025$, $\eta_p^2 = 0.44$). This latter result is extremely important given that the literature points to research findings in which the presence of passengers increased the risk of accidents two-fold (described in Chapter 4, section 4–2.4). Therefore, perhaps such hazards can be somewhat offset in a form of a self-mediated intervention – like listening to an alternative music background.

6–3. Authenticating '*In-Car Music*'

Considering the findings of the above two field studies, more than anything else we might view such results as a validation of the structural template employed in designing the experimental music background. That is, when people drove with the experimental music background, their positive affect and perceived level of driver caution remained stable throughout, while their music-associated cognitive attentive processes decreased. The truth is that the concept of providing self-mediated methods with which drivers could strategically modify their behaviour in an attempt to maintain their driving performance is not new. For example, ten years ago Oron-Gilad (2003; Oron-Gilad, Ronen, & Shinar, 2008) developed a model by which drivers could recruit tasks as the means for adaptive change to undesirable driving circumstances (see Figure 6.3). Accordingly, drivers can increase situational demands when they find themselves beneath the optimum, as well as decrease situational demands when they find themselves above and

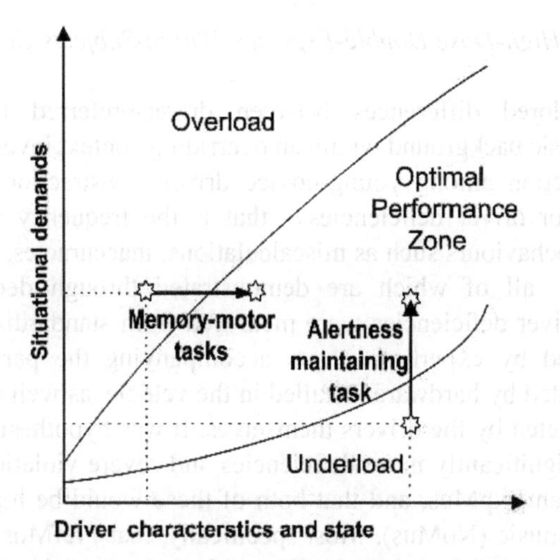

Figure 6.3 Application of countermeasures for underload and overload driving conditions

Source: Oren-Gilad (2003). Reprinted with permission of the author.

beyond. Oron-Gilad et al. suggest two forms of undertaking: memory-motor tasks to decrease demands, and alertness-maintaining tasks to increase demands. In either case, the notion is that these actions would contribute to a more appropriate driver fitness, thus enabling them to achieve vehicular performances in an optimal zone.

Having said the above, and refocussing the discussion on background music, such notions have already been conceptualized within the form of the Affective Music Player (see Figure 5.4). As described previously in Section 5–2.1.1, when the situational demands reach levels outside an optimal zone, and especially if the background of music consists of inappropriate features, there would be an immediate need to reduce/increase the dynamism by introducing music of an alternate complexity. Namely, AMP selects pieces based on the greatest probability of modifying the listener's energy to a more functional level. To be fair, Oron-Gilad et al. did explore the application of music as a possible task to increase low arousal and fatigue from driving underload; however these attempts were not effective. Perhaps there is more utility in applying music designed with an aural architecture that from the *onset* targets the most optimal performance zone rather than wait and only introduce pieces to transform driver behaviour to more functional levels when needed. With this in mind, it seems prudent to undertake more precisely controlled explorations with larger samples of participants driving on the road while listening to the experimental music specifically developed for driving a car. Such investigations should employ in-vehicle data recorders (IVDRs) that can more objectively measure the effects of driving with music, and therefore provide an autonomous means of quantifying vehicular performances variegated by music conditions.

6–3.1. On-Road High-Dose Double-Exposure Within-Subjects Clinical-Trial

The study explored differences between driver-preferred music and the experimental music background within an overriding context investigating music-generated distraction among young-novice drivers. Distraction was measured by accounting for driver deficiencies – that is the frequency and severity of deficient driver behaviours such as miscalculations, inaccuracies, aggressiveness, and violations – all of which are demonstrated through decreased vehicle performance. Driver deficiencies were measured from standardized behavioural assessments rated by expert observers accompanying the participant-drivers, from data generated by hardware installed in the vehicle, as well as from journey log-books completed by the drivers themselves. It was hypothesized that drivers would commit significantly more deficiencies and severe violations during trips with PrefMus than ExpMus, and that both of these would be higher than when driving without music (NoMus). Most specifically, that PrefMus would escalate risks for distraction leading to more traffic accidents compared to NoMus, and that ExpMus would cut-down risks leading to distraction and traffic accidents therefore increasing driver safety. To reiterate, employing driver-preferred music brought from home is a strategy that has been shown to increase *ecological validity* as music stimuli are based on familiarity of the music tracks the participants usually listen to when driving everyday (Unal et al., 2012).

In the 16-month clinical trial, 85 young novice drivers were recruited through local high schools and social media from a 650 square-kilometre (403 square-mile) residential area. The participants were on average 17.5 years old ($SD = 0.41$, 58% male) with a valid licence for at least seven months ($SD = 2.64$). None of the young drivers had been prosecuted in a traffic court, and 92% claimed that they had never been in a collision. Of the participants, 86% reported to listen to music *all the time* while driving; 94% claimed they reproduce in-cabin music at an intensity volume described as *moderately-loud to loud*; 99% reported listening to music portrayed as *moderately-fast to very-fast*. The study employed two driving instructors as on-site clinical monitors and expert observers; each one was allocated to half of the sample. The instructors were both males in their early 60s with roughly 39 years certification. Each instructor brought his own automatic-gear Learner's Vehicle with double accelerator and braking pedals to the study. In each automobile, an identical in-vehicle data recorder was installed and calibrated by an automotive engineer,[5] and both cars came with a manufacturer-fitted built-in dashboard-mounted CD player and two pairs of stereo speakers. Subsequent analyses found no meaningful differences between the two driving instructors, or for the two vehicles. The clinical

5 The study used a *Traffilog* in-vehicle data recorder connected directly to the controller area network (CANBUS) and on-board diagnostics. The IVDR measured applied g-force during driving (quantifying acceleration). Proprietary software calculated g-sensor triggered events, and analysed the online stream (at a sampling rate of 100 milliseconds) into meaningful manoeuvres. Subsequently, scores were codified as driving profiles.

trial required each driver to complete six trips within a 2-week period, in the same learner's vehicle, with the same accompanying driving instructor. The trial implemented two trips, each of three driving conditions in a randomized order: PrefMus, ExpMus, and NoMus (for empirical control). In total there were 510 trips, covering 20,400 kilometres (12,676 miles), with an average trip covering 39 kilometres (25 miles) during 42 minutes, for a total 357 hours of on-road driving (for the entire sample). It is important to note here that driving with music consisted of a *high-dose* 40-minute exposure per trial; this level of saturation is 10-times more potent than the maximum 4-minute exposure employed in many studies reported in the literature (described in Chapter 5).

The driving instructors independently scheduled each session with the participants. For the most part, the participants came to the first session with several CDs in hand and a written playlist; the CDs remained in the learner's vehicles throughout all six trips. On average, each participant brought 12 pieces to the study (that is, a total of 1,035 music tracks); the majority (70%) consisted of international tracks. A post-study musicological analysis indicated that six overriding music genres accounted for more than 90% of the music heard by the participants (see Figure 6.4). As can be seen in the figure, these were: Dance/Trance/House, Hebrew-Middle Eastern, Rock/Hard-Rock/Punk, Pop, Hebrew-Pop, and Hip-Hop/Rap music styles.[6] Before each session, the driving instructors registered trip details in a logbook (date, time, starting kilometre), after which participants inserted their first CD into the disk-player; they independently chose track numbers, switched CDs, and adjusted intensity volumes. All six trips were structured in a similar fashion: (1) From Home to Highway (10 minutes) via limited-access residential roads to the closest divided boulevard leading to a highway; (2) Highway Driving (15 minutes

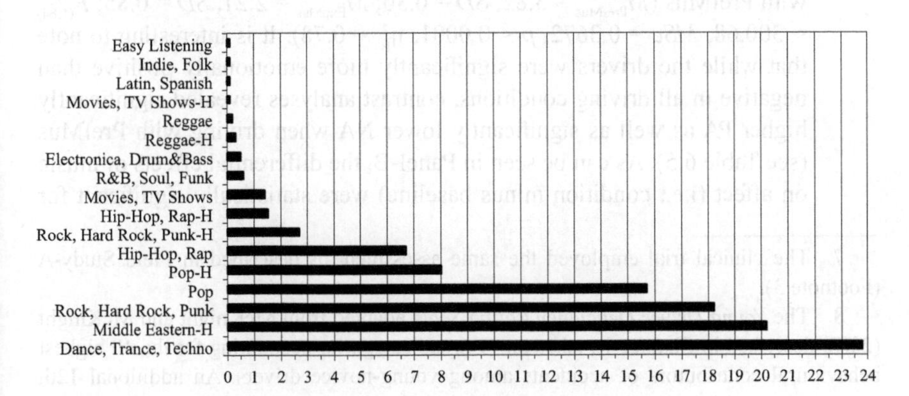

Figure 6.4 Music genres of young novice drivers

Note: H = Hebrew language lyrics.

Source: Adapted from Brodsky and Slor (2013).

6 A list of the 20 most popular International English-language tunes can be seen in Section 3–1.2.

in each direction, total 30 minutes) along an urban interstate freeway with 2- or 3-lane traffic in each direction; and (3) From Highway to Home (10 minutes) via a divided boulevard leading to limited-access residential roads returning home. Upon conclusion, each participant completed a post-trip questionnaire including an evaluation of their mood state,[7] as well as an assessment of their self-perceived awareness and enjoyment of the music. After each participant left the vehicle, the accompanying driving instructor registered trip details (time, ending kilometre), wrote a brief text commentary describing the overall trip circumstances and driver behaviour, and then rated the driving performance on a standardized assessment.[8] All observed driver deficiencies per trip were later summed, and the weighted total scores emulated a *trip-risk profile* reflecting a ranking in which the highest scores were coded as the most deficient at-risk driving.

The results of the clinical trial relate to four overriding areas: The Driving Experience, Observed Driver Behaviour, Mechanical Events (as recorded from the automobile), and Driver Profiles.

1. *The Driving Experience.* For the most part, the drivers reported a high-level of concurrence between the empirical trips and their everyday driving ($M = 3.40$, $SD = 0.58$). Namely, the music brought to the study was *highly-similar* ($M = 3.82$, $SD = 0.47$) to what they usually listen to in their own vehicle, and the intensity volumes they adjusted during the study were *quite-similar* ($M = 3.40$, $SD = 0.58$) to how they usually hear music in their own vehicle ($M = 85.5$dBls, $SD = 4.04$). In general, the drivers reported to be significantly more aware of the music background when driving with PrefMus ($M_{PrefMus} = 3.65$, $SD = 0.44$; $M_{ExpMus} = 2.89$, $SD = 0.66$; $F_{(1,84)} = 107.33$, $MSe = 0.2280$, $p < 0.0001$, $\eta_p^2 = 0.56$), and enjoyed driving more with PrefMus ($M_{PrefMus} = 3.82$, $SD = 0.30$; $M_{ExpMus} = 2.21$, $SD = 0.85$; $F_{(1,84)} = 300.68$, $MSe = 0.3672$, $p < 0.0001$, $\eta_p^2 = 0.78$). It is interesting to note that while the drivers were significantly more emotionally positive than negative in all driving conditions, contrast analyses revealed significantly higher PA as well as significantly lower NA when driving with PrefMus (see Table 6.5). As can be seen in Panel-B, the differential effects of music on affect (i.e., condition minus baseline) were statistically significant for

7 The clinical trial employed the same assessment as described in Field Study-A (Footnote 3).

8 The *Young Driver Deficiency Rating Scale* adapted from McKnight and McKnight (2003, described in Chapter 4), takes on 11 broad categories accounting for the 40 highest behavioural contributors of accidents among young-novice drivers. An additional 12th category was added to account for manoeuvres that had to be executed by the driving instructor to avoid an impending accident. All 40 behaviours were rated on 4-point scale (1 = Least Severe, 4 = Most Severe), and supplementary scores were added per frequency of implemented verbal warnings (x 5) and physical interventions (x 6). A square-number sequence was used as a weighting method to transform event frequency to event severity; numerical frequencies (1, 2, 3, 4, etc.) became value strengths (1, 4, 9, 16, etc.). Reliability of both 11-item and 12-item versions was seen to be acceptable (α_{11}=0.70; α_{12}=0.71).

Table 6.5 Clinical trial: Mood states

	Positive Affect *M*	*(SD)*	Negative Affect *M*	*(SD)*	*p*
A. Driving Condition					
I. No Music	2.78	(0.59)	1.62	(0.39)	< 0.0001
II. Preferred Music	3.27	(0.44)	1.38	(0.20)	< 0.0001
III. Experimental Music	2.85	(0.52)	1.62	(0.39)	< 0.0001
B. Effects of Music					
II. Preferred Music	0.49	(0.55)	-0.23	(0.41)	< 0.0001
III. Experimental Music	0.07	(0.49)	0.01	(0.42)	0.482

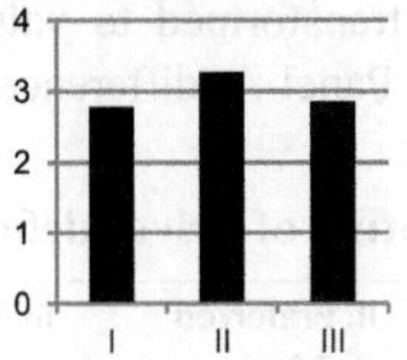

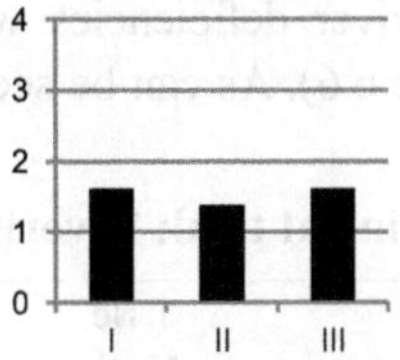

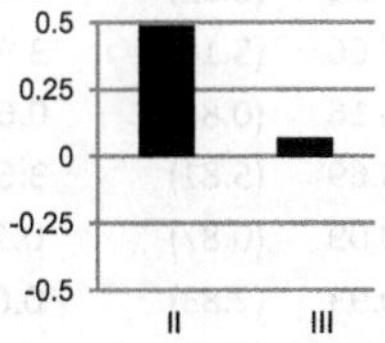

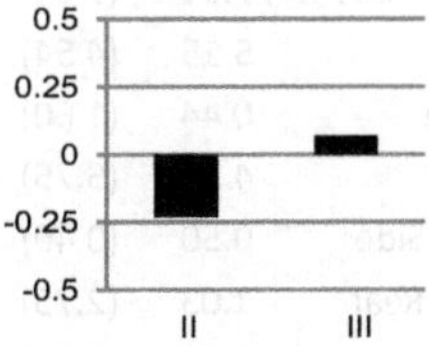

Source: Adapted from Brodsky and Slor (2013).

trips with PrefMus. That is, PrefMus was accountable for both increased PA and decreased NA, whereas the experimental music seems to have had little effect on mood at all. Finally, the study found that all drivers perceived a high level of caution throughout, regardless of the driving condition.

2. *Observed Driver Behaviour*. Out of the total 510 trips, the accompanying driving instructors considered only 61 (12%) as totally efficient without violations. This level of incidence indicates an average three driver deficiencies in at least 1-out-of-6 trips for all participants. The most frequently observed violations were deficiencies of speed (20%), maintaining attention (14%), following distance (11%), lane use (11%), hands on the steering wheel (10%), lateral control (5%), and overtaking vehicles (2%). Together, these seven categories account for 73% of all driver deficiencies in the study. It is interesting to note that six out of these seven deficiencies have also been listed by Braitman et al. (2008) as characteristic deviations of young novice drivers. Further, 32% of the drivers committed at least one violation that required an emergent warning from the accompanying driving instructor, while 20% (i.e., one-in-five) committed at least one violation that required an actual physical intervention (steering wheel alignment or braking manoeuvre) in order to avoid an imminent crash. In general, 90% of the drivers committed

at least one violation when driving with ExpMus, 92% committed at least one violation when driving without music, and 98% committed at least one violation when driving with PrefMus. Nonetheless, *how frequent* a driver commits violations is far less as important as *how severe* are the violations they commit. Certainly we all go about our business everyday committing a wide range of driver deficiencies with little concern, and yet the unfortunate truth is that it only takes one brutal event to cause us to be in harm's way. Hence, driver deficiencies were transformed to values of event severity (see Table 6.6). As can be seen in Panel-A, differences between the driving

Table 6.6 Clinical trial: Severity rating of driver deficiencies

	I. No Music		II. Preferred Music		III. Experimental Music		
	M	*(SD)*	*M*	*(SD)*	*M*	*(SD)*	*p*
A. Deficiencies							
1. Vehicle control	2.99	(4.62)	4.60	(5.58)	3.44	(4.71)	< 0.05
2. Traffic Controls Use	7.91	(7.08)	7.23	(6.02)	5.82	(5.95)	< 0.05
3. Attention	5.15	(4.54)	6.66	(5.14)	3.94	(4.17)	< 0.001
4. Driver Fatigue	0.44	(1.60)	0.16	(0.84)	0.65	(2.03)	
5. Search Ahead	4.20	(5.75)	5.69	(5.81)	3.62	(4.62)	< 0.05
6. Search to the Side	0.50	(0.49)	0.09	(0.87)	0.19	(1.22)	
7. Search to the Rear	1.03	(2.75)	0.93	(2.83)	0.65	(2.08)	
8. Adjusting Speed	7.91	(5.67)	8.82	(5.44)	6.96	(5.50)	< 0.01
9. Maintaining Space	2.34	(5.05)	2.05	(4.06)	1.74	(3.11)	
10. Signals	1.01	(3.23)	0.92	(2.56)	0.48	(1.82)	
11. Emergencies	4.89	(6.96)	5.76	(7.00)	4.65	(6.50)	< 0.01
12. Interventions	12.76	(15.3)	10.92	(13.6)	8.29	(12.7)	0.06
B. Total Score	50.68	(37.3)	53.89	(31.4)	40.42	(25.8)	< 0.001
C. Effect of Music			3.21	(34.3)	-49.13	(37.2)	< 0.001

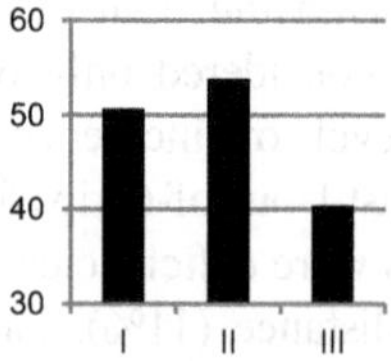

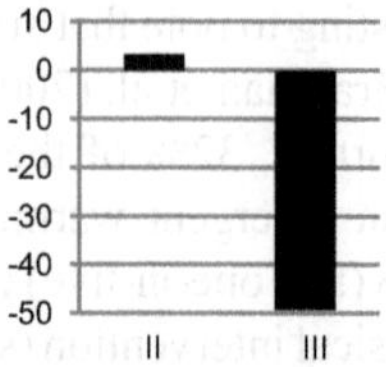

Source: Adapted from Brodsky and Slor (2013).

conditions surfaced for 6-out-of-11 categories, and these deficiencies were most severe when driving with PrefMus. Panel-B indicates that when all values for deficiency severity were summated into an overall total score, the trips that were significantly more efficient – and hence safer – were when drivers listened to ExpMus. Finally, Panel-C shows that the differential effects of music on event severity were statistically significant. Namely, the alternative experimental background significantly decreased the severity of driving deficiencies as rated by highly experienced expert observers.

3. *Mechanical Events Recorded from the Automobile.* The IVDR continuously logged 27 event-behaviours from five independent event-categories and the combinations thereof, including: speeding, acceleration, breaking, turning, and motor revs (i.e., engine RPMs). These parameters and associated g-forces have been employed in studies such as Simons-Morton et al. (2011) who assessed elevated g-force events, including longitudinal deceleration/hard-braking ($\leq$ -0.45g), longitudinal acceleration/rapid-starts ($\geq$ 0.35g), hard left/right turns ($\leq$ -0.50g / $\geq$ 0.50g) and yaw (± 6° within 3 seconds). In a similar fashion, the clinical trial calculated specific potencies, and tabulated total g-force event thresholds per driving condition (see Table 6.7). Panel-A denotes that both the *rate of incidence* and the *severity of events* were significantly higher with PrefMus. Then, Panel-B indicates that the differential effects of music on event-frequency and event-severity were statistically significant. Contrast analysis revealed that PrefMus caused significant increases in both rate and severity of deficient events. This finding strengthens the observational data (above) indicating that music effects of ExpMus were unlike those of PrefMus, and in direct contrast, actually increased driver safety.
4. *Driver Profiles.* Most industrial manufacturers of in-vehicle data recorders have invested in application features that engage in some type of categorical ranking of drivers. In general, these profiles are referred to as *Driver-DNA*. In a similar fashion the clinical trial employed event-severity scores (0–350) to designate participants into one-of-three overriding driver types: Cautious ($\leq$ 49), Moderate (50 – 249), and Aggressive ($\geq$ 250) (see Table 6.8). As can be seen in Panel-A, although the majority (64%) were designated as Moderate drivers, there were still a substantial number designated as Cautious (16%) and Aggressive (20%) drivers. But, then when removing the Moderate driver sub-group from the analyses, important proportional differences between Cautious and Aggressive drivers surfaced. Panel-B clearly shows a *between-condition* analysis for proportional ratios between Cautious *versus* Aggressive driver sub-groups; these results indicate near meaningful differences ($z_{I\text{-}II} = 1.61$, $p = 0.054$; $z_{II\text{-}III} = 1.39$, $p = 0.082$). However, a *within-condition* analysis examining the same ratios found statistically significant differences for participants driving with PrefMus

($x^2_{(1)} = 4.50$, $p < 0.05$) whereas no differences surfaced for ExpMus and NoMus conditions. That is, driving with PrefMus caused a proportion of Cautious drivers to behave in a more reckless manner, with far less control over their vehicle, and commit more deficient driving behaviour; as a result these participants were reclassified as Aggressive drivers.

Table 6.7 Clinical trial: In-vehicle data recorder

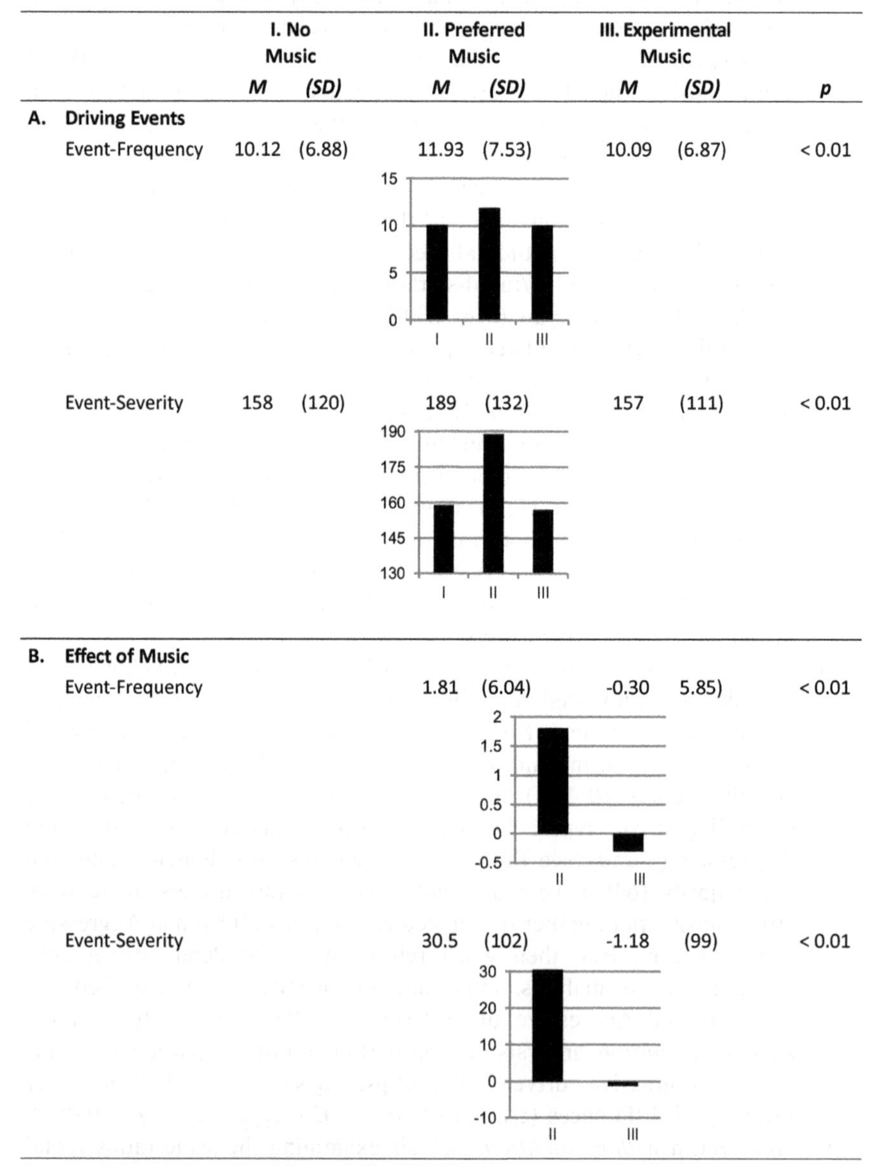

	I. No Music		II. Preferred Music		III. Experimental Music		
	M	*(SD)*	*M*	*(SD)*	*M*	*(SD)*	*p*
A. Driving Events							
Event-Frequency	10.12	(6.88)	11.93	(7.53)	10.09	(6.87)	< 0.01
Event-Severity	158	(120)	189	(132)	157	(111)	< 0.01
B. Effect of Music							
Event-Frequency			1.81	(6.04)	-0.30	5.85)	< 0.01
Event-Severity			30.5	(102)	-1.18	(99)	< 0.01

Source: Adapted from Brodsky and Slor (2013).

Table 6.8 Clinical trial: IVDR Driver-DNA

		I. No Music		II. Preferred Music		III. Experimental Music	
		N	*%*	*N*	*%*	*N*	*%*
A. Driver Profiles (Full)							
Cautious	Solid Grey	14	17	10	12	15	18
Moderate	Broken Grey	58	68	53	62	54	63
Aggressive	Solid Black	13	15	22	26	16	19
B. Driver Profiles (Condensed)							
Cautious	Solid Grey	14	52	10	31	15	48
Aggressive	Solid Black	13	48	22	69	16	52

Source: Adapted from Brodsky And Slor (2013).

Finally, the clinical trial explored possible individual differences based on gender (male/female) and temperament (Impulsivity/Sensation-Seeking [*Imp-SS*]). These are important because studies (for example, Cestac, Paran, & Delhomme, 2010) have shown that risk-taking motivations evolve with gender and perceived behavioural control.

1. *Differences of Gender.* Foremost, female drivers reported higher levels of perceived driver caution (NoMus: $M_{\text{Female}} = 3.90$, $SD = 0.29$; $M_{\text{Male}} = 3.66$, $SD = 0.50$; $F_{(1,\,83)} = 6.97$, $MSe = 0.170$, $p < 01$, $\eta_p^2 = 0.07$; PrefMus: $M_{\text{Female}} = 3.87$, $SD = 0.25$; $M_{\text{Male}} = 3.67$, $SD = 0.42$; $F_{(1,\,83)} = 6.17$, $MSe = 0.129$, $p < 0.05$, $\eta_p^2 = 0.07$). Albeit, no differences between the genders for perceived caution surfaced for ExpMus. Further, females reported to be more aware of the ExpMus background ($M_{\text{Female}} = 3.91$, $SD = 0.19$; $M_{\text{Male}} = 3.76$, $SD = 0.34$; $F_{(1,\,83)} = 5.93$, $MSe = 0.083$, $p < 0.05$, $\eta_p^2 = 0.07$). However, the most important difference between the genders relates to driving deficiencies. Namely, male drivers were involved in significantly more events at significantly greater severities in every driving condition (see Table 6.9).

Table 6.9 Individual differences: Gender with IVDR scores

	Female		Male		
	M	(*SD*)	*M*	(*SD*)	*p*
I. No Music					
Event-Frequency	8	(3.99)	12	(7.95)	< 0.01
Event-Severity	113	(62.2)	190	(140)	< 0.01
II. Preferred Music					
Event-Frequency	10	(6.25)	13	(8.03)	< 0.05
Event-Severity	150	(107)	216	(141)	< 0.05
III. Experimental Music					
Event-Frequency	8	(4.90)	12	(7.49)	< 0.01
Event-Severity	113	(78.8)	189	(124)	< 0.01

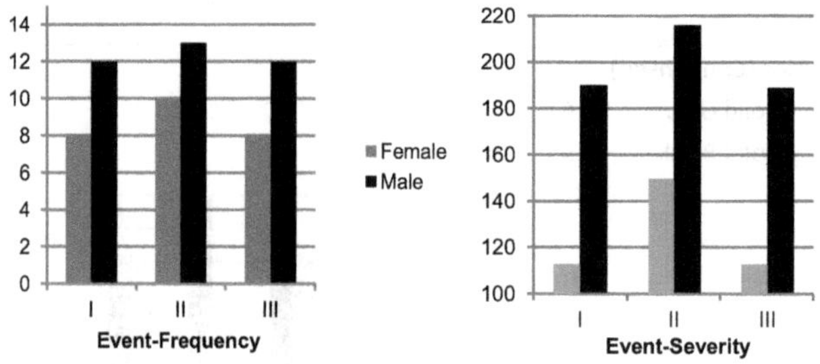

Source: Adapted from Brodsky and Slor (2013).

2. *Differences of Temperament.* Foremost, low-*ImpSS* participants reported a higher concurrence between their preferred music brought to the study and their driving experiences ($M_{\text{Low-ImpSS}}$ = 3.61, *SD* = 0.50; $M_{\text{High-ImpSS}}$ = 3.11, *SD* = 0.58; $F_{(1,\ 34)}$ = 7.61, *MSe* = 0.30, $p < 0.01$, $\eta_p^2 = 0.18$), as well as concerning the volume reproduced in the cabin ($M_{\text{Low-ImpSS}}$ = 3.56, *SD* = 0.51; $M_{\text{High-ImpSS}}$ = 2.89, SD = 0.58; $F_{(1,\ 34)}$ = 13.30, *MSe* = 0.30, $p < 0.001$, $\eta_p^2 = 0.28$). Further, for trips accompanied by ExpMus, low-*ImpSS* participants reported higher sensorial enjoyment ($M_{\text{Low-ImpSS}}$ = 2.42, *SD* = 0.88; $M_{\text{High-ImpSS}}$ = 1.81, *SD* = 0.67; $F_{(1,34)}$ = 5.52, *MSe* = 0.61, $p < 0.05$, $\eta_p^2 = 0.14$), lower NA ($M_{\text{Low-ImpSS}}$ = 1.57, *SD* = 0.37; $M_{\text{High-ImpSS}}$ = 1.94, *SD* = 0.42; $F_{(1,34)}$ = 8.01, *MSe* = 0.16, $p < 0.01$, $\eta_p^2 = 0.19$), and increased perceived caution ($M_{\text{Low-ImpSS}}$ = 3.91, *SD* = 0.19; $M_{\text{High-ImpSS}}$ = 3.64, *SD* = 0.41; $F_{(1,\ 34)}$ = 6.69, *MSe* = 0.104, $p < 0.05$, $\eta_p^2 = 0.17$). However, the most important difference between the *ImpSS* subgroups relates to driving deficiencies. Namely, high-*ImpSS* drivers were involved in significantly more events at significantly greater severities in every driving condition (see Table 6.10).

Table 6.10 Individual differences: Impulsivity sensation-seeking with IVDR scores

	Low-ImpSS *M* (*SD*)		High-ImpSS *M* (*SD*)		*p*
I. No Music					
Event-Frequency	7	(3.72)	13	(8.83)	< 0.05
Event-Severity	100	(59.6)	203	(172)	< 0.05
II. Preferred Music					
Event-Frequency	9	(6.79)	14	(6.92)	< 0.06
Event-Severity	145	(117)	221	(128)	< 0.07
III. Experimental Music					
Event-Frequency	6	(3.60)	12	(6.25)	< 0.01
Event-Severity	83	(50.1)	186	(98.0)	< 0.01

14
12
10
8
6
4
2
0
I II III
Event-Frequency

Low-ImpSS
High-ImpSS

230
210
190
170
150
130
110
90
70
I II III
Event-Severity

Source: Adapted from Brodsky and Slor (2013).

Unlike other studies employing simulated driving in a virtual vehicle on an animated road that can only approximate real driving tasks, Brodsky and Slor (2013) carried out the current study in the real world offering an authentic three-dimensional visual field incorporating distinct behaviours related to survival. Although some researchers (Just et al., 2008; Horrey & Wickens, 2006; Santos et al., 2005) formulate an argument for simulated driving being comparable to those conducted in the field with instrumented automobiles (i.e., as both seem to produce similar combined effect sizes of distraction on driving performance), these sentiments are not held dear to the heart. It should be pointed out that real-world on-road investigations guarantee that the attentional requirements tested are equivalent to genuine valid driving, and do engage the perception of risk when it comes to time-sharing strategies (especially as regards in-car music listening). Moreover, on-the-road driving-related stimuli presented via a computer screen are perhaps easier overwhelmed by music than perhaps tangible perceptual cues from engine noise and revs, vibrations, motor feedback, pedestrian density, weather conditions, and additional traffic – all of which can be significantly different when presented in the laboratory as they are either controlled, tuned-down, or eliminated altogether. On the other hand, many studies carried outside of controlled laboratory settings have been implemented on sterile test tracks. In this connection, Stutts et al. (2005) professed that largely absent from the literature are reports on drivers' exposure to various potentially distracting events while engaged in everyday driving.

Perhaps we might understand that *in-cabin listening* provides prime conditions and great potential for distraction that can result in driver miscalculation, inaccuracy, driver error, traffic violations, and driver aggressiveness. Along these lines, a recent review and meta-analysis by Young and Salmon (2012) found that driver error and distraction were highly related to one another, with overall estimates that these are a causal factor in as much as 75–95% of road crashes. They further suggest that secondary task distraction is a contributing factor in at least 23% of all accidents. The findings of the above clinical-trial suggest that listening to music background is such as secondary task. Nevertheless, these findings also demonstrated that not all musics cause the same *ill-effects*, and some musics may even contribute towards improved well-being and driver welfare. Accordingly, the experimental music programme developed by Brodsky (Brodsky & Kizner, 2012) seems to be such a background, and therefore we might view the programme as a form of self-mediated intervention – to be used as an alternative when drivers perceive it to be most fitting and adaptive in circumstances of higher risk such as when driving home from work at the day's end or under after-party duress in the wee-hours.

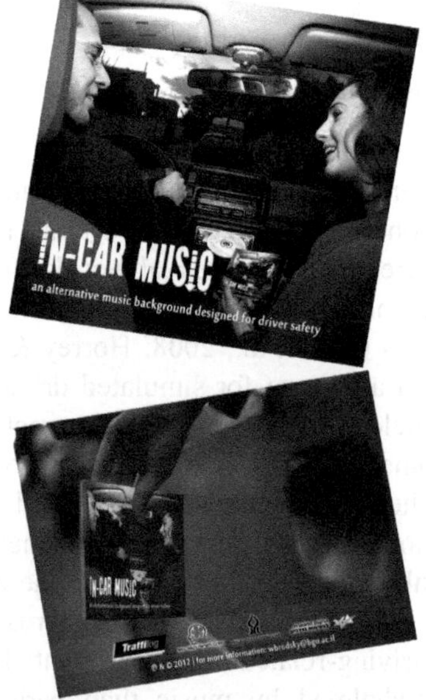

Figure 6.5 ***In-Car Music*** **CD covers: Music composed, arranged, and produced by Micha Kizner**

Source: Cover photography and graphic design by Elisha Brodsky and Raviv Kraemer. Reprinted with permission.

Chapter 7
Postscript

In the book *Driving with Music* I have attempted to present a comprehensive picture of the cognitive-behavioural implications that might surface from circumstances when driving an automobile while listening to music. From the onset it was important to relate to the globalization of the motorcar in modern Westernized societies, and focus on 'the humanity of the car' in the sense that the automobile has become an "integral part of the cultural environment with which we see ourselves as human" (Miller, 2001, p. 2). Indeed, some postmodernists view the car as a product that is no longer produced, purchased, or used to express class distinctions, but rather the car is the badge of membership in one of many lifestyle groups – none of which are necessarily superior to the other (Gartman, 2004). Automobility is today one of the most taken for granted undercurrents of our lives, and the emotions attached to the car, as well as those that are interjected into the cabin and projected onto the roads, must be seen as paramount. While traffic research has made great strides in exploring and explaining a host of causes for driver distraction that can lead to incidents, crashes, and fatalities, for the most part driver emotion has been taken out of the equation. To illustrate the point, almost every conceptual definition outlining the co-variables and processes of driver distraction list three main facets. These are: structural mechanical interference, perceptual masking, and cognitive exhaustion or capacity interference. Albeit, a few of these also mention social diversion (i.e., passengers) as a fourth component. Nonetheless, it is rare that *emotional dissonance* is considered – with the exception of a passing remark about the adverse circumstances that occur when the driver is engaged in a passionate emotive conversation or dispute (for example, see McKnight & McKnight, 1993). Therefore, we might wonder about driving situations in which environmental conditions incite change of emotion among drivers – for better or worse. Certainly, music has the power to provoke and inflame sentiment.

Although cooperative ventures between international agencies may have closed gaps in forging globally accepted uniform codes of driving, there is still great diversity. Undoubtedly, we all experience quite distinct driver behaviours in one country that are not seen in others – and perhaps these deviations are more often than not viewed as unacceptable forms of conduct. Such sets of behaviour surrounding the automobile – whether related to ownership and maintenance or to operation and performance – are all unique aspects of localized car-cultures (Featherstone, 2004). Yet, among all this diversity, one thing that is true for all drivers everywhere is the fact that we all listen to music while we drive. Therefore, *Driving With Music* has weaved the development of car-audio into the social fabric of automobility and has placed the interface of *cars-&-music* at the forefront of popular culture as it had taken command of the visual, literary, and the sonic art forms already from the 1940s. To this end, the book provides seven appendixes as resources for readers; these list the names of full feature films about the road, as well as those that highlight car movies, car video games, car songs, and CD compilations of driving tracks. Although the lists cannot cover every exemplar in the each category, these resources act as a useful springboard for anyone interested in the topic.

Today, we not only listen to music via standard broadcast signal radio and platforms based on Internet or satellite, but we bring CDs and portable music players into the vehicle in increasing numbers (Bayly, Young, & Regan, 2009; Regan, Young, & Lee 2009a). Some nomadic devices, such as the *iPod*, have already been considered by the auto industry in their designs of entertainment systems in few high-market car models; these feature docking cradles in centre consoles anchored beneath the dashboard employing customized electronic interfaces. However, for the most part, a host of music gadgets are simply plugged into the car-audio system. Certainly, this was the case with the Sony *Walkman* cassette tape, the Sony *Discman* compact disk player, and now occurs with numerous MP3 file digital music players (especially the Apple *iPod* and *iPhone*). Today's infotainment systems allow for wireless connection via *Bluetooth* technology, which is especially effective to couple mobile smartphones – as these have taken over the mobile music market. According to Bayly et al.:

> These devices usually allow for a wider range of personalized music than fixed auditory entertainment systems (e.g. car radios), and provide the option of skipping and searching through extensive libraries of songs, pictures, and videos… Portable music players may be listened to through headphones, or car kits that allow the output to be played through the vehicle speakers (p. 193).

Throughout the book I have attempted to place just one overriding concept onto the table (so to speak). The one deduction I bring forward is:

> IT IS NOT ONLY THE PHYSICAL ACTIVITY WHEN LISTENING TO MUSIC WHILE DRIVING THAT INFLUENCES DRIVING PERFORMANCE, BUT RATHER IT IS THE MUSIC – THE SOUNDS THEMSELVES – THAT AFFECT DRIVERS' BEHAVIOUR.

I would also like to emphasize the fact that part and parcel to this view is that behavioural effects clearly result from *certain kinds of music*, and that not all musics cause the same response. This was made very clear in Chapter 6, where driver-preferred musics brought to the car by young novice drivers demonstrated increased driver deficiencies (both in frequency and severity) compared to a newly composed aural background, which was found to be music that increases driver safety. In this connection, Dibben and Williamson (2007) feel that such notions lead to another more peripheral exploration. They query: Are different listening preferences associated with driving performance? Actually, for quite some time most drivers have noted particular driving behaviours that seem to correlate with specific in-car listening practices. Perhaps we have experienced these ourselves or have observed them among other drivers – especially when a car pulls up alongside us at an intersection. More than anything else, we have all simply come to assume that driving performance *is* somewhat associated with our musical preferences. Namely, vehicular control is directly linked to *what* we hear and *how* we play it. Using the popular Anthelme Brillat-Savarin (1826) saying, 'Tell me what you eat and I will tell you what you are' as an analogy, we might transfer the idea to the driving arena as follows: *Tell me what music you listen to (i.e., your playlist) and I will tell you how you drive.* This was exactly what several surveys, such as those by Allianz and Kanetix, attempted to discover (both described in Chapter 5). More than anything else, such observations mean that listening to music is not just simply a casual factor of influence on the driver, but rather like musical preferences themselves, there are a host of interacting variables that weigh on our inclinations – such as age, gender, persona, lifestyle, and the sensorial acoustic qualia of the listening context. Yet, most importantly, the effects of in-car music on the driver are moderated by music complexity. As pointed out in Chapter 2, *complexity* is a more general term used when accounting for the typical characteristics within music that are recognized as the identifying features pertaining to a specific genre. Therefore, it is imperative we consider that different genres may affect driving safety to different extents and in different ways – precisely because of their distinct musical characteristics and features. Hence, in Chapter 2, I have brought forth the relevant material to widen readers' horizons about music listening, listeners, and response theories. I do believe that drivers bring music to the car that has meaning to them. But what gives music meaning? And, does music have to have meaning to be found among our musical preferences? I have discussed both of these issues to lead to answering the ultimate question: Do we bring specific music to the car as an accompaniment when we drive? That is, do we select meaningful preferred pieces for driving? Or, do we essentially just bring to the car the same music we generally prefer when engaged in other activities (such as dancing, eating, exercising, learning, loving, and relaxing)?

Driving With Music was written with several reader populations in mind. This is important, especially as the implications of driving with music go far beyond the simulators of traffic researchers, the laboratories of music psychologists, and the clipboards of accident investigators. The subject matter herein is found in the minds and hearts of driving instructors, parents of young drivers, and in fact every driver. Further, while professional and academic readers may see some of material presented as far too elementary and based on concepts that certainly fall well within their domain of expertize, this same material is equally presented for everyday layman readers to widen their outlook and knowledge. In both cases, I have chosen and presented the materials in an effort to demonstrate that music *is* a principal variable that dovetails at-risk behaviour subsequent to distraction. All through the text I have provided several definitions of driver distraction, and indeed each has a coveted place and contribution towards uncovering the cognitive-behavioural implications of music on driver behaviour. Nevertheless, the definition I prefer to bring forward in conclusion, is:

> Distraction reflects a mismatch between the attention demanded by the road environment and the attention devoted to it... The degree to which the driver's engagement in a competing activity poses a threat of distraction depends on the combined demands of the roadway and the competing activity relative to the available capacity of the driver (Lee, Regan, & Young, 2009, p. 35).

I chose this definition here specifically because it is highly fitting for our discourse on in-car music, and I view this definition as one with which we might be able to resolve disputes that have surfaced between the various research findings (or null findings) that have been reported in the literature concerning driving with music. I now add a second parable to my overriding conceptual message of the book:

> IT IS NOT ONLY THE PHYSICAL ACTIVITY WHEN LISTENING TO MUSIC WHILE DRIVING THAT INFLUENCES DRIVING PERFORMANCE, BUT RATHER IT IS THE MUSIC – THE SOUNDS THEMSELVES – THAT AFFECT DRIVERS' BEHAVIOUR. IN-CAR MUSIC AFFECTS DRIVER SAFETY, TO DIFFERENT EXTENTS AND IN DIFFERENT WAYS, DUE TO THE MUSICAL CHARACTERISTICS AND FEATURES FOUND IN THE AURAL BACKGROUND RELATIVE TO THE AVAILABLE CAPACITY OF EACH INDIVIDUAL DRIVER.

Thus, we can understand that: (1) the same music style or piece may cause no harm to one driver while causing havoc to another driver; and (2) the same music style or piece may cause no harm to a driver on one day/trip while causing havoc to the same driver on a different day/trip.

Considering the above, driving with music may be highly beneficial. We have read in Chapter 3 that drivers the world over enjoy listening to music while cruising the countryside, just going from home to office, or running errands in the neighbourhood. The car actually provides a space – the autosphere – for us to experiment with repertoire that we otherwise might not ever hear; to sing our hearts out, to pretend we are in the band, and even dance in our seats – all with the security of anonymity. In addition, music may keep us out of harms way by combating fatigue. Then again, these benefits can crumble without a moments notice. We also know there to be some more general contraindications to in-car listening. That is, driving a car in traffic is certainly not like the other activities we so often engage in when listening to music (such as dancing, eating, exercising, learning, loving, and relaxing), and therefore the music we usually choose to listen to as a background in our everyday lives might not be very appropriate for driving a car. Chapter 4 outlines the circumstances that should be considered when choosing music. I have detailed many occurrences of driver distraction and inattention that could result from music listening based on structural mechanical interference, perceptual masking, capacity interference, and social diversion. In other words, perhaps we might have to consider that sometimes listening to music may not be in the best interests of driver safety. I do believe that the crux of the matter is not the *use* of music per se, but rather the *misuse* of music by selecting exemplars that are highly inappropriate for driving a vehicle. Therefore, there is much to be gained by raising awareness among drivers about the most appropriate music choices for driving to begin with and how to tailor these for each and every one of us based on our personalogical make up. We may even need to think of such competencies as a component in driver education courses. And most definitely, we all need some experience in learning how to adapt and modify our personalised playlists based on the *ill-effects* that some musics cause while driving. Among those described in Chapter 5, are musics that evoke inappropriate levels of arousal, music tempos that generate distraction, and music genres that induce aggressive driving. Then, a third parable must be added to the overriding conceptual message of the book:

> IT IS NOT ONLY THE PHYSICAL ACTIVITY WHEN LISTENING TO MUSIC WHILE DRIVING THAT INFLUENCES DRIVING PERFORMANCE, BUT RATHER IT IS THE MUSIC – THE SOUNDS THEMSELVES – THAT AFFECT DRIVERS' BEHAVIOUR. IN-CAR MUSIC AFFECTS DRIVER SAFETY, TO DIFFERENT EXTENTS AND IN DIFFERENT WAYS, DUE TO THE MUSICAL CHARACTERISTICS AND FEATURES FOUND IN THE AURAL BACKGROUND RELATIVE TO THE AVAILABLE CAPACITY OF EACH INDIVIDUAL DRIVER. DRIVER DEFICIENCIES ARE THE RESULT OF SELECTING MUSICS THAT ARE HIGHLY INAPPROPRIATE FOR DRIVING A VEHICLE.

In this book I have taken on a highly critical stance when it came to reviewing the empirical efforts of the many researchers who had investigated traffic-related driver behaviour whereby music was entered into the empirical platform. Among these are a few music studies that chose driving-game simulations to investigate listener responses to music. I would hope that given the fact that the book is the very first full-length volume on the topic, the text not only reflects the current state of the scientific field, but that the analyses and reports can be seen as having been carried out honestly. Unfortunately, one cannot but be impressed by the underdevelopment of the topic, and some might even say the current situation is rather pitiful. These circumstances are reflected in the following summary by Bayly et al. (2009), who attempt to bring forth some statement about our current knowledge (or perhaps 'gap in knowledge') regarding in-car music listening.

> The in-vehicle radio has been commercially available since the 1930s, and is now a standard feature in most vehicles. Despite this, relatively little research concerning the distracting effects of these systems has occurred to date. The auditory output from the radio programs (or CD players) is rarely cited, or even investigated, as a potential source of distraction, despite evidence showing that audio entertainment is used by over 70% of driving time. Of the studies conducted regarding the effects of audio entertainment, findings such as increases in lane position deviations have been observed, whereas other studies have found no or minor effects. Stutts found no increases in the proportion of time spent driving with eyes off the road or both hands off the steering wheel when auditory stimuli (from either the radio or CD player) were present (p. 204).

First, it is certainly an embarrassment for traffic research when a leading research team can claim: (1) there is a prevalent human behaviour covering 70% of incidence of all those who engage in driving that has been acknowledged for at least 80 years; (2) this behaviour is notorious for affecting drivers; but (3) we know almost nothing about the consequences simply because little to no empirical investment has been made. Second, while the above is an ill-fated situation in its own right, when looking at research protocols employed by the few who *have* attempted to explore the effects of listening to music while driving, we are faced with empirical trials that were implemented with less rigor towards the music (that is, the actual stimuli itself, the format of reproduction, and the methods of exposure) than we would expect for high level scientific investigations. Is it no surprise, then, that studies examining how music listening affects driving behaviour have found either 'no' or 'minor' outcomes. Third, the traffic-research community themselves seem to have been less than thorough in their assessment of music-related findings, and continue to cite falsehoods in their own reports. Moreover, editors in traffic science have been less than systematic in reviewing manuscripts on music when submitted for publication, the impression being that all music is music, it is something we have been familiar with since we were born, and thus there is no further need for detailed descriptions. Subsequently, the scientific literature itself is full of misrepresentations (*False Positives* and *Type II Errors*) about the effects of music on driving behaviour, and by default traffic scientists continue to perpetuate such distortions by replicating methodical errors. For example, in the summary above, Bayly et al. refer to the infamous findings by Stutts et al. (2003) who claim to have demonstrated that *music causes no effect on drivers*. As pointed out in Chapter 4, this conclusion was made on the basis that they found no increases in the proportion of time drivers spent driving with eyes off the road or both hands off the steering wheel when listening to music. Nevertheless, as stated previously (Brodsky & Kizner, 2012; Brodsky & Slor, 2013) such conclusions are illusive simply because these behavioural measures – though highly relevant to secondary tasks of a visual nature, or for tasks requiring physical manipulation – are not in the least ecologically valid or reliable measures for evaluating the effects of *music listening* on mental resources and vehicular performance.

So what is it about music that is so vague and intangible, that it causes so many to misperceive the fit for utility (i.e., the appropriateness of function) when it comes to investigating driving? Bruner (1990) astutely asserts that while music may effectively be used to serve one purpose, it may also be inappropriate for another. By integrating 50 years of musicological findings related to the characteristic features that produce emotional expressions, Bruner produced a highly illustrative model (see Figure 7.1). As can be seen in the table, the sonic structure of eight affective states (i.e., Serious, Sad, Sentimental, Serene, Happy, Exciting, Majestic, Frightening) are characterised by six music components (Mode, Tempo, Pitch, Rhythm, Harmony, Volume) whose interactive fusion engenders each unique emotional expression. I would like to point out that no two emotions have the same architectural structure.

The overall assumption that empirical driving conditions can be based on music that has been selected as an expression of a specific affect, is quite presumptuous even for highly celebrated traffic researchers. Although there may be no doubt that all music expresses some form of emotion, the ability to design music with the appropriate structural qualia (i.e., complexity) to express a specific emotion, or perceptive process, or cognitive state, is a far beyond those without intricate life-long musical training and advanced music-research experience.

	Mode		Tempo			Pitch			Rhythm			Harmony		Volume			
	Major	Minor	Slow	Medium	Fast	Low	Medium	High	Uneven	Flowing	Firm	Dissonant	Consonant	Soft	Medium	Loud	Varied
Serious	✓		✓			✓					✓		✓		✓		
Sad		✓	✓			✓					✓	✓		✓			
Sentimental		✓	✓				✓			✓			✓	✓			
Serene	✓		✓				✓			✓			✓	✓			
Happy		✓			✓			✓		✓			✓		✓		
Exciting		✓			✓		✓		✓			✓				✓	
Majestic		✓		✓			✓				✓	✓				✓	
Frightening	✓		✓			✓			✓			✓					✓

Figure 7.1 Musical characteristics for producing emotional expressions
Source: Adapted from Bruner (1990).

The truth is that when I designed the new experimental music background for in-car listening (described in Chapter 6), and had to account for the functional congruence between the aural conditions and the perceptual-motor tasks necessary to safely operate a car, I realized that several other music components besides the six listed by Bruner were crucial. These were: Arrangement, Instrumentation, Instrument Range (Timbre, Tone Frequency), and Orchestral Texture. Further, in order not to express any of the specific emotional expressions as detailed in Figure 7.1, it seemed paramount that the music would have to remain somewhere in the middle range of each musical component – as much as possible reflecting an overall optimal well-balanced sonic architecture. Hence, to complete the overriding conceptual message of the book, I add here the fourth and final parable:

> It is not only the physical activity when listening to music while driving that influences driving performance, but rather it is the music – the sounds themselves – that affect drivers' behaviour. In-car music affects driver safety, to different extents and in different ways, due to the musical characteristics and features found in the aural background relative to the available capacity of each individual driver. Driver deficiencies are the result of selecting musics that are highly inappropriate for driving a vehicle. The fit between driving and music accompaniment, which is essential for improved vehicular performance and increased driver safety, is moderated by aural backgrounds structurally designed to generate moderate levels of perceptual complexity that do not express specific formulaic emotional qualities.

Appendix A
Road Movies

Compiled by Richard F. Weingroff
Reprinted with permission from
www.fhwa.dot.gov/interstate/roadmovies.htm

This listing of 72 movies (presented in ascending alphabetical order) was compiled by the information liaison specialist with the Federal Highway Administration (United States Department of Transportation). The list covers a 62-year period (between 1934–1996). The listing was completed for the celebratory '50th Anniversary of the Eisenhower Interstate Highway System' in 1996. Most of the films on the list were previously referenced, among 19,000 other titles, in the *Movie & Video Guide 1996* (Maltin, 1995).

Big Bus, The	1976
Boys on the Side	1995
Breaking the Rules	1992
Cannonball	1976
Cannonball II	1984
Cannonball Run, The	1981
Chase, The	1994
Coast to Coast	1980
Convoy	1978
Duel	1971
Easy Rider	1969
Eat My Dust	1976
Electric Glide in Blue	1973
Even Cowgirls Get the Blues	1994
From Dusk to Dawn	1996
Grapes of Wrath, The	1940
Great Dynamite Chase, The	1977
Great Race, The	1965
Great Smokey Roadblock, The	1976
Gumball Rally, The	1976
Hell's Angels on Wheels	1967
High-Ballin'	1978
Highway 61	1991
Highway 301	1950
Highway to Hell	1992
Hitch-Hiker, The	1953

Honeysuckle Rose	1980
Honky Tonk Freeway	1981
Honkytonk Man	1982
Intersection	1994
It Happened One Night	1934
It's a Gift	1934
Joyride	1977
Kalifornia	1993
Last Chase, The	1981
Long, Long Trailer, The	1954
Lost in America	1985
Miracle on I-880	1994
National Lampoon's Vacation	1983
Out	1982
Pickup on 101	1972
Planes, Trains, and Automobiles	1987
Psycho	1960
Road Killers, The	1995
Road Scholar	1993
Roadgames	1981
Roadhouse 66	1984
Roadside Prophets	1990
Six of a Kind	1934
Smokey and the Bandit	1977
Something Wild	1986
Song of the Open Road	1944
Speed	1994
Speed Zone!	1989
Sugarland Express, The	1974
Sure Thing, The	1985
Thelma and Louise	1991
They Drive by Night	1940
Thieves' Highway	1949
Those Daring Young Men in Their Jaunty Jalopies	1969
Thunder Road	1958
To Wong Foo, Thanks for Everything! Julie Newmar	1995
Tommy Boy	1995
Truck Stop Women	1974
Two-Lane Blacktop	1971
Vanishing, The	1993
Vanishing Point	1971
Wayward Bus, The	1957
Wheels of Terror	1990
White Line Fever	1975
Who Framed Roger Rabbit	1988
Wild One, The	1954

Appendix B
Car Movies

Compiled by Erez Ein Dor

This listing of 145 movies (presented in ascending alphabetical order) is a compilation of English-language internationally distributed films on car themes. The movie genres include: action, animation, comedy, crime, drama, fantasy, fiction, horror, musical, mystery, romance, sport, spy, and thriller. The list covers a 78-year period (between 1926–2014). The list is assembled from various public access sources, including: 'Top 40 Car Movies' from *IMDB* (Internet Movie Data Bank, www.imdb.com), 'Best Car Movies' as rated by viewers on *Ranker* (www.ranker.com), as well as from several pieces found in the literature (Eyerman & Lofgren, 1995; Lutz & Fernandez, 2012; Mottram, 2002; Pascoe, 2002; Rees, 2002; Wollen, 2002).

2 Fast 2 Furious	John Singleton	2003	Action, Crime, Thriller
A View to a Kill (007)	John Glen	1985	Action, Spy
American Graffiti	George Lucas	1973	Comedy
Assault on Precinct 13	John Carpenter	1976	Action, Thriller
At War with the Army	Hal Walker	1950	Comedy, Musical, Romance
Bad Boys	Michael Bay	1995	Action, Comedy
Bad Boys II	Michael Bay	2003	Action, Comedy
Black Dog	Kevin Hooks	1998	Action
Blood Simple	Joel Coen	1984	Crime, Film Noir
Blues Brothers, The	John Landis	1980	Action, Comedy, Musical
Blues Brothers 2000	John Landis	1998	Action, Comedy, Musical
Bonny And Clyde	Arthur Penn	1967	Crime, Drama
Boulevard Nights	Michael Pressman	1979	Crime, Drama
Breathless	Jean-Luc Godard	1960	Crime, Drama
Bullitt	Peter Yates	1968	Action, Mystery, Thriller
Cannonball Run II	Hal Needham	1984	Action, Comedy
Cannonball Run, The	Hal Needham	1981	Action, Comedy
Cars	Joe Ranft	2006	Adventure, Animation, Comedy
Cars 2	John Lasseter	2011	Adventure, Animation, Comedy
Casino Royale	Martin Campbell	2006	Action, Spy
Chitty Chitty Bang Bang	Ken Hughes	1968	Comedy
Christine	John Carpenter	1983	Horror
Convoy	Sam Peckinpah	1978	Action, Comedy, Drama
Corvette Summer	Matthew Robbins	1978	Action

Crowd Roars, The	Howard Hawks	1932	Drama, Action
Days of Thunder	Tony Scott	1990	Action, Drama, Romance, Sport
Death Proof	Quentin Tarantino	2007	Action, Thrille
Death Race	Paul W.S. Anderson	2008	Action, Sci-Fi, Thriller
Death Race 2	Roel Reiné	2010	Action, Sci-Fi, Thriller
Death Race 2000	Paul Bartel	1975	Action, Sci-Fi, Sport
Detective, The	Gordon Douglas	1968	Crime, Drama
Devil on Wheels	Crane Wilbur	1947	Drama
Diamonds Are Forever (007)	Guy Hamilton	1971	Action, Spy
Die Another Day (007)	Lee Tamahori	2002	Action, Spy
Dirty Mary, Crazy Larry	John Hough	1974	Action, Adventure
Dr. No (007)	Terence Young	1962	Action, Spy
Drive	Nicolas Winding Refn	2011	Crime, Drama
Drive Angry	Patrick Lussier	2011	Action, Crime, Fantasy, Thriller
Driven	Renny Harlin	2001	Action, Drama, Sport
Driver, The	Walter Hill	1978	Action, Crime, Drama, Thriller
Duel	Steven Spielberg	1971	Action, Thriller
Each Dawn I Die	William Keighley	1939	Crime
Fast and Furious	Melville W. Brown	1927	Comedy
Fast and Furious	Busby Berkeley	1939	Comedy, Crime
Fast and Furious	Justin Lin	2009	Action, Crime, Drama, Thriller
Fast and the Furious, The	Rob Cohen	2001	Action, Crime, Thriller
Fast and the Furious: Tokyo Drift, The	Justin Lin	2006	Action, Crime, Drama, Thriller
Fast Five	Justin Lin	2011	Action, Crime, Thriller
For Your Eyes Only (007)	John Glen	1981	Action, Spy
Four Jills in a Jeep	William A. Seiter	1944	Musical, Romance
Four Men in a Jeep	Leopold Lindtberg	1951	Drama
French Connection, The	William Friedkin	1971	Action, Crime, Thriller
From Russia with Love (007)	Terence Young	1963	Action, Spy
Getaway, The	Roger Donaldson	1972	Action, Crime, Thriller,
Goldeneye (007)	Martin Campbell	1995	Action, Spy
Goldfinger (007)	Guy Hamilton	1964	Action, Spy
Gone In 60 Seconds	H. B. Halicki	1974	Action, Crime, Drama
Gone In Sixty Seconds	Dominic Sena	2000	Action, Crime, Thriller
Grand Prix	John Frankenheimer	1966	Drama, Sport
Grapes of Wrath, The	John Ford	1940	Drama
Grease	Randal Kleiser	1978	Musical, Romance
Great Race, The	Blake Edwards	1965	Slapstick
Green Helmit, The	Michael Forlong	1961	Drama
Green Hornet, The	Ford Beebe	1940	Action, Crime
Green Hornet, The	Michel Gondry	2011	Action, Comedy
Gumball Rally, The	Chuck Bail	1976	Comedy
Herbie Goes Bananas	Vincent Mceveety	1980	Comedy
Herbie Goes to Monte Carlo	Vincent Mceveety	1977	Comedy

Herbie Rides Again	Robert Stevenson	1974	Comedy
Herbie: Fully Loaded	Angela Robinson	2005	Comedy
It's a Gift	Norman Z. Mcleod	1934	Comedy
It's the Old Army Game	Edward Sutherland	1926	Comedy, Romance
Italian Job, The	Peter Collinson	1969	Action, Comedy, Crime, Thriller
Italian Job, The	F. Gary Gray	2003	Action, Crime, Thriller
Killers, The	Robert Siodmak	1946	Crime, Drama
Kiss Me Deadly	Robert Aldrich	1955	Film Noir
Knight Rider	Steve Shill	2008	Action, Drama
Kustom Kar Kommandos	Kenneth Anger	1970	Short
Last Stand, The	Jee-woon Kim	2013	Action, Crime
Le Mans	Lee H. Katzin	1971	Action, Adventure, Sport
License to Kill (007)	John Glen	1989	Action, Spy
Live and Let Die (007)	Guy Hamilton	1973	Action, Spy
Living Daylights, The (007)	John Glen	1987	Action, Spy
Lost Highway	David Lynch	1997	Thriller, Neo Noir
Love Bug, The	Robert Stevenson	1968	Comedy
Mad Max	George Miller	1979	Dystopian Fiction, Action
Mad Max 2	George Miller	1981	Dystopian Fiction, Action
Mad Max Beyond Thunderdome	George Miller	1985	Post-Apocalyptic
Mad Max: Fury Road	George Miller	2014	Post-Apocalyptic
Magnificent Ambersons, The	Orsen Wells	1942	Drama
Man with the Golden Gun, The (007)	Guy Hamilton	1974	Action, Spy
Mission: Impossible	Brian De Palma	1996	Action, Spy
Mission: Impossible II	John Woo	2000	Action, Spy
Mission: Impossible III	J. J. Abrams	2006	Action, Spy
Mission: Impossible – Ghost Protocol	Brad Bird	2011	Action, Spy
Moonraker (007)	Lewis Gilbert	1979	Action, Spy
National Lampoon's Vacation	Harold Ramis	1983	Comedy
North by Northwest	Alfred Hitchcock	1959	Thriller
Octopussy (007)	John Glen	1983	Action, Spy
On Her Majesty's Secret Service (007)	Peter R. Hunt	1969	Action, Spy
Playtime	Jaques Tati	1967	Comedy
Quantum of Solace (007)	Marc Forster	2008	Action, Spy
Racers, The	Henry Hathaway	1955	Drama, Sport
Rebel without a Cause	Nicholas Ray	1955	Drama
Roadracers	Robert Rodriguez	1994	Action, Drama
Ronin	John Frankenheimer	1998	Action, Crime, Drama, Thriller
Senna	Asif Kapadia	2010	Documentary, Biography, Sport
Skyfall (007)	Sam Mendes	2012	Action, Spy
Smokey and the Bandit	Hal Needham	1977	Action, Comedy, Crime, Romance

Smokey and the Bandit II	Hal Needham	1980	Comedy
Smokey and the Bandit Part 3	Dick Lowry	1983	Comedy
Solid Gold Cadillac, The	Richard Quine	1956	Comedy, Romance
Speed	Jan De Bont	1994	Action, Thriller
Spy Who Loved Me, The (007)	Lewis Gilbert	1977	Action, Spy
St Valentines' Day Massacre	Roger Corman	1967	Crime, Drama
Steel Cowboy	Harvey S. Laidman	1976	Drama
Sugarland Express	Steven Spielberg	1974	Adventure, Comedy, Crime
Sunset Boulevard	Billy Wilder	1950	Film Noir
Talladega Nights: The Ballad of Ricky Bobby	Adam Mckay	2006	Action, Comedy, Sport
Taxi	Gérard Pirès	1998	Action, Comedy, Crime
Taxi 2	Gérard Krawczyk	2000	Action, Comedy, Crime
Taxi 3	Gérard Krawczyk	2003	Action, Comedy
Taxi 4	Gérard Krawczyk	2007	Action, Comedy
Taxi Dancer, The	Harry F. Millarde	1927	Silent
Taxi Driver	Martin Scorcese	1976	Drama
They Drive by Night	Raoul Walsh	1940	Crime, Drama, Thriller
Thunder Road	Arthur Ripley	1958	Crime, Drama, Thriller
Thunderball (007)	Terence Young	1965	Action, Spy
To Please a Lady	Clarence Brown	1950	Action, Romance, Sport
Tomorrow Never Dies (007)	Roger Spottiswoode	1997	Action, Spy
Transporter, The	Louis Leterrier	2002	Action
Transporter 2, The	Louis Leterrier	2005	Action
Transporter 3, The	Olivier Megaton	2008	Action
Two Tars	James Parrott	1928	Comedy, Short
Two-Lane Blacktop	Monte Hellman	1971	Drama
Used Cars	Robert Zemeckis	1980	Comedy
Vanilla Sky	Cameron Crowe	2001	Action, Sci-Fi
Vanishing Point	Richard C. Sarafian	1971	Action, Drama
Week-End	Jean-Luc Godard	1967	Comedy, Drama
Winning	James Goldstone	1969	Action, Drama, Sport
World Is Not Enough, The (007)	Michael Apted	1999	Action, Spy
XXX	Rob Cohen	2002	Action, Adventure
XXX: State Of The Union	Lee Tamahori	2005	Action, Adventure
You Only Live Twice (007)	Lewis Gilbert	1967	Action, Spy
Young Racers, The	Roger Corman	1965	Action, Drama

Appendix C
Car Games for Video and PC

Compiled by Erez Ein Dor

This listing of 101 PC computer games (presented in ascending alphabetical order) is a compilation of internationally distributed video and personal computer games on car themes. In general, these are the 'best sellers' of racing simulators (however, a few action and adventure games are also listed). The list covers a period from when the very first 'popular' car video game appeared on the market until current times (between 1987–2013). The list is assembled from various public access web sites for games and gamers, including: *GameSpot* (www.gamespot.com), and *IGN* (www.ign.com). The entries on the listing are exclusive to automobiles (that is, there are no games highlighting trucks or motorcycles, which could have easily doubled the number of entries). Further, the list does not embrace games in which the car is used for any other purpose than driving – including drama, theft, or violent manslaughter (such as the trendy *Grand Theft Auto*, *Saints Row*, and *True Crime*). Finally, all games listed allow for the user to drive in the 'first-person cockpit-view mode' (opposed to controlling the vehicle in the 'third person aerial-view mode').

Auto Club Revolution	Eutechnyx	2012	Racing Simulator
Burnout	Acclaim Entertainment	2001	Racing Simulator
Burnout 2: Point of Impact	Acclaim Entertainment	2003	Racing Simulator
Burnout 3: Takedown	EA Games	2004	Racing Simulator
Burnout Crash	EA Games	2011	Racing Simulator
Burnout Dominator	EA Games	2007	Racing Simulator
Burnout Legends	EA Games	2005	Racing Simulator
Burnout Paradise	EA Games	2008	Racing Simulator
Burnout Revenge	EA Games	2005	Racing Simulator
Colin Mcrae Rally	Codemasters	1998	Racing Simulator
Colin Mcrae Rally 04	Codemasters	2003	Racing Simulator
Colin Mcrae Rally 2.0	Codemasters	2002	Racing Simulator
Colin Mcrae Rally 2005	Codemasters	2004	Racing Simulator
Colin Mcrae: Dirt	Codemasters	2007	Racing Simulator
Colin Mcrae: Dirt 2	Codemasters	2009	Racing Simulator
Dirt 3	Codemasters	2011	Racing Simulator
Dirt: Showdown	Codemasters	2012	Racing Simulator
Driv3r	Ubisoft Reflections	2004	Action, Adventure
Driver	Ubisoft Reflections	1999	Action, Adventure

Driver 2: Back on the Streets	Ubisoft Reflections	2000	Action, Adventure
Driver 76	Ubisoft Reflections	2007	Action, Adventure
Driver Renegade	Ubisoft Reflections	2011	Action, Adventure
Driver: Parallel Lines	Ubisoft Reflections	2007	Action, Adventure
Driver: San Francisco	Ubisoft Reflections	2011	Action, Adventure
EA Sports Nascar Team Racing	EA Sport	2007	Arcade Racing
F1 2010	Codemasters	2010	Racing Simulator
F1 2011	Codemasters	2011	Racing Simulator
F1 2012	Codemasters	2011	Racing Simulator
Forza Horizon	Microsoft Studios	2012	Racing Simulator
Forza Motorsport	Microsoft Studios	2005	Racing Simulator
Forza Motorsport 2	Microsoft Studios	2007	Racing Simulator
Forza Motorsport 3	Microsoft Studios	2009	Racing Simulator
Forza Motorsport 4	Microsoft Studios	2011	Racing Simulator
Forza Motorsport 5	Microsoft Studios	2013	Racing Simulator
Gran Turismo	Sony Computer Entertainment	1997	Racing Simulator
Gran Turismo 2	Sony Computer Entertainment	1999	Racing Simulator
Gran Turismo 3: A-Spec	Sony Computer Entertainment	2001	Racing Simulator
Gran Turismo 4	Sony Computer Entertainment	2005	Racing Simulator
Gran Turismo 4 Prologue	Sony Computer Entertainment	2003	Racing Simulator
Gran Turismo 5	Sony Computer Entertainment	2011	Racing Simulator
Gran Turismo 5 Prologue	Sony Computer Entertainment	2008	Racing Simulator
Gran Turismo Concept: 2001 Tokyo	Sony Computer Entertainment	2001	Racing Simulator
Gran Turismo Concept: 2002 Tokyo-Geneva	Sony Computer Entertainment	2002	Racing Simulator
Gran Turismo Concept: 2002 Tokyo-Seoul	Sony Computer Entertainment	2002	Racing Simulator
Gran Turismo Hd Concept	Sony Computer Entertainment	2006	Racing Simulator
Midnight Club 3: Dub Edition	Rockstar San Diego	2005	Racing Simulator
Midnight Club II	Rockstar San Diego	2003	Racing Simulator
Midnight Club: Los Angeles	Rockstar San Diego	2008	Racing Simulator
Midnight Club: Street Racing	Rockstar San Diego	2000	Racing Simulator
Midtown Madness	Microsoft Studios	1999	Racing Simulator
Midtown Madness 2	Microsoft Studios	2000	Racing Simulator
Midtown Madness 3	Microsoft Studios	2002	Racing Simulator
Nascar 06: Total Team Control	EA Sport	2005	Racing Simulator
Nascar 07	EA Sport	2006	Racing Simulator
Nascar 08	EA Sport	2007	Racing Simulator
Nascar 09	EA Sport	2008	Racing Simulator
Nascar 2000	EA Games	1999	Racing Simulator
Nascar 2001	EA Games	2001	Racing Simulator
Nascar 99: Legacy	EA Sport	1999	Racing Simulator
Nascar 2005: Chase for the Cup	EA Games	2004	Racing Simulator
Nascar Racers	Hasbro Interactive	2000	Racing Simulator
Nascar Racing 2	Sierra Entertainment	1996	Racing Simulator
Nascar Racing 3	Sierra Entertainment	1999	Racing Simulator
Nascar Racing 4	Sierra Entertainment	2001	Racing Simulator

Nascar Revolution	EA Games	1999	Racing Simulator
Nascar Rumble	EA Games	2000	Racing Simulator
Nascar Simracing	EA Games	2005	Racing Simulator
Nascar The Game: Inside Line	Activison	2012	Racing Simulator
Nascar Thunder 2002	EA Games	2001	Racing Simulator
Nascar Thunder 2003	EA Games	2002	Racing Simulator
Nascar Thunder 2004	EA Games	2003	Racing Simulator
Need for Speed II	EA Canada	1997	Racing Simulator
Need for Speed III: Hot Pursuit	EA Canada	1998	Racing Simulator
Need for Speed World	EA Singapore	2010	Racing Simulator
Need for Speed: Carbon	EA Canada	2006	Racing Simulator
Need for Speed: High Stakes	EA Canada	1999	Racing Simulator
Need for Speed: Hot Pursuit	EA Singapore	2010	Racing Simulator
Need for Speed: Hot Pursuit 2	EA Canada	2002	Racing Simulator
Need for Speed: Most Wanted	EA Canada	2005	Racing Simulator
Need for Speed: Most Wanted 2012	Criterion Games	2012	Racing Simulator
Need for Speed: Nitro	EA Canada	2009	Racing Simulator
Need for Speed: Porsche Unleashed	EA Canada	2000	Racing Simulator
Need for Speed: Prostreet	EA Canada	2007	Racing Simulator
Need for Speed: Shift	EA Canada	2009	Racing Simulator
Need for Speed: The Run	EA Black Box	2011	Racing Simulator
Need for Speed: Undercover	EA Canada	2009	Racing Simulator
Need for Speed: Underground	EA Canada	2003	Racing Simulator
Need for Speed: Underground 2	EA Canada	2004	Racing Simulator
Shift 2: Unleashed	Slightly Mad Studios	2011	Racing Simulator
Td Overdrive: The Brotherhood of Speed	Midway Studios – Newcastle	2002	Racing Simulator
Test Drive	Accolade	1987	Racing Simulator
Test Drive 4	Accolade	1997	Racing Simulator
Test Drive 5	Accolade	1998	Racing Simulator
Test Drive 6	Midway Studios – Newcastle	1999	Racing Simulator
Test Drive III: The Passion	Accolade	1990	Racing Simulator
Test Drive Unlimited	Atari	2006	Racing Simulator
Test Drive Unlimited 2	Atari	2011	Racing Simulator
Test Drive: Eve of Destruction	Atari	2004	Racing Simulator
Test Drive: Ferrari Racing Legends	Atari	2012	Racing Simulator
The Duel: Test Drive II	Accolade	1989	Racing Simulator
The Need for Speed	EA Canada	1994	Racing Simulator

Appendix D
The Ultimate List of Car Songs

Compiled by David Louis Harter

Reprinted with permission from
www.calif-tech.com/blog/carsongs.html

This list is indeed the ultimate list of car songs (presented in ascending alphabetical order). There are over 520 titles. The criteria for entry to the listing have been as wide as possible. Among the songs are those that feature themes related to cars, motor fatality, the road and highway, travelling, and loving (with automobility serving the lyrics as a metaphor for human behaviour). Several titles appear more than once, indicating coverage of the same tune by different performing artists.

1000 Dollar Car	The Bottle Rockets
18 Wheels and a Dozen Roses	Kathy Mattea
1965 GTO	The Amazing Royal Crowns
2468 Motorway	Tom Robinson
409	The Beach Boys
426 Super Stock	Dick Dale
426 Super Stock	Gary Usher
455 Rocket	Kathy Mattea
455 SD	The Hellacopters
455 SD	Radio Birdmen
5 Years, 4 Months, 3 Days	Brian Setzer
500 Miles	Bobby Bare
57 Chevrolet	Billie Jo Spears
'59	Brian Setzer
68 Pontiac	Mario Panacci
8 Miles a Gallon	Scott Miller And Commonwealth
8 Miles a Gallon	Shotgun Mccoy
8 More Miles to Louisville	Boys Of The Fort
8 Track	Brian Setzer
90 Miles an Hour	Bo Ladner
928	Keith Sykes
A Bone	The Trashmen
Ain't No Getting Around Getting Around	Julian Cope
All I Want	Joni Mitchel
All I Wanted Was a Car	Brad Paisley
All The Roadrunning	Mark Knopfler & Emmylou Harris

Always Crashing in the Same Car	David Bowie
American Idle	John Rewind
American Pie	Don Mclean
Aneheim Azuza and Cucamonga Sewing Circle Book Review and Timing Association, The	Jan And Dean
Angel in a Cadillac	Bill Carter
Another Sad Love Song	Toni Braxton
Antique 32 Studebaker Dictator Coupe	Ronny And The Daytonas
Automatic Woman	Joe Hill Louis
Automobile	John Prine
Automobile	NWA
Automobile Girl	Sidney Chapman
Automobile Song, The	Benny Bell
Automobile Song, The	Luke Mcdaniel
Automobile Song, The	Tennessee Buck
Automobiles	Ken Clinger
Baby Driver	Simon And Garfunkel
Baby, Let's Play House	Elvis Presley
Ballad of the General Lee, The	Doug Kershaw And The Hazzard County Boys
Ballad of Thunder Road, The	The Charlie Daniels Band
Ballad of Thunder Road, The	Robert Mitchum
Battery to My Heart, The	Billy Briggs
Be Thankful for What You've Got	Massive Attack
Be Thankful for What You've Got	William Devaughn
Beep, Beep	The Playmates
Behind the Wheel	Depeche Mode
Belvedere	Andy Swindell
Betsy	The Beach Boys
Big Road	Bonnie Raitt
Billricay Dickie	Ian Dury
Bitchin' Camaro	The Dead Milkmen
Black Bear Road	C.W. Mccall
Black Cadillac	Roseanne Cash
Black Limousine	The Rolling Stones
Black Sunshine	White Zombie
Blonde in the 406	The Challengers
Blondes in Black Cars	Autograph
Blue Café	Brian Setzer
Bonneville Bonnie	The Rip Cords
Born to Be Wild	Hinder
Born to Be Wild	Steppenwolf
Born to Run	Bruce Springsteen
Born to Take The Highway	Joni Mitchel
Boss Barracuda	Joanie Sommers
Boss Barracuda	The Safaris

Boss Hoss	The Sonics
Bowtie Hunters	John Rewind
Brand New Cadillac	Brian Setzer
Brand New Cadillac	The Clash
Brand New Cadillac	Vince Taylor
Brand New Car	The Rolling Stones
Bucket T	Ronny And The Daytonas
Buick Electra	Black Helicopter
Buildin' a Hot Rod	Firebird Trio
Bump in the Trunk	Bone Thugs-n-Harmony
Cadillac	Bo Diddley
Cadillac Assembly Line	Albert King
Cadillac Blues	Southern Comfort On The Skids
Cadillac Boogie	Jimmy Liggins
Cadillac Daddy	Howlin' Wolf
Cadillac Girls	Ludacris
Cadillac in Model A	Chuck Mead
Cadillac in Model A	Bob Wills & The Texas Playboys
Cadillac Ranch	Bruce Springsteen
Cadillac Walk	Mink Deville
California Dreamin'	Mamas And The Papas
Camaro	The Cyrkle
Camaro	Pernod Fils
Car Carrier Blues	Leo Kottke And Mike Gordon
Car Jamming	The Clash
Car Song	Elastica
Car Song	Woody Guthrie
Car Song, The	Cat Empire
Car Trouble	Adam Ant
Car Wash	Rose Royce
Car Wheels on a Gravel Road	Lucinda Williams
Car, The	Jeff Carson
Carefree Highway	Gordon Lightfoot
Cars	Gary Numan
Cars with the Boom	L'Trimm
Chevette	Audio Adrenaline
Chevrolet	The Black Crowes
Chevrolet	Luke Bryan
Chevrolet	Foghat
Chevrolet	Robben Ford
Chevrolet	ZZ Top
Chevy Van	Sammy Johns
Chevys and Fords	Bryant Keith
Chevys and Fords	Mac Dre
Chicken Fried	Zac Brown Band

Chitty Chitty Bang Bang	The Sherman Brothers
Convertibles	Whiteheart
Convoy	C.W. Mccall
Cool Blue Corvette	Black Shep
Corvette Song, The	George Jones
Country Road	James Taylor
Country Roads	John Denver
Coupe de Ville	Neil Young
Crawling from The Wreckage	Dave Edmunds
Crazy 'Bout an Automobile	Ry Cooder
Crossroads	Cream
Crosstown Traffic	Jimi Hendrix
Cruisin'	Smokey Robinson
Crusin' and Boozin'	Sammy Hagar
Custom Machine	The Beach Boys
Custom Machine	Bruce And Terry
D B Blues	Blind Lemon Jefferson
Daddy Let Me Drive	Alan Jackson
Daddy's Going to Pay for Your Crashed Car	U2
Dartin' Around	John Rewind
Dead and Gone	T.I.
Dead Man's Curve	Jan And Dean
Dead Skunk	Louden Wainwright III
Dear Dad	Chuck Berry
Deuce and a Quarter	New Power Generation
Devil in My Car	The B 52's
Distance, The	Cake
Do You Know the Way to San Jose	Dionne Warwick
Don't It Make You Wanna Go Home	Kedron Taylor
Don't Think about Her When You're Trying to Drive	Little Village
Don't Worry Baby	The Beach Boys
Drag City	Jan And Dean
Dragula	Rob Zombie
Dreamsville	Brian Setzer
Drive	The Cars
Drive	Incubus
Drive	R.E.M.
Drive (Theme from Hardcastle and Mccormick)	David Morgan
Drive In	The Beach Boy
Drive It All Over Me	My Bloody Valentine
Drive Like I Never Been Hurt	Ry Cooder
Drive Like Lightning	Brian Setzer
Drive My Car	The Beatles
Drive My Car	Billy Thorpe

Drive South	Suzy Bogguss
Drive South	John Hiatt
Driver's Seat	Sniff 'N' The Tears
Drivin' Down the Wrong Side of the Road	Ricky Riddle
Drivin' My Life Away	Eddie Rabbit
Drivin' Slow	Johnny London
Driving in My Car	Madness
Driving with Private Malone	David Ball
Drop Top	Billy Love
East Bound and Down	Jerry Reed
El Camino	Ween
Eliminator	ZZ Top
Even the Man in the Moon Is Crying	Mark Collie
Fast Car	Tracy Chapman
Fast Cars and Freedom	Rascal Flatts
Faster	George Harrison
Fifteen Kisses a Gallon	David Mack
Flying Low	Don Johnson
Ford Econoline	Nanci Griffith
Ford Mustang	Tim Workman & John Mcclellan
Forty Miles of Bad Road	Duane Eddy
Four Wheel Drive	C.W. Mccall
Free Ride	Edgar Winter
Free Ride	Foghat
Freeway of Love	Aretha Franklin
From a Buick 6	Bob Dylan
Fuel	Metallica
Full Blown 426 Hemi	The Untamed Youth
Fun, Fun, Fun	The Beach Boys
Galaxy 500	Reverend Horton Heat
Gallon of Gas, A	The Kinks
Gasoline	Seether
General Lee	Johnny Cash
Geronimo's Cadillac	Michael Martin Murphey
Getaway Car	Audioslave
Get 'Em on the Ropes	Brian Setzer
Get In the Car	Echo & The Bunnymen
Get Out of My Dreams, Get into My Car	Billy Ocean
Get out of the Car	Richard Berry
Get Your Daddy's Car Tonight	The Petites
Get Your Kicks on Route 66	Nat King Cole
Goin' Mobile	The Who
Goin' Up The Country	Canned Heat
Going Up around the Bend	Creedence Clearwater Revival
Grandpappy's Hot Rod Blues	Grandpappy Earl Davis

Greased Lightning	John Travolta
Greased Lightning	Liz Phair
Grey Cortina	Tom Robinson
Guitars, Cadillacs	Dwight Yoakam
Hard to Rule Woman Blues	Ramblin' Thomas
Hardtop Race	George Stogner
Heaven's in the Back Seat of My Cadillac	Hot Chocolate
Heavy Chevy	Grin
Hell Bent	Brian Setzer
Hell on Wheels	The Clarks
Hemi Barracuda	Dan Olsen
Hemi Charger	Kill Switch
Hemi Cuda	Hemi Cuda
Hemi Head	John Rewind
Hey Good Lookin'	Hank Williams
Hey Little Cobra	The Rip Cords
Highway 23	Ry Cooder
Highway 40 Blues	Merle Haggard
Highway 40 Blues	Ricky Skaggs
Highway Junkie	Randy Travis
Highway Patrol	Junior Brown
Highway Star	Deep Purple
Highway to Hell	AC/DC
Highyway '61 Revisited	Bob Dylan
Hitchin' A Ride	Vanity Fair
Holiday Road	Lindsay Buckingham
Hollywood Nights	Bob Seager
Honk Your Horn	Jimmie Heap
Horsepower	Chris Ledoux
Hot Rod	Larry Collins
Hot Rod	Ruby Short
Hot Rod Blues	The Southernairess
Hot Rod City	The Super Stocks
Hot Rod Girl	Brian Setzer
Hot Rod Hades	Charlie Ryan
Hot Rod Heart	John Fogerty
Hot Rod Hearts	Robbie Dupree
Hot Rod Holiday	The Rip Chords
Hot Rod Lincoln	Johnny Bond
Hot Rod Lincoln	Commander Cody
Hot Rod Lincoln	Charlie Ryan
Hot Rod Race	Arthur Guitar Boogie Smith
Hot Rod Race No. 2	Bob Williams
Hot Rod Race No. 3	Bob Williams
Hot Rod Rag	Paul Westmoreland

Hot Rod Shotgun Boogie No. 2	Tillman Franks
Hot Rod USA	The Ripcords
Hot Rods & My Old Man	Rod Bob Lopez
How We Do	The Game
Hubbin' It	Bob Wills & The Texas Playboys
I Can't Drive 55	Sammy Hagar
I Drove All Night	Cindy Lauper
I Drove All Night	Roy Orbison
I Drove All Night	Pinmonkey
I Fahr' Daimler	Wolle Kriwanek
I Get Around	The Beach Boys
I Got My Ragtop Down	Neil W. Young
I Like Driving in My Car	Madness
I Love My Car	Belle And Sebastian
I Saw A Cop	Jill Sobule
I Watch the Cars	Robyn Hitchcock
I'm Changing My Name to Chrysler	Arlo Guthrie
I'm Going to Park Myself in Your Arms	Ted Weems And His Orchestra, With Dusty Rhoades
I'm in Love with My Car	Mike Ness
I'm in Love with My Car	Queen
I'm Trading You in on a Later Model	Shot Jackson
I'm Your Vehicle, Baby	Ides Of March
I've Been Everywhere	Johnny Cash
I've Got A Ford Engine Movements In My Hips	Cleo Gibson
I've Been Everywhere In Texas	Brian Burns
If You Drink, Don't Drive	Johnny Rector
Ignition	Brian Setzer
In My Car	The Beach Boys
In My Car	Shinia Twain
In My Merry Oldsmobile	Bonzi Buddy, G. Edwards/V.P. Bryan
In the Driver's Seat	John Schnieder
In the Parking Lot	The Beach Boys
Inner City Suburbs 4WD Association, The	Kedron Taylor
It Was a Good Day	Ice Cube
Jaguar	The Who
Jaguar and the Thunderbird	Chuck Berry
James Dean	The Eagles
Jenny Take a Ride	Mitch Ryder
Jerry Was a Racecar Driver	Primus
Jesus Built My Hot Rod	Ministry
Johnny Kool Pt. 2 (The Legend of)	Brian Setzer
Judy's Got a Stick Shift	The Hot Rods
Jump in My Car	Ted Mulry
Keep Between Them Ditches	Doug Kershaw And The Hazzard County Boys

Keep It Between the Lines	Ricky Van Shelten
Keep on Truckin'	Eddie Kendricks
Keys to the Car, The	Michael Nesmith
Keys To The Highway	Eric Clapton
Killer Cars	Radiohead
King Kong	Jibbs
Kustom Kar Show	Davis Marks And The Marksmen
Last Kiss	J. Frank Wilson
Leader of the Pack, The	Shangri Las
Let Me Be Your Car	Rod Stewart
Let Me Ride	Dr Dre
Let Sally Drive	Sammy Hagar
Let's Take Some Drugs and Drive Around	Silos
Life in the Fast Lane	The Eagles
Life Is a Highway	Tom Cochrane
Life Is a Highway	Rascal Flatts
Life Is a Highway	Chris Ledoux
Lights on the Hill	Slim Dusty
Like a Racecar	Hawk Nelson
Little Deuce Coupe	The Beach Boys
Little G.T.O.	Ronny And The Daytonas
Little Honda	The Beach Boys
Little Honda	The Hondells
Little Old Lady from Pasadena	Jan And Dean
Little Red Corvette	Prince
Little Scrambler	Ronny And The Daytonas
Little Woody, This	The Rip Chords
Long Black Limousine	Merle Haggard
Long Black Limousine	Gram Parsons
Long Black Limousine	Elvis Presley
Long Gone	Monde Yeux
Long May You Run	Neil Young
Long Tall Sally	Elvis Presley
Long Tall Sally	Little Richard
Love in an Automobile	A.R. Dixon
Low Rider	War
Lucille	Fred Eaglesmith
Mabelene	Elaine Britt
Mag Wheels	Dick Dale
Magic Bus	The Who
Makin' Thunderbirds	Bob Seger
Malagueña	Brian Setzer
Manic Mechanic	ZZ Top
Many Fast Cars	Pearl Jam
Maybelline	Chuck Berry

Maybelline	Paul Revere And The Raiders
Meet Virginia	Train
Mercedes Benz	Janis Joplin
Mercury Blues	Ry Cooder & David Lindley
Mercury Blues	K.C. Douglas
Mercury Blues	El Rayo X
Mercury Blues	Alan Jackson
Mercury Blues	David Lindley
Mercury Blues	Steve Miller
Mercury Boogie	K.C. Douglas
Midget Auto Blues	Paul Tutmark
Mighty Big Car	Fred Eaglesmith
Model T Baby	Jack Turner
Motor Away	Guided By Voices
Motor City	Neil Young
Motorcity Madhouse	Ted Nugent
Move Along	The All American Rejects
Mr. Limousine Driver	Grand Funk Railroad
Mr. Policeman	Brad Paisley
Mustang Sally	Buddy Guy
Mustang Sally	Wilson Pickett
Mustang Sally Bought a GTO	John Lee Hooker
My Automobile	Parliament
My Chevrolet	Phil Vassar
My Hooptie	Sir Mix A Lot
My System	Daz Dillinger
Nadine	Chuck Berry
Nationwide	ZZ Top
Nifty 50	The Customs
Night Moves	Bob Seger
No Money Down	Chuck Berry
No Parking Here	Jimmy Littlejohn
No Particular Place to Go	Chuck Berry
No Wheels	Ronny And The Daytonas
Not Fade Away	Buddy Holly
Ode to My Car	Adam Sandler
Ol' '55	The Eagles
Ol' '55	Tom Waits
Old Holden Waltz, The	Kedron Taylor
Old Yellow Car	Dan Seals
On the Road Again	Canned Heat
On the Road Again	Willie Nelson
One Piece at a Time	Johnny Cash
Our Car Club	The Beach Boys
Out with the Wrong Woman	Washboard Sam

Panama	Van Halen
Paradise by the Dashboard Lights	Meat Loaf
Peak Hour	The Moody Blues
Pink Cadillac	Bruce Springsteen
Plymouth Belvedere	Trish Lester
Pontiac	Fred Eaglesmith
Pontiac Blues	Eric Clapton
Pontiac Blues	Sonny Boy Williamson
Poor Man's Friend	Sleepy John Estes
POW 369	Darryl Worley
Racing In The Streets	Bruce Springsteen
Radar Gun	The Bottle Rockets
Radar Love	Golden Earring
Radar Love	Ministry
Ragtop Cadillac	Lonestar
Ragtop Day	Jimmy Buffett
Rambling Man	Allman Brothers
Rapid Roy (The Stock Car Boy)	Jim Croce
Rapture	Blondie
Real Gone Rocket	Jackie Brenstone
Red Barchetta	Rush
Red Corvette, The	John Mccutcheon
Red Hot Roadster	The Rip Cords
Red Hot Rod	Vincent Razorbacks
Red Ragtop	Alan Jackson
Rehab	Amy Winehouse
Rehab	Rihanna
Repossession Blues	Roger Christian
Rev It on the Red Line	Foreigner
Ride, The	David Allen Coe
Ridin' Dirty	Chamillionaire
Ridin' High	The Dixie Dregs
Riding in My Car	NRBQ
Riding with Private Malone	David Bell
Riding with The Blues	Ry Cooder
Riding With The King	John Hiatt
Road Hog	John D. Loudermilk
Road Runner	The Gants
Road to Hell, The	Chris Rea
Road You Leave Behind, The	David Lee Murphy
Road's My Middle Name, The	Bonnie Raitt
Roadhouse Blues	The Doors
Roadrunner	Bo Diddley
Roadrunner	The Modern Lovers
Rocket 88	Bill Haley And The Saddlemen

Rocket 88	Jackie Brenstone & The Kings Of Rhythm With Ike Turner
Rockin' Down the Highway	The Doobie Brothers
Rolls Royce Papa	Virginia Liston
Rooster Rock	Brian Setzer
Route 66	Asleep At The Wheel
Route 66	Bobby Troup
Route 66	Bob Dylan
Route 66	Chuck Berry
Route 66	The Cramps
Route 66	The Rolling Stones
Route 66	Van Morrison
Running on Empty	Jackson Browne
Rusty Chevrolet	Da Yoopers
Santa Rosa Rita	Brian Setzer
Sausalito Summernight	Diesel
Schlock Rod	Jan And Dean
Service Station Blues	Drifting Johnny Smith
Setting the Woods on Fire	Hank Williams
Seven Little Girls Sittin' in the Back Seat	Paul Evans
She Loves My Automobile	Willie Nelson
She Loves My Automobile	ZZ Top
She Won't Turn Over For Me	Floyd Compton
She's My Chevy	Hotrod Hillbillies
Shirley	L7
Shut Down	The Beach Boys
Silver Thunderbird	Marc Cohn
Silver Thunderbird	Jo Dee Messina
Six Days on the Road	Dave Dudley
Six Days on the Road	George Thorogood
Six Days on the Road	Sawyer Brown
Six Days On The Road	Taj Mahal
Six White Cadillacs	Emmylou Harris
Sixteen Days under the Hood	The Paladins
Sixty Days	Slim Rhodes
Skidmarks on My Heart	The Go Go's
Slick Black Cadillac	Quiet Riot
Slick Black Limousine	Alice Cooper
Slow Ride	Foghat
Slow Ride	Kenny Wayne Shepherd
Solid Gold Cadillac	Pearl Bailey
Son of Mustang Ford	Swervedriver
Spacegrass	Clutch
Speed	Montgomery Gentry
Speed Kills	Ten Years After

Speed King	Deep Purple
Speeding	The Go Go's
Spirit of America	The Beach Boys
Sports Model Mama	Bertha Chippie Hill
SS396	Paul Revere And The Raiders
Stand on It	Bruce Springstein
Steven Mcqueen	Sheryl Crow
Stick Shift	The Duals
Stick Shift	The Trashmen
Stick Shift	The Ventures
Stick Shifts and Safety Belts	Cake
Still Cruisin'	The Beach Boys
Street Machine	Sammy Hagar
Stripped Gears and a Broken Heart	Tim Ponzek
Stuck in My Car	The Go Go's
Surf City	Jan And Dean
Swing Low Sweet Cadillac	Dizzy Gillespie
Switchblade 327	Brian Setzer Orchestra
T-Model Boogie	Roscoe Gordon
Take a Little Ride	Jason Aldean
Take it Easy	The Eagles
Teen Angel	Mark Dinning
Tell Laura I Love Her	Ray Peterson
Tennessee Plates	John Hiatt
Terraplane Blues	Eric Clapton
Terraplane Blues	Foghat
Terraplane Blues	John Lee Hooker
Terraplane Blues	Robert Johnson
Texas Bound and Flyin'	Jerry Reed
Texas Plates	Kelly Coffey
Theme From 'The Dukes of Hazzard'	Waylon Jennings
Theme Song from 'Smokey and the Bandit, The	Jerry Reed
This Car of Mine	The Beach Boys
Three Window Coupe	The Rip Cords
Thunder Road	Bruce Springsteen
Traffic Jam	James Taylor
Trampled Underfoot	Led Zeppelin
Trans Am	Neil Young
Trans Am	Sammy Hagar
Transfusion	Nervous Norvus
Trophy Machine	The Rip Cords
Truckin'	The Grateful Dead
Turbo Lover	Judas Priest
Turn the Page	Bob Seger
Turn the Page	Metallica

Two Lane Blacktop	Rob Zombie
Uneasy Rider	Charlie Daniels Band
Up around the Bend	John Fogerty
Vehicle	Ides Of March
Ventura Highway	America
Wake Up Little Suzie	Everly Brothers
Wanderer, The	Dion
Welfare Cadillac	Guy Drake
Wheel, The	Grateful Dead
When Horsepower Meant What It Said	Sandi Thom
Whiskey Girl	Toby Keith
White Line Fever	John Mayall And The Bluesbreakers
Who Would Love This Car But Me?	Brian Setzer
Woody Walk	Shutdown Douglas
Wordplay	Jason Mraz
Wrecked T Bird	Ruby Short
You Can Sleep While I Drive	Melissa Etheridge
You Can Sleep while I Drive	Trisha Yearwood
You're Gonna Get Yours	Public Enemy
Z28	Static X

Appendix E
100 Greatest Car Songs

Compiled by Gup Gascoigne – The 'Golden Gup' –
Cadillac of Mobile Disc-Jockeys
For Lewis Carlock, DigitalDreamDoor, 2010
Reprinted with permission from
www.digitaldreamdoor.com/pages/best_songs-car.html

This list highlights the greatest car songs ever written (presented in ascending alphabetical order). Each example is highly distinct from all others; they are the 100 *best of the best* car songs. DJ Gup Cascoigne used very strict criteria for songs to be placed on an initial 'short-list', and then by calculating an overall score using a rating procedure, songs were accepted for the final cut of the 100-song listing. Foremost, the songs were selected by subject matter: songs about cars were rated higher than songs that merely referenced one. For example, 'Rocket 88' and 'Maybelline', two great tunes that made it to the list, but the former is about a car (an Oldsmobile), while the other is about a girl (Maybelline). Second, good quality automotive references had to be present in the text. For example, "I got a fuel injected engine sittin' under my hood" gets a higher rating than "ridin' in the hood" (as DJ Gup Cascoigne claims that composers who have grease under the fingernails generally write a better *car* song). Finally, the song has to have high popularity; besides killer musicianship, and lyrics with auto references, if the song has not made a mark on pop culture, then it would be rated lower and subsequently denied entry to the list. Titles do not appears more than once, as only the best version (artist/performance) is listed.

'49 Mercury Blues	Brian Setzer Orchestra
409	The Beach Boys
455 Rocket	Kathy Mattea
Ain't Nothing Wrong with the Radio	Aaron Tippen
Be Thankful for What You Got	William DeVaughn
Beep Beep	Playmates
Black and White Thunderbird	Delicates
Black Sunshine	White Zombie
Buick 59	Medallions
Built for Speed	Stray Cats
Bump in the Trunk	Bone Thugs-n-Harmony
Cadillac Jack	Andre Williams
Cadillac Ranch	Bruce Springsteen

Cadillac Red	Judds
Car Crazy Cutie	Beach Boys
Car Wash	Rose Royce
Cars	Gary Numan
Cars with the Boom	L'Trimm
Crown Victoria Custom '51	Jerry Lee Lewis
Dead Man's Curve	Jan & Dean
Dear Dad	Chuck Berry
Deuce and a Quarter	All The Kings Men
Don't Worry Baby	The Beach Boys
Drag City	Jan & Dean
Drive It Home	Little Caesar
Drive My Car	Beatles
Fireball Rolled a Seven	Dave Dudley
Fishtail Blues	Wynonie Harris
Fun Fun Fun	Beach Boys
Getaway Car	Audioslave
G.T.O.	Ronny and the Daytonas
Goin' Mobile	The Who
Greased Lightnin'	John Travolta 'Grease' Soundtrack
Heavy Chevy	Grin
Hey Little Cobra	Rip Chords
Highway Star	Deep Purple
Hot Rod Gang	Stray Cats
Hot Rod Lincoln	Commander Cody/Charlie Ryan/Johnny Bond/ Bill Kirchen
Hot Rod Man	Tex Rubinowitz
Hot Rod Race	Arkie Shibley & Ramblin' Jimmy Dolan
I Gotta New Car	Big Boy Groves
I'm in Love with My Car	Queen
In My Merry Oldsmobile	Billy Murray
It Was a Good Day	Ice Cube
Jaguar and the Thunderbird	Chuck Berry
Jeepster	T-Rex
King Kong	Jibbs
Last Kiss	J. Frank Wilson and the Cavaliers
Let It Roll	Little Feat
Little Deuce Coupe	Beach Boys
Little Red Corvette	Prince
Little Rivi	Airhead
Long May You Run	Neil Young
Long White Cadillac	Dwight Yoakam
Look at That Cadillac	Stray Cats
Lord Mr. Ford	Jerry Reed
Low Rider	War

Makin' Thunderbirds	Bob Seger
Maybelline	Chuck Berry
Mercedes Benz	Janis Joplin
Mercury Blues	K.C. Douglas/Alan Jackson/Steve Miller Band
Motoring	Martha and the Vandellas
Mustang Sally	Wilson Pickett/Sir Mack Rice
My Hooptie	Sir Mix
My Old Car	Lee Dorsey
My System	Daz Dillinger
No Money Down	Chuck Berry
No Particular Place to Go	Chuck Berry
Ol '55	Eagles
Old Betsy Goes Boing, Boing, Boing	Hummers
One Piece at a Time	Johnny Cash and the Tennessee Three
Panama	Van Halen
Pink Cadillac	Bruce Springsteen
Pontiac Blues	Sonny Boy Williamson
Race With the Devil	Gene Vincent
Racing in the Street	Bruce Springsteen
Radar Love	Golden Earring
Rapid Roy	Jim Croce
Red Barchetta	Rush
Red Cadillac and a Black Moustache	Bob Luman/Warren Smith
Rev on the Red Line	Foreigner
Ride on Josephine	George Thorogood/Bo Diddley
Rocket 88	Jackie Brenston and his Delta Cats
Rockin' Down the Highway	Doobie Brothers
Roll On Down the Highway	B.T.O.
Rollin' in My Rolls	Moon Martin
Rusty Chevrolet	Da Yoopers
See the U.S.A. in Your Chevrolet	Dinah Shore
(Seven Little Girls) Sitting in the Back Seat	Paul Evans
Shut Down	Beach Boys
SS 396	Paul Revere and the Raiders
Still Cruisin'	The Beach Boys
'T' Model Blues	Lightnin' Hopkins
Tell Laura I Love Her	Ray Peterson
Terraplane Blues	Robert Johnson
The Ballad Of Thunder Road	Robert Mitchum
The Little Old Lady (from Pasadena)	Jan & Dean
The One I Loved Back Then	George Jones
Transfusion	Nervous Norvus
V8 Ford (Going to Your Funeral)	Buddy Moss
Vehicle	Ides of March
Voodoo Cadillac	Southern Culture on the Skids

White Cadillac	The Band
Whose Cadillac Is That	War
You Can't Catch Me	Chuck Berry
You're Not Safe in a Japanese Car	John Goldsmith

Appendix F
Driving Tracks

Compiled by Erez Ein Dor

This listing of 24 CDs (that include 655 tracks), presented according to musical style genre, is a compilation of English language internationally distributed CD collections dedicated to accompanied driving. In general, these collections are the 'Top-10 Best Sellers' assembled from Amazon online shopping (www.amazon.com) in the category of 'music for driving'. The criteria for CDs to be placed on this listing, are: CDs specifically collected, labelled with title, and marketed as Driving Music; and music tracks were performed by *bona fide* original or well-known performing artists (that is, CDs with cover bands, instrumental versions, and copy artists were omitted). The entries on the listing are exclusive for everyday automobile and truck driving (as CDs highlighting motorcycling were not listed). Further, the list presents a wide diversity of music genres, including: Pop and Rock, Country, Funk and Jazz, and Classical music styles.

I. Rock and Pop	*101 Driving Songs* (Phantom Sound and Vision)
	Best of Driving Rock (EMI Gold Imports)
	Cruisin' – Driving Songs (Collecting Records OMP)
	Driving in the Rain 3 AM: Songs to Get Lost with (Bongo Beat Records)
	Driving Songs (EMI Operations)
	Hot Wheels and Highways: Great Driving Songs (Golden Stars Holland)
	Late Night Driving Songs (Ectypal Music OMP)
	Music for Drunk Driving (Asphodel Records)
	Top Gear (Universal UK)
	Top Gear: Seriously Cool Driving Music (EMI Operations/Ceva Logistics)
	Ultimate Driving Collection: California (Polygram Records)
II. Country Music	*Best of the Truck Driving Songs* (Hollywood)
	Classic Country: Road Songs (Shamrock-n-Roll)
	Eighteen Truck Driving Classics (Double Play)
	Road Music: Truckin' Favorites – Vol. 1 (Tee Vee Records)
	Truckers Jukebox (Time Life Records)
	What a Truckin' Life! Country Music Road Driving Feelgood Songs (Mediaworld)
III. Dance and DJ	*A Little Driving Music* (Allosonic)
	Best of Driving Forces Vol.1 (Driving Forces Recordings)

IV. Funk and Jazz	*Funkengruven: The Joy Of Driving a B3* (Allegro Chicken Coup)
	Ultimate Driving Collection: Smooth Ride (Polygram Records)

V. Classical Music	*My Classical Life, 40 Classical Songs for Driving* (U-5)
	Music for Driving (Sony/BMG Int.)
	Must Have Driving Classics (Gut Active)

References

AAA. (2010a). 86 Per cent of Teens Have Driven while Distracted, According to AAA and *Seventeen* Magazine. *AAA NewsRoom.* http://usatoday30.usatoday.com/news/nation/2010–08–02-teendrivers02_ST_N.htm?loc=interstitialskip.

AAA. (2010b). AAA, *Seventeen* and DOT Recognize National Two-Second Turnoff Day. *AAA NewsRoom.* http://newsroom.aaa.com/2010/09/aaa-seventeen-and-dot-recognize-national-two-second-turnoff-day/.

AASHTO. (2010). Survey Finds 86% Teens Have Driven while Distracted. *AASHTO Journal: Weekly Transportation Report. Journal of the American Association of State Highway and Transportation Officials.* http://www.aashtojournal.org/Pages/080610distracted.aspx.

Abeles, H.F. (1980). Responses to Music. *In* D.A. Hodges (ed.), *Handbook of Music Psychology* (1st ed., pp. 105–140). Lawrence, KS: National Association for Music Therapy.

Abeles, H.F., & Chung, J.W. (1996). Responses to Music. *In* D.A. Hodges (ed.), *Handbook of Music Psychology* (2nd ed., pp. 285–342). San Antonio: IMR Press.

ACF. (2009). Duffy Drives Us Home Safely. *ACF Car Finance.* http://www.acfcarfinance.co.uk/news/duffy-drives-us-home.

Akesson, K.-P., & Nilsson, A. (2002). Designing Leisure Applications for the Mundane Car-Commute. *Personal and Ubiquitous Computing* 6, 176–187.

Allianz Insurance (2014). Personal Communication, March 10, to W. Brodsky from M. Bishop – Music Story Data Broken Down.

Alpert, M.I., Alpert, J.I., & Maltz, E.N. (2005). Purchase Occasion Influence on the Role of Music in Advertising. *Journal Of Business Research* 58, 369–376.

Anderson, C.A., Carnagey, N.L., & Eubanks, J. (2003). Exposure to Violent Media: The Effects of Songs with Violent Lyrics on Aggressive Thoughts and Feelings. *Journal of Personality and Social Psychology* 84(5), 960–971.

Andringa, T.C., & Lanser, J.J.L. (n.d.). *Pleasure and Annoying Sounds and How these Impact on Life*. Unpublished paper, University of Groningen. Accessed January 2014. www.ai.rug.nl/~tjeerd/publications/Andringa_2011S1.pdf.

Andringa, T.C., & Lanser, J.J.L. (2013). How Pleasant Sounds Promote and Annoying Sounds Impede Health: A Cognitive Approach. *International Journal of Environmental Research and Public Health* 10, 1439–1461.

Angell, L., Auflick, J., Austria, P.A., Kochhar, D., Tijerina, L., Biever, W., Diptiman, T., Hogsett, J., & Kiger, S. (2006). *Driver Workload Metrics Task 2 Final Report (DOT HS 810 635)*. Washington, DC: National Highway Traffic Safety Administration, U.S. Department of Transportation.

Anund, A., Kecklund, G., Peters, B., & Akerstedt, T. (2008). Driver Sleepiness and Individual Differences in Preferences for Countermeasures. *Journal of Sleep Research* 17, 16–22.

Arbitron/Edison. (1999). *Los Angeles In-Car Listening Study: Arbitron and Edison Media Research.* http://asaha.com/ebook/UNDY4NzI-/The-Los-Angeles-In-Car-Listening-Study.pdf.

Arbitron/Edison. (2003a). *The Arbitron National In-Car Study: Arbitron and Edison Media Research.* www.vehiclewrappricing.com/LiteratureRetrieve.aspx?ID=82711.

Arbitron/Edison. (2003b). *Shifting Gears: The UK In-Car Study: Arbitron and Edison Media Research.* www.edisonresearch.com/homeimg/archives/NAB%20UK%20Final%20Draft.pdf.

Arbitron/Edison. (2011). *The Road Ahead: Media and Entertainment in the Car: Arbitron and Edison Research Media Group.* www.arbitron.com/home/studies_chron.htm and www.edisonresearch.cm/?s=in-car+listening and www.slideshare.net/webby2001/the-road-ahead-in-car-entertainment-2011-from-edison-research and www.arbitron.com/downloads.the_road_ahead_2011.pdf.

Arbitron/Edison. (2013). *The Infinite Dial 2013: Navigating Digital Platforms: Arbitron/Edison, Report.* www.edisonresearch.com/wp-content/uploads/2012/04/2012_infirnate_dial_companiaon_report.pdf. Slides at: www.edisonresearch.com/wp-content/uploads/2012/04/Edison_Arbitron_Infinate_Dial_2012.pdf.

Areni, C.S. (2003). Exploring Managers' Implicit Theories of Atmospheric Music: Comparing Academic Analysis to Industry Insight. *The Journal of Services Marketing* 17(2/3), 161–184.

Areni, C.S., & Kim, D. (1993). The Influence of Background Music on Shopping Behavior: Classical versus Top-Forty Music in a Wine Store. *Advances in Consumer Research* 20, 336–340.

Arnett, J. (1991a). Adolescents and Heavy Metal Music: From the Mouth of Metalheads. *Youth and Society* 23(1), 76–98.

Arnett, J. (1991b). Heavy Metal Music and Reckless Behavior among Adolescents. *Journal of Youth and Adolescence* 20(6), 573–591.

Arnett, J. (2002). Developmental Sources of Crash Risk in Young Drivers. *Injury Prevention* 8(Suppl. II), ii17–ii23.

Atchley, P., & Dressel, J. (2004). Conversation Limits the Functional Field of View. *Human Factors* 46(4), 664–673.

AutosCom. (n.d.). Car Radios: A History of Mobile Audio. *Autos.com.* Retrieved 9.2.2013. www.autos.com/aftermarket-parts/car-radios-a-history-of-mobile-audio.

AutoTraderBlog. (2012). Dire Straits Beats Queen in Auto Trader's Top 10 Driving Songs. *Auto Trader Blog.* http://autotraderblog.co.uk/2012/07/12/dire-straits-beats-queen-in-auto-trader-top-ten-driving-songs/.

Ayres, T.J., & Hughes, P. (1986). Visual Acuity with Noise and Music at 107 dbA. *The Journal of Auditory Research* 26(1), 65–74.

Bach, K.M., Jaeger, M.G., Skov, M.B., & Thomassen, N.G. (2009). Interacting with In-Vehicle Systems: Understanding, Measuring, and Evaluating Attention. *HCI2009 – People and Computers XXIII – Celebrating People and Technology, Proceedings of the 23rd British HCI Group Annual Conference* (pp. 453–462). Liverpool, UK: British Computer Society.

Bailey, N., & Areni, C.S. (2006). When a Few Minutes Sounds Like a Lifetime: Does Atmosphere Music Expand or Contract Perceived Time? *Journal of Retailing* 82(3), 189–202.

Baltrunas, L., Kaminskas, M., Ludwig, B., Moling, O., Ricci, F., Aydin, A., Luke, K.-H., & Schwaiger, R. (2011). InCarMusic: Context-Aware Music Recommendations in a Car. *In* C. Huemer & T. Setzer (eds.). *Proceedings of the 12th International Conference of E-Commerce and Web Technologies, EC-Web 2011* (pp. 89–100). Toulouse, France, August 30–September 1, 2011.

Barari, A. (2010). SEAT Studies Driving Songs! *MOTORWARD (MW).* http://www.motorward.com/2010/08/seat-studies-driving-songs/.

Barnet, R.D., & Burris, L.L. (2001). *Controversies of the Music Industry*. Westport, CT: Greenwood Press.

Barretts. (2010). SEAT Study – Driving Song Secrets. *Barretts Seat Kent.* http://seat.barrettskent.co.uk/2010/08/seat-study-driving-song-secrets.html.

Bartlett, D.L. (1996). Physiological Responses to Music and Sound Stimuli. *In* D.A. Hodges (ed.), *Handbook of Music Psychology*. 2nd ed. (pp. 343–386). San Antonio: IMR Press.

Bayly, M., Young Kristies, L., & Regan, M.A. (2009). Sources of Distraction inside the Vehicle and Their Effects on Driving Performance. *In* M. Regan, J.D. Lee, & K.L. Young (eds). *Driver Distraction: Theory, Effects, and Mitigation* (pp. 191–214). Boca Raton, FL: CRCPress, Taylor and Francis Group.

BBC. (2003). Fast Music – Dangerous Driving: High Speed Music Is Linked to High Speed Driving. *BBC News, Health On-line.* http://news.bbc.co.uk/2/hi/health/1870853.stm.

BBC. (2004). Drivers Warned against Loud Music: Listening to Loud Music while Driving Can Seriously Hamper Reaction Times and Cause Accidents, New Research Suggests. *BBC News.* http://news.bbc.co.uk/2/hi/uk_news/3623237.stm.

Beckmann, J. (2004). Mobility and Safety. *Theory, Culture and Society* 21(4/5), 81–100.

Beh, H.C., & Hirst, R. (1999). Performance on Driving-Related Tasks During Music. *Ergonomics* 42(8), 1087–1098.

Behne, K.-E. (1997). The Development of 'Musikerleben' in Adolescence: How and Why Young People Listen to Music. *In* I. Deliege & J. Sloboda (eds), *Perception and Cognition of Music* (pp. 143–160). East Sussex, UK: Psychology Press.

Bellamy, J. (2013). Bad Driver Blames Music. *The Daily Star.* http://www.dailystar.co.uk/posts/view/307334Bad-driver-blames-music.

Bellinger, D.B., Budde, B.M., Machida, M., Richardson, G.B., & Berg, W.P. (2009). The Effect of Cellular Telephone Conversation and Music Listening on Response Time in Braking. *Transportation Research Part F,* 12, 441–451.

Belojevic, G., Slepcevic, V., & Jarkovljevic, B. (2001). Mental Performance in Noise: The Role of Introversion. *Journal of Environmental Psychology* 21(209–213).

Berger, I. (2002). Car as Concert Hall: Audio on the Road. *New York Times,* 12 April 2002.

Berkowitz, J. (2010). The History of Car Radios: Car Tunes; Life before Satellite Radio. *Car and Driver: Intelligence, Independence, Irreverence.* www.caranddriver.com/features/the-history-of-car-radios.

Bermudez, P., & Zotorre, R.J. (2005). Difference in Gray Matter between Musicians and Non-Musicians. *Annals of The New York Academy of Sciences* 1060 (The Neurosciences and Music II: From Perception to Performance), 395–399.

Berti, S., Munzer, S., Schroger, E., & Pechmann, T. (2006). Different Interference Effects in Musicians and Control Groups. *Experimental Psychology* 53(2), 111–116.

Besson, M., & Faita, F. (1995). An Event Related Potential (ERP) Study of Musical Expectancy: Comparison of Musicians with Non-Musicians. *Journal of Experimental Psychology: Human Perception and Performance,* 21(6), 1278–1296.

Betts, S.L. (2009). Taylor Swift's 'Love Story' Encourages Safe Driving? *The Boot.* http://www.theboot.com/2009/03/12/taylor-swifts-love-story-encourages-safe-driving/.

Bever, T.G., & Chiarello, R.J. (1977). Cerebral Dominance in Musicians and Non-Musicians. *Science* 185, 537–539.

Blattner, M.M., Sumikawa, D.A., & Greenberg, R.M. (1989). Earcons and Icons: Their Structure and Common Design Principles. *Human-Computer Interaction* 4, 11–44.

Blesser, B. (2007). The Seductive (Yet Destructive) Appeal of Loud Music *eContact! 9.4 – Hearing (Loss) and Related Issues* (June). Montreal: Canadian Electroacoustic Community.

Blesser, B., & Salter, L.-R. (2007). *Spaces Speak, Are You Listening? Experiencing Aural Architecture*. Cambridge, MA: MIT Press.

Bouvard, P., Rosin, L., Snyder, J., & Noel, J. (2003). *The Arbitron National In-Car Study*. New York: Arbirtron.

Boxberger, R. (1963). Historical Bases for the Use of Music in Therapy. *In* E.H. Schneider (ed.). *Music Therapy 1961* 11, 125–168). Lawrence, KS: National Association for Music Therapy.

Braitman, K.A., Kirley, B.B., McCartt, A.T., & Chaudhary, N.K. (2008). Crashes of Novice Teenage Drivers: Characteristics and Contributing Factors. *Journal of Safety Research* 39, 47–54.

Brice, M. (2013). Black Eyed Peas Most Dangerous Music to Listen to while Driving, Coldplay Least. *Medical Daily.* http://www.medicaldaily.com/articles/13940/20130117/black-eyed-peas-dangerous-music-listen-driving.htm.

Brodsky, W. (2002). The Effects of Music Tempo on Simulated Driving Performance and Vehicular Control. *Transportation Research Part F* 4, 219–241.

Brodsky, W. (2013). *Background Music As a Risk Factor for Distraction among Young Drivers: An IVDR Study*. (Translated into Hebrew by Zack Slor.) Jerusalem, Israel: Israel National Road Safety Authority. www.ras.gov.il/meidamechkar/MechkarimSkarim/Documents/Music2013.pdf.

Brodsky, W., Henik, A., Rubinstein, B.S., & Zorman, M. (2003). Auditory Imagery from Musical Notation in Expert Musicians. *Perception and Psychophysics* 65(4), 602–612.

Brodsky, W., Kessler, Y., Rubinstein, B.-S., Ginsborg, J., & Henik, A. (2008). The Mental Representation of Music Notation: Notational Audiation. *Journal of Experimental Psychology: Human Perception and Performance*.

Brodsky, W., & Kizner, M. (2012). Exploring an Alternative In-Car Music Background Designed for Driver Safety. *Transportation Research, Part F: Traffic Psychology and Behaviour* 15(3), 162–173.

Brodsky, W., & Slor, Z. (2013). Background Music as a Risk Factor for Distraction among Young-Novice Drivers. *Accident Analysis and Prevention* 59, 382–393.

Broekemier, G., Marquardt, R., & Gentry, J.W. (2008). An Exploration of Happy/Sad and Liked/Disliked Music Effects on Shopping Intentions in a Women's Clothing Store Service Setting. *Journal of Service Marketing* 22(1), 59–67.

Brookhuis, K.A., & de Waard, D. (1993). The Use of Psychophysiology to Assess Driver Status. *Ergonomics* 36, 1099–1110.

Brown, I.D. (1965). Effect of Car Radio on Driving in Traffic. *Ergonomics* 8(4), 475–479.

Brown, I.D. (1994). Driver Fatigue. *Human Factors* 36(2), 298–314.

Brown, I.D., Tickner, A.H., & Simmonds, D.C.V. (1969). Interference between Concurrent Tasks of Driving and Telephoning. *Journal of Applied Psychology* 55(5), 419–424.

Bruce, B. (2010). Police: Loud Music Leads to Crash Involving Ambulance (21 October 2010). *KSL.com.* http://www.ksl.com/?nid=148andsid=12911899.

Brumby, D.P., Salvucci, D.D., Mankowski, W., & Howes, A. (2007). A Cognitive Constraint Model of the Effects of Portable Music-Player Use on Driver Performance *Proceedings of the Human Factors and Ergonomic Society 51st Annual Meeting*. Santa Monica, CA: Human Factors and Ergonomic Society.

Bruner II, G.C. (1990). Music, Mood, and Marketing. *Journal Of Marketing* 54(4), 94–104.

Budiansky, S. (2002). Missing Pieces: The Strange Case of the Disappearing Arias and Adagios. *Atlantic Monthly*.

Bull, M. (2001). Soundscapes of the Car: A Critical Study of Automobile Habituation. *In* D. Miller (ed.), *Car Cultures* (pp. 185–202). Oxford, UK: Oxford University Press.

Bull, M. (2004). Automobility and the Power of Sound. *Theory, Culture and Society* 21(4/5), 243–259.

Bull, M. (2005). No Dead Air! The iPod and the Culture of Mobile Listening. *Leisure Studies* 24(4), 343–355.

Burns, J.E., & Sawyer, P.R. (2010). The Portable Music Player as a Defense Mechanism. *Journal of Radio and Audio Media* 17(1), 2012.

Button, D.C., Behm, D.G., Holmes, M., & Mackinnon, S.N. (2004). Noise and Muscle Contraction Affects Vigilance Task Performance. *Occupational Ergonomics* 4, 157–171.

Caldwell, C., & Hibbert, S.A. (2002). The Influence of Music Tempo and Musical Preference on Restaurant Behavior. *Psychology and Marketing* 19(11), 895–917.

Cameron, M.A., Baker, J., Peterson, M., & Braunsberger, K. (2003). The Effects of Music, Wait-Length Evaluation, and Mood on a Low-Cost Wait Experience. *Journal of Business Research* 56, 421–430.

Campbell, K. (2002). If You Drive, Don't Groove: While Driving, Crank with Care. *Popular Science* 260 (no. 6), 38.

CanadianUnderwriter. (2013). Talk Radio, Music Genres Have Effect on Driving Behaviour, Survey Says. *Canadian Underwriter: Canada's Insurance and Risk Magazine.* www.canadianunderwriter.ca/news/talk-radio-music-genres-have-effect-on-driving-behaviour-survey-says/1002507379/.

Carpentier, F.D., Knobloch, S., & Zillmann, D. (2003). Rock, Rap, and Rebellion: Comparisons of Traits Predicting Selective Exposure to Defiant Music. *Personality and Individual Differences* 33, 1643–1655.

Caspy, T., Peleg, E., Schlam, D., & Goldberg, J. (1988). Sedative and Stimulative Music Effects: Differential Effects on Performance Impairment following Frustration. *Motivation and Emotion* 12(2), 123–137.

Cassidy, G., & MacDonald, R.A.R. (2007). The Effect of Background Music and Background Noise on the Task Performance of Introverts and Extroverts. *Psychology of Music* 35(3), 517–537.

Cassidy, G., & MacDonald, R.A.R. (2010). The Effects of Music on Time Perception and Performance of a Driving Game. *Scandinavian Journal of Psychology* 51, 455–464.

Cattell, R.B., & Anderson, J.C. (1953). The Measurement of Personality and Behavior Disorders by the I.P.A.T. Music Preference Test. *The Journal of Applied Psychology* 37(6), 446–454.

Cestac, J., Oaran, F., & Delhomme, P. (2010). Young Drivers' Sensation Seeking, Subjective Norms, and Perceived Behavioral Control and their Roles in Predicting Speeding Intention: How Risk-Taking Motivations Evolve with Gender and Driving Experience. *Safety Science* 49(3), 424–432.

Chafin, S., Roy, M., Gerin, W., & Christenfeld, N. (2004). Music Can Facilitate Blood Pressure Recovery from Stress. *British Journal of Health Psychology* 9, 393–403.

Chamorro-Premuzic, T., & Furnham, A. (2007). Personality and Music: Can Traits Explain How People Use Music in Everyday Life? *British Journal of Psychology* 98, 175–185.

Chisholm, S.L., Caird, J.K., & Lockhart, J. (2008). The Effects of Practice with MP3 Players on Driving Performance. *Accident Analysis and Prevention* 40, 700–713.

Christman, S.D. (2013). Handedness and 'Open-Eardness': Strong Right-Handers Are Less Likely to Prefer Less Popular Musical Genres. *Psychology of Music* 41(1), 89–96.

Claims Journal. (2012). Survey Finds Music Adds to Distracted Driving. *Claims Journal*. www.claimsjournal.com/news/international/2012/08/31/212934.htm.

Clarke, E., Dibben, N., & Pitts, S. (2010). *Music and Mind in Everyday Life*. Oxford: Oxford University Press.

clickondetroit.com. (2006a). Four Detroit Firefighters Injured When Fire Truck Crashes into Home (27 June 2006). *FireRescue1.com*. www.firerescue1.com/fire-products/fire-apparatus/articles/107617-Four-Detroit-firefighters-injured-when-fire-truck-crashes-into-home/.

clickondetroit.com. (2006b). Detroit Fire Truck Crashes into House (27 June 2006). *Firehouse*. www.firehouse.com/news/10505320/detroit-fire-truck-crashes-into-house.

CNW-Newswire. (2013). Driving the Beat of the Music Can Lead to Higher Insurance Rates. *CNW A PR Newswire Company*. www.newswire.ca/en/story/1205443/driving-to-the-beat-of-the-music-can-lead-to-higher-insurance-rates.

Colley, A. (2008). Young People's Music Taste: Relationship with Gender and Gender-Related Traits. *Journal of Applied Social Psychology* 38(8), 2039–2055.

Confused.com. (2013a). The Ultimate Driving Songs Playlist – Download the Safest Songs to Drive To. *Confused.com*. www.confused.com/car-insurance/ultimate-driving-songs-playlist.

Confused.com. (2013b). Ultimate Driving Playlist. *Spotify*. http://open.spotify.com/user/www.confused.com/playlist/5jpkSit5EmTBtqOQwYRas5.

Connolly, T., & Alberg, L. (1993). Some Contagion Models of Speeding. *Accident Analysis and Prevention* 25, 57–66.

Consiglio, W., Driscoll, P., Witte, M., & Berg, W.P. (2003). Effect of Cellular Telephone Conversations and Other Potential Interference on Reaction Time in a Braking Response. *Accident Analysis and Prevention* 35, 495–500.

Copeland, L. (2010). Most Teens Still Driving while Distracted. *USA Today*.

Corhan, C.M., & Roberts-Gounard, B. (1976). Types of Music, Schedules of Background Stimulation, and Visual Vigilance Performance. *Perceptual and Motor Skills* 42, 662.

Cosper, A. (n.d.). The History of Car Audio. *eHow tech*. Retrieved 9/2/2013, from www.ehow.com/about_5380379_history-car-audio.html.

Crundall, D., & Underwood, G. (1998). The Effects of Experience and Processing Demands on Visual Information in Drivers. *Ergonomics* 41, 448–458.

Crundall, D., & Underwood, G. (2001). The Priming Function of Road Signs. *Traffic Research, Part F* 4, 187–200.

Crundall, D., Bains, M., Chapman, P., & Underwood, G. (2005). Regulating Conversation during Driving: A Problem for Mobile Phones. *Transportation Research Part F,* 8, 197–211.

Crutchfield. (n.d.). Multichannel Music in Your Car. *Crutchfield.* Retrieved 10.2013 from www.crutchfield.com/S-dPUWfRFjW9P/learn/learningcenter/car/video_surround.html.

Cummings, P., Koepsell, T.D., Moffat, J.M., & Rivara, F.P. (2001). Drowsiness, Counter-Measures to Drowsiness, and the Risk of a Motor Vehicle Crash. *Injury Prevention, 7*(3), 194–199.

Dalton, B.H., & Behm, D.G. (2007). Effects of Noise and Music on Human and Task Performance: A Systematic Review. *Occupational Ergonomics* 7, 143–152.

Dalton, B.H., Behm, D.G., & Kibele, A. (2007). Effects of Sound Types and Volumes on Simulated Driving, Vigilance Tasks, and Heart Rate. *Occupational Ergonomics* 7, 153–168.

Dant, T. (2004). The Driver-Car. *Theory, Culture and Society* 21(4/5), 61–79.

DavisLawGroup. (2010). Teens Still Engaging in Distracted Driving: Seattle Injury. *Davis Law Group, Injury Trial Lawyer.* www.injurytriallawyer.com/news/teens-still-engaging-in-distracted-driving-seattle-injury-lawyer20100803.cfm.

Delsing, M.J.M.H., Bogt, T.F.M.T., Engles, R.C.M.E., & Meeus, W.H.J. (2008). Adolescents' Music Preferences and Personality Characteristics. *European Journal of Personality* 22, 109–130.

DeMain, B. (2012). When the Car Radio Was Introduced, People Freaked Out. *Mental_Floss.* www.mentalfloss.com/article/29631/when-the-car-was-introduced-people-freaked-out.

DeNora, T. (2000). *Music in Everyday Life.* Cambridge: Cambridge University Press.

DeNora, T. (2003). The Sociology of Music Listening in Everyday Life. *In* R. Kopiz, A.C. Lehmann, I. Wolther, & C. Wolf (eds.), *Proceedings of the 5th Triennial ESCOM Conference, Hanover, Germany, 8–13 September 2003.* Hanover, Germany: Hanover University of Music and Drama, Germany.

de Waard, D., & Brookhuis, K.A. (1991). Assessing Driver Status: A Demonstration Experiment on the Road. *Accident Analysis and Prevention* 23, 297–307.

Dibben, N., & Williamson, V.J. (2007). An Exploratory Survey of In-Vehicle Music Listening. *Psychology of Music* 35(4), 571–590.

Dingus, T.A., Klauer, S.G., Neale, V.L., Petersen, A., Lee, S.E., Sudweeks, J., Perez, M.E., Hankey, J., Ramsey, D., Gupta, S., Bucher, C., Doerzaph, Z.R., Jermeland, J., Knipling, R.R. (2006). *The 100-Car Naturalistic Driving Study Phase II-Results of the 100-Car Field Experiment.* Washington, DC: National Highway Traffic Safety Administration.

Direct Line Insurance. (n.d.). Music to Drive By: Music Effects on Drivers and Driving Safety. *Direct Line Insurance.* www.directline.com/car-insurance/advice/motoring/driving-music#.VCzrluc2arc.

Dittmar, H. (2004). Are You What You Have? *The Psychologist* 17(4), 206–210.

Dobbs, S., Furnham, A., & McClelleand, A. (2011). The Effects of Background Music and Noise on Cognitive Test Performance of Introverts and Extroverts. *Applied Cognitive Psychology* 25, 307–313.

Dolak, K. (2013a). Want to Drive Safer? Switch the Radio to Coldplay. *ABC News, Music.* http://abcnews.go.com/blogs/entertainment/2013/01/want-to-driver-safer-switch-the-radio-to-coldplay/.

Dolak, K. (2013b). Want to Drive Safer? Listen to Coldplay. *ABC News Blogs, Good Morning America.* http://news.yahoo.com/blogs/abc-blogs/want-driver-safer-switch-radio-coldplay-132826386--abc-news-music.html.

Drews, F.A., Pasupathi, M., & Strayer, D.L. (2008). Passenger and Cell Phone Conversations in Simulated Driving. *Journal of Experimental Psychology: Applied* 14(4), 392–400.

Dube, L., & Morin, S. (2001). Background Music Pleasure and Store Evaluation Intensity Effects and Psychlogical Mechanisms. *Journal of Business Research* 54, 107–113.

Du Lac, J.F. (2008). Rollin on Empty. *Washington Post*, 7 September 2008.

Eby, D.W., & Kostyniuk, L.P. (2003). Driver Distraction and Crashes: An Assessment of Crash Databases and Review of the Literature. Technical Report UMTRI-2003-12. Ann Arbor, MI: University of Michigan, Transportation Research Institute. http://deepblue.lib.umich.edu/bitstream/2027.42/1533/2/97314.0001.001.pdf.

Edison. (2013). The Streaming Audio Task Force: The New Mainstream; Edison Research. http://www/edisonresearch.com/wp-content/uploads/2013/09/The-New_Mainstream-2013-from-Edison-Research.pptx.pdf.

EMS1. (2010). Loud Music Leads to Crash Involving Utah Ambulance: The Driver of a Car Failed to Hear the Sirens because His Music Was So Loud, Say Police (22 October 2010). *EMS1.com.* www.ems1.com/ems-news/898652-loud-music-leads-to-crash-involving-utah-ambulance.

Engström, J., Johansson, E., & Östlund, J. (2005). Effects of Visual and Cognitive Load in Real and Simulated Motorway Driving. *Transportation Research Part F,* 8, 97–120.

Erb, E., & Stichling, J. (2012). First Car Radios: History and Development of Early Car Radios. *Radio Museum.* www.radiomeseum.org/forum/first_car_radios_history.

Eyerman, R., & Lofgren, O. (1995). Romancing the Road: Road Movies and Images of Mobility. *Theory, Culture and Society* 12(1), 53–79.

Fagerstrom, K.O., & Lisper, H.O. (1977). Effects of Listening to Car Radio, Experience, and Personality of the Driver on Subsidiary Reaction Time and Heart Rate in a Long-Term Driving Task. *In* R.R. Mackie (ed.), *Vigilance: Theory, Operational Performance, and Physiological Correlates* (pp. 73–85). New York: Plenum Press.

Fagioli, S., & Ferlazzo, F. (2006). Shifting Attention across Spaces While Driving: Are Hands-Free Mobile Phones Really Safe? *Cognitive Processes* 7 (Supplement 1), S147.

Fairclough, S., van der Zwaag, M.D., Spiridon, E., & Westerink, J.H.D.M. (2014). Effects of Mood Induction via Music on Cardiovascular Measures of Negative Emotion during Simulated Driving. *Physiology and Behavior 129*, 173–180.

Farrier, J. (2013). Man Arrested for Driving under the Influence of Awesome Music. *Neatorama.* www.neatorama.com/2013/04/04/Man-Arrested-for-Driving-under-the-Influence-of-Awesome-Music/.

Featherstone, M. (2004). Automobiles: An Introduction. *Theory, Culture and Society* 21(4/5), 1–24.

Ferlazzo, F., Fagioli, S., Di Nocera, F., & Sdola, S. (2008). Shifting Attention Across Near and Far Spaces: Implications for Use of Hands-Free Cell Phones While Driving. *Accident Analysis and Prevention* 40(6), 1859–1864.

Field, P. (2002). No Particular Place to Go. *In* P. Wollen & J. Keer (eds), *Autopia: Cars and Culture* (pp. 59–64). London: Reakton Books.

Fischer, P., & Greitemeyer, T. (2006). Music and Aggression: The Impact of Sexual-Aggressive Song Lyrics on Aggression-Related Thoughts, Emotions, and Behavior toward the Same and the Opposite Sex. *Personality and Social Psychology Bulletin* 32(9), 1165–1176.

Fraine, G., Smith, S.G., Zinkiewicz, L., Chapman, R., & Sheehan, M. (2007). At Home on the Road? Can Drivers' Relationships with Their Cars Be Associated with Territoriality? *Journal of Environmentl Psychology* 27, 204–214.

Furnham, A., & Allass, K. (1999). The Influence of Musical Distraction of Varying Complexity on the Cognitive Performance of Extroverts and Introverts. *European Journal of Personality* 13, 27–38.

Furnham, A., & Bradley, A. (1997). Music While You Work: The Differential Distraction of Background Music on the Cognitive Test Performance of Introverts and Extroverts. *Applied Clinical Psychology* 11, 445–255.

Gaab, N., & Schlaug, G. (2003). Musicians Differ from Non-Musicians in Brain Activation Despite Performance Matching. *In* G. Avanzini, C. Faienza, D. Minciacchi, L. Lopez, & M. Majno (eds), *The Neurosciences and Music* 999, 385–388. New York: The New York Academics of Sciences.

Garay-Vega, L., Pradhan, A.K., Werinberg, G., Schmidt-Nielsen, B., Harsham, B., Shen, Y., Dvekar, G., Romoser, M., Knoder, M., & Fisher, D.L. (2010). Evaluation of Different Speech and Touch Interfaces to In-Vehicle Music Retrieval Systems. *Accident Analysis and Prevention* 42, 913–920.

Garlin, F.V., & Owen, K. (2006). Setting the Tone with the Tune: A Meta-Analytic Review of the Effects of Background Music in Retail Settings. *Journal of Business Research* 59, 755–764.

Gartman, D. (2004). Three Ages of the Automobile: The Cultural Logics of the Car. *Theory, Culture and Society* 21(4/5), 169–195.

Gascoigne, G. (2010). 100 Greatest Car Songs. *DigitalDreamDoor.* www.digitaldreamdoor.com/pages/best_songs-car.html.

Gaston, E.T. (1968). *Music in Therapy*. New York: Macmillian.

Gatersleben, B. (2012). The Psychology of Sustainable Transport. *The Psychologist* 25(9), 676–679.

Gaver, W.W. (1993a). What in the World Do We Hear?: An Ecological Approach to Auditory Event Perception. *Ecological Psychology* 5(1), 1–29.

Gaver, W.W. (1993b). How Do We Hear in the World? Explorations in Ecological Acoustics. *Ecological Psychology* 5(4), 285–313.

Gilroy, P. (2001). Driving While Black. *In* D. Miller (ed.), *Car Cultures* (pp. 84–104). Oxford: Oxford University Press.

Godwin, J. (1985). *The Devils Disciples: The Truth about Rock.* Chino, CA: Chick.

Goodwin, A.H., Foss, R.D., Harell, S.S., & O'Brian, N.P. (2012). *Distracted Driving among Newly Licensed Teen Drivers.* Washington, DC: AAA Foundation for Traffic Safety. www.distraction.gov/download/DistractedDrivingAmongNewlyLicensedTeenDrivers.pdf.

Goodwin, B. (2012). History of the Car Radio in Motor Cars/Automobiles. *Carhistory4u.* www.carhistory4u.com/the-last-100-years/parts-of-the-car/car-radio.

Gopinath, S., & Stanyek, J. (eds). (2014). *The Oxford Handbook of Mobile Music.* New York: Oxford University Press.

Gratton, L. (2013). The Three Paradoxes of Generation Y. *Forbes. Leadership. 6/06/2013.* http://www.forbes.com/sites/lyndagratton/2013/06/06/the-three-paradoxes-of-generation-y/.

Greasley, A.E., & Lamont, A. (2011). Exploring Engagement with Music in Everyday Life Using Experience Sampling Methodology. *Musicae Scientiae* 15(1), 45–71.

Green, P. (2001). *Variations in Task Performance between Younger and Older Drivers: UMTRI Research on Telematics*. Paper presented at the Association for the Advancement of Automotive Medicine Conference on Aging and Driving, Southfield, Michigan, 19–20 February 2001.

Gregersen, N.P., & Berg, H.Y. (1994). Lifestyle and Accidents among Young Drivers. *Accident Analysis and Prevention* 26(3), 297–303.

Gregory, A.H. (1997). The Roles of Music in Society: The Ethnomusicological Perspective. *In* D. Hargreaves, J., & A. North, C. (eds), *The Social Psychology of Music*, 123–140. Oxford: Oxford University Press.

Groeger, J.A. (2000). *Understanding Driving: Applying Cognitive Psychology to a Complex Everyday Task.* Hove, East Sussex, UK: Psychology Press, Taylor and Francis Group.

GroningenUniversity. (2013). Listening to Music While Driving Has Very Little Effect on Driving Performance. *News and Events.* www.rug.nl/news-and-events/news/archief2013/nieuwsberichten/muziek-luisteren-tijdens-autorijden-beinvloedt-rijprestatie-nauwelijks?lang=en.

Gueguen, N., & Jacob, C. (2002). The Influence of Music on Temporal Perceptions in an On-Hold Waiting Situation. *Psychology of Music* 30(2), 210–214.

Gueguen, N., Jacob, C., Lourel, M., & La Guellec, H. (2007). Effect of Background Music on Consumer's Behavior: A Field Experiment in an Open-Air Market. *European Journal of Scientific Research* 16(2), 268–272.

Haack, P.A. (1980). The Behavior of Music Listeners. *In* D.A. Hodges (ed.), *Handbook of Music Psychology* (1st ed., pp. 141–182). Lawrence, KS: National Association for Music Therapy.

Haake, A. (2011). Individual Music Listening in Workplace Settings: An Exploratory Survey of Offices in the UK. *Musicae Scientiae* 15(1), 107–129.

Hamer, M. (2002). Death by Music: Be Careful What You Play on Your Car Stereo. *New Scientist* 2334, 8.

Hancock, P.A., Lesh, M., & Simmons, L. (2003). The Distraction Effects of Phone Use during a Crucial Driving Maneuver. *Accident Analysis and Prevention* 35, 501–514.

Hancock, P.A., Simmons, L., Hashemi, L., Howarth, H., & Ranney, T. (1999). The Effects of In-Vehicle Distraction on Driver Response during a Crucial Driving Maneuver. *Transportation Human Factors* 1(4), 295–309.

Hanowski, R. (2006). *Overview of the 100-Car Naturalistic Driving Study.* Paper presented at the PATH Driver Behavior and Safety Workshop. 23–24 October 2006. www.techtransfer.berkeley.edu/humanfactors/hundred_car.pdf?, Richmond, CA.

Hanowski, R.J., Olson, R.L., Hickman, J.S., & Dingus, T.A. (2006). *The 100-Car Naturalistic Driving Study: A Descriptive Analysis of Light Vehicle–Heavy Vehicle Interactions from the Light Vehicle Driver's Perspective* (FMCSA-RRR-06–004). Washington, DC: National Highway Traffic Safety Administration (NHTSA). www.fmcsa.dot.gov/facts-research/research-technology/report/100-car-naturalistic-study/100-car-naturalistic-study.pdf.

Harbluck, J.L., Noy, Y.I., Trbovich, P.L., & Eizenman, M. (2007). An On-Road Assessment of Cognitive Distraction: Impacts on Drivers' Visual Behavior and Braking Performance. *Accident Analysis and Prevention* 39, 372–379.

Hargreaves, D.J. (1987–88). Verbal and Behavioral Responses to Familiar and Unfamiliar Music. *Current Psychological Research and Reviews* 6(4), 323–330.

Hargreaves, D.J. (2012). Musical Imagination: Perception and Production, Beauty and Creativity. *Psychology of Music* 40(5), 539–557.

Hargreaves, D.J., MacDonald, R., & Miell, D. (2005). How Do People Communicate Using Music? *In* D. Miell, MacDonald, R., & Hargreaves, D.J. (eds), *Musical Communication* (pp. 1–26). Oxford: Oxford University Press.

Hargreaves, D.J., & North, A.C. (1999). The Functions of Music in Everyday Life: Redefining the Social in Music Psychology. *Psychology of Music* 27(1), 71–83.

Hargreaves, D.J., & North, A.C. (2010). Experimental Aesthetics and Liking for Music. *In* Juslin, P.N. & Sloboda, J.A. (eds), *Handbook of Music and Emotion: Theory, Research, Applications* (pp. 515–546). Oxford: Oxford University Press.

Hargreaves, D.J., North, A.C., & Tarrant, M. (2006). Musical Preference and Taste in Childhood and Adolescence. *In* G.E. McPherson (ed.), *A Handbook of Musical Development* (pp. 135–154). Oxford: Oxford University Press.

Harter, D.L. (n.d.). The Ultimate List of CarSongs! Retrieved 21.02.13, from www.calif-tech.com/blog/carsongs.html.

Hasegawa, C., & Oguri, K. (2006). The Effects of Specific Musical Stimuli on Driver's Drowsiness. *Proceedings of the 2006 IEEE Intelligent Transportation Systems Conference, 17–20 September 2006, Toronto Canada* (pp. 817–822). Toronto, Canada.

Hawkins, R. (2012). 'Travelling at the Speed of Sound': Top Gear Compilations as (British) Musical Expressions of Driving. *In* C. Hart, M. Duffett, & B. Peter (eds), *Proceedings of the Popular Music and Automobile Culture Conference*. University of Chester, England: www.hartchester.blogspot.com.

He, J., Becic, E., Lee, Y.-C., & McCarley, J.S. (2011). Mind Wandering behind the Wheel: Performance and Oculomotor Correlates. *Human Factors* 53(13), 13–21.

Heck, K.E., & Carlos, R.M. (2008). Passenger Distractions among Adolescent Drivers. *Journal of Safety Research* 39, 437–443.

Henry, S., & Lehrman, S. (1997). Hit the Pedal with Heavy Metal. *Health*.

Herrington, J., & Capella, L.M. (1996). Effects of Music in Service Environments: A Field Study. *The Journal of Services Management* 10(2), 26–41.

Hevner, K. (1936). Experimental Studies in the Elements of Expression in Music. *American Journal of Psychology* 48, 246–268.

Hevner, K. (1937). The Affective Value of Pitch and Tempo in Music. *American Journal of Psychology* 49, 621–630.

Heye, A., & Lamont, A. (2010). Mobile Listening Situations in Everyday Life: The Use of MP3 Players while Traveling. *Musicae Scientiae* 14(1), 95–120.

History.com. (n.d.). September 26, 1928: First Day of Work at the Galvin Manufacturing Corporation. *This Day in History.* Retrieved 9.2.2013, from www.history.com/this-day-in-history/first-day-of-work-at-the-galvin-manufacturing-corporation.

Ho, C., Reed, N., & Spence, C. (2007). Multi-Sensory Warning Signals for Collision Avoidance. *Human Factors* 49(6), 1107–1114.

Ho, C., & Spence, C. (2005). Assessing the Effectiveness of Various Auditory Cues in Capturing a Driver's Visual Attention. *Journal of Experimental Psychology: Applied* 11(3), 157–174.

Ho, C., & Spence, C. (2008). *The Multisensory Driver: Implications for Ergonomic Car Interface Design*. Aldershot, Hampshire: Ashgate.

Hodges, D.A. (2010). Psychophysiological Measures. In Juslin, P.N. & Sloboda, J.A. (eds), *Handbook of Music and Emotion: Theory, Research, Applications* (pp. 279–312). Oxford: Oxford University Press.

Hodges, D.A., & Haack, P.A. (1996). The Influence of Music on Human Behavior. *In* D.A. Hodges (ed.), *Handbook of Music Psychology* (2nd ed., pp. 469–556). San Antonio: IMR Press.

Hodges, D.A., & Sebald, D.C. (2011). *Music in the Human Experience: An Introduction to Music Psychology*. New York: Rutledge, Taylor and Francis.

Horberry, T., Anderson, J., Regan, M.A., Triggs, T.J., & Brown, J. (2006). Driver Distraction: The Effects of Concurrent In-Vehicle Tasks, Road Environment and Age on Driving Performance. *Accident Analysis and Prevention* 38, 185–191.

Horberry, T., & Edquist, J. (2009). Distractions Outside the Vehicle. *In* Regan, M., Lee, J. D., & Young, K.L. (eds), *Driver Distraction: Theory, Effects, and Mitigation* (pp. 215–228). Boca Raton, FL: CRCPress, Taylor and Francis.

Horrey, W.J. (2009). On Allocating the Eyes: Visual Attention and In-Vehicle Technologies. *In* C. Castro (ed.), *Human Factors of Visual and Cognitive Performance in Driving* (pp. 151–166). Boca Raton, FL: CRC Press, Taylor and Francis.

Horrey, W.J. (2011). Assessing the Effects of In-Vehicle Tasks on Driving Performance. *Ergonomics in Design: The Quarterly of Human Factors Applications* 19(4), 4–7.

Horrey, W.J., Lesch, M.F., & Garabet, A. (2009). Dissociation between Driving Performance and Drivers' Subjective Estimates of Performance and Workload in Dual-Task Conditions. *Journal of Safety Research* 40, 7–12.

Horrey, W.J., & Wickens, C.D. (2006). Examining the Impact of Cell Phone Conversations on Driving Using Meta-Analytic Techniques. *Human Factors* 48(1), 196–205.

Hughes, G.M., Rudin-Brown, C.M., & Young, K.L. (2013). A Simulator Study of the Effects of Singing on Driving Performance. *Accident Analysis and Prevention* 50, 787–792.

IASCA. (n.d.). International Auto Sound Challenge Association: The Standard by Which Great Mobile Electronics Performance Is Measured. *IASCA Worldwide.* Retrieved 05.2013, from www.iasca.com.

Inglis, D. (2004). Auto Couture: Thinking the Car in Post-War France. *Theorem Culture and Society* 21(4/5), 197–219.

Iwamiya, S.-I., & Sugiomoto, M. (1996). Interaction between Auditory and Visual Processing In Car Audio: Simulation Experiment Using Video Reproduction. *In* B. Pennycook & E. Costa-Giomi (eds), *Proceedings of the 4th ICMPC – Fourth International Conference on Music Perception and Cognition, McGill University, 11–15 August 1996* (pp. 309–314). Montreal, Quebec: Faculty of Music, McGill University.

Jacob, C. (2006). Styles of Background Music and Consumption in a Bar: An Empirical Evaluation. *Hospitality Management* 25, 716–720.

Jacob, C., Gueguen, N., & Boulbry, G. (2010). Effects of Songs with Prosocial Lyrics on Tipping behavior in a restaurant. *International Journal Of Hospitality Management, 29*, 761–763.

Jamson, A.H., & Merat, N. (2005). Surrogate in-Vehicle Information Systems and Driver Behavior: Effects of Visual and Cognitive Load in Simulated Lural Driving. *Transportation Research Part F, 8*, 79–96.

Jancke, L., Brunner, B., & Esslen, M. (2008). Brain Activation during Fast Driving in a Driving Simulator: The Role of the Lateral Prefrontal Cortex. *NeruroReport* 19(11), 1127–1130.

Janssen, J.H., van den Broek, E.L., & Westerink, J.H.D.M. (2009). Personalised Affective Music Player. *Proceedings of the 3rd International Conference of Affective Computing and Intelligent Interaction, 10–12 September, Amsterdam.* The Netherlands: University of Twente. http://purl.utwente.nl/publications/73203.

Janssen, J.H., van den Broek, E.L., & Westerink, J.H.D.M. (2012). Tune In to Your Emotions: A Robust Personalised Affective Music Player. *User Modeling and User-Adapted Interaction* 22(3), 255–270.

Johansson, S. (2003). Most Great Car Songs Are About Volvos, Expert Reports (ID 723). *Volvo Cars of Canada Corp.* www.media.volvocars.com/download/media/articles/pdf/723_3_3.aspx.

Johnson, P. (2012). Moving Sounds: Hearing Route 66 on the Car Radio Then and Now. *In* C. Hart, M. Duffett, & B. Peter (eds), *Proceedings of the Popular Music and Automobile Culture Conference.* University of Chester, England: www.hartchester.blogspot.com.

Jolley, A. (2013). The 10 Most Dangerous Driving Songs. *Confussed.com.* www.confused.com/car-insurance/articles/top-ten-most-dangerous-driving-songs.

Joo, J. (2007). The Impact of the Automobile and Its Culture in the U.S. *International Area Review* 10(1), 39–54.

Jordan, P.W., & Johnson, G.I. (1993). Exploring Mental Workload via TLX: The Case of Operating a Car Stereo Whilst Driving. *In* A. Gale, I.D. Brown, C.M. Haselgrave, H. Kruysse, W. & S.P. Tayler (eds.), *Vision in Vehicles IV*, 255–262. Amsterdam: Elsevier Science.

Just, M.A., Keller, T.A., & Cynkar, J. (2008). A Decrease in Brain Activation Associated with Driving When Listening to Someone Speak. *Brain Research* 1205, 70–80.

Kallinen, K. (2002). Reading News from a Pocket Computer in a Distracting Environment: Effects of the Tempo of Background Music. *Computers in Human Behavior* 18, 537–551.

Kallinen, K. (2004). The Effects of Background Music on Using a Pocket Computer in a Cafeteria: Immersion, Emotional Responses, and Social Richness of Medium. *In* E. Dykstar-Erickson & M. Tscheligi (eds), *Proceedings of the Conference on Human Factors in Computing Systems, 24–29 April 2004* (pp. 1227–1230). Vienna and Paris: ACM Press.

Kampfe, J., Sedlmeier, P., & Renkewitz, F. (2011). The Impact of Background Music on Listeners: A Meta-Analysis. *Psychology of Music* 39(4), 424–448.

Kanetix. (2013a). How Music Impacts Your Driving – Research Findings. *Kanetix, Advocate for Choice.* https://www.kanetix.ca/how-music-impacts-your-driving.

Kanetix. (2013b). How Music Impacts Your Driving – Infographics. *Kanetix, Advocate for Choice.* https://www.kanetix.ca/how-music-impacts-your-driving-infographic.

Kanetix (2013c). *Revised Kinetix Music Survey Data.* Personal Communication, September 17, to W. Brodsky from N. Carr.

Katic, M. (2013). Study: Music You Listen to Affects Driving Speed. *CBS Boston WBZO.* http://boston.cbslocal.com/2013/01/10/study-music-you-listen-to-affects-driving-speed/.

Kellaris, J.J., & Kent, R.J. (1992). The Influence of Music on Consumers' Temporal Perceptions: Does Time Fly When You're Having Fun? *Journal of Consumer Psychology* 1(4), 365–376.
Kemp, A.E. (1996). *The Musical Temperament: Psychology and Personality of Musicians*. Oxford: Oxford University Press.
Kemp, A.E. (1997). Individual Differences in Musical Behaviour. *In* D. Hargreaves, J. & A.C. North (eds), *The Social Psychology of Music* (pp. 25–45). Oxford: Oxford University Press.
Klauer, S.G., Dingus, T.A., Neale, V.L., Sudweeks, J.D., & Ramsey, D.J. (2006a). *The Impact of Driver Inattention on Near Crash/Crash Risk: An Analysis Using the 100 Car Naturalistic Driving Study Data.* Washington, DC: National Highway Traffic Safety Administration.
Klauer, S.G., Sudweeks, J., Hickman, J.S., & Neale, V.L. (2006b). *How Risky Is It? An Assessment of the Relative Risk of Engaging in Potential Unsafe Driving Behaviors.* Blacksburg, VA: Virginia Tech Transportation Institute (VTI), Virginia Polytech Institute and State University.
Konecni, V.J. (1982). Social Interaction and Musical Preference. *In* D. Deutch (ed.), *Psychology of Music* (1st ed., pp. 497–516). Orlando, FL: Academic Press.
Konz, S., & McDougal, D. (1968). The Effect of Background Music on Control Activity of an Automobile Driver. *Human Factors* 10(3), 233–244.
Kortum, P., Bias, R.G., Knott, B.A., & Bushey, R.G. (2008). The Effect of Choice and Announcement Duration on the Estimation of Telephone Hold Time. *International Journal of Technology and Human Interaction* 4(4).
Krahe, B., & Bieneck, S. (2012). The Effect of Music-Induced Mood on Aggressive Affect, Cognition, and Behavior. *Journal of Applied Psychology* 42(2), 271–290.
KristinS. (2009). Volvo Tunes. *Official Blog of Volvo Cars of North America.* www.volvoblog.us/2009/09/09/volvo-tunes/.
Kujala, T., & Saariluoma, P. (2011). Effects of Menu Structure and Touch Screen Style on the Variability of Glance Durations during In-Vehicle Visual Search Tasks. *Ergonomics* 54(8), 716–732.
Lacourse, E., Claes, M., & Villenuve, M. (2001). Heavy Metal Music and Adolescent Suicide Risk. *Journal of Youth and Adolescence* 30(3), 321–332.
Lamb, B. (2012). Top 10 Best Driving Songs. *About.com.* http://top40.about.com/od/top10lists/tp/drivingsongs.htm.
Lamont, A., & Greasley, A. (2009). Musical Preferences. *In* S. Hallam, I. Cross, & M. Thaut (eds), *The Oxford Handbook of Music Psychology* (pp. 160–168). Oxford: Oxford University Press.
Lanza, J. (1994). *Elevator Music: A Surreal History of Muzak, Easy-Listening, and Other Moodsong*. New York: Picador USA.
Laukkanen, R., & Virtanen, P. (1998). Heat Rate Monitors – State of the Art. *Journal of Sports Sciences* 16, S3-S7.
Laurier, E. (2004). Doing Office Work on the Motorway. *Theory, Culture and Society* 21(4/5), 261–277.

Lee, J.D. (2007). Technology and Teen Drivers. *Journal of Safety Research* 38, 203–213.

Lee, J.D., Regan, M.A., & Young, K.L. (2009). What Drives Distraction? Distraction as Breakdown of Multi-Level Control. *In* M. Regan, J.D. Lee & K.L. Young (eds), *Driver Distraction: Theory, Effects, and Mitigation* (pp. 41–56). Boca Raton, Fl: CRCPress, Taylor and Francis.

Lee, J.D., Roberts, S.C., Hoffman, J.D., & Angell, L.S. (2012). Scrolling and Driving: How an MP3 Player and Its Aftermarket Controller Affect Driving Performance and Visual Behavior. *Human Factors* 54(2), 250–263.

Lee, J.D., Young, K.L., & Regan, M.A. (2009). Defining Driver Distraction. *In* M. Regan, J.D. Lee, & K.L. Young (eds), *Driver Distraction: Theory, Effects, and Mitigation* (pp. 31–40). Boca Raton, FL: CRCPress, Taylor and Francis.

Lees-Maffei, G. (2002). Men, Motors, Markets, and Women. *In* P. Wollen & J. Keer (eds), *Autopia; Cars and Culture* (pp. 363–370). London: Reakton Books.

Lendino, J. (2012). The History of the Car Stereo. *PC Magazine PCMAG. COM.* www.pcmag.com/article2/0,2817,2399878,00.asp.

Lesiuk, T. (2005). The Effect of Music Listening on Work Performance. *Psychology of Music* 33(2), 173–192.

Lewis, B.E., & Schmidt, C.P. (1991). Listeners' Responses to Music as a Function of Personality Type. *Journal of Research in Music Education* 59(4), 311–321.

Lezotte, C. (2012). Born to Take the Highway: The Automobile, Women and Rock-n-Roll. *In* C. Hart, M. Duffett, & B. Peter (eds), *Proceedings of the Popular Music and Automobile Culture Conference*. University of Chester, England: www.hartchester.blogspot.com.

Lezotte, C. (2013). Born to Take the Highway: Women, the Automobile, and Rock 'n' Roll. *Journal of American Culture* 36(3), 161–176.

Litle, P., & Zuckerman, M. (1986). Sensation Seeking and Music Preference. *Personality and Individual Differences* 7(4), 575–577.

Liu, N.-H., Chiang, C.-Y., & Hsu, H.-M. (2013). Improving Driver Alertness through Music Selection Using a Mobile EEG to Detect Brainwaves. *Sensors* 13, 8199–8221.

Lopez, L., Jurgens, R., Diekmann, V., Becker, W., Ried, S., Grozinger, B., & Erne, S.N. (2003). Musicians Versus Non-Musicians: A Neurophysiological Approach. *In* G. Avanzini, C. Faienza, D. Minciacchi, L. Lopez, & M. Majno (eds), *The Neurosciences and Music*, 999, 124–130. New York: The New York Academics of Sciences.

LowerTheBoom. (n.d.). Is There Really a Problem. *National Alliance against Loud Car Stereo Assault.* Retrieved 05.2013, from www.lowertheboom.com.

Lutz, C., & Fernandez, A.L. (2010). *Carjacked: The Culture of the Automobile and Its Effect on Our Lives*. New York: Palgrave Macmillian.

Madsen, C.K. (1987). Background Music: Competition for Focus of Attention. *In* C.K. Madsen & C.A. Prickett (eds.), *Applications of Research in Music Behavior* (pp. 315–326). Tuscaloosa, AL: University of Alabama Press.

Magnini, V.P., & Parker, E.E. (2009). The Psychological Effect of Music: Implications for Hotel Firms. *Journal of Hospitality and Tourism Research* 37(1), 281–299.

Maltin, L. (1995). *Movie and Video Guide 1996*. New York: Penguin.

Manni, M. (2005). Honey, They're Playing Our Song… in an Ad. *Fox News.com*. October 17 2005.

Marling, K.A. (2002). America's Love Affair with the Automobile in the Television Age. In P. Wollen & J. Keer (eds), *Autopia: Cars and Culture* (pp. 354–362). London: Reakton Books.

Marsui, T., Tanaka, S., Kazai, K., & Tsuzaki, M. (2013). Activation of the Left Superios Temporal Gyrus of Musicians by Music-Derived Sounds. *NeuroReport* 24, 41–45.

Matthews, G., Quinn, C.E.J., & Mitchell, K.J. (1998). Rock Music, Task-Induced Stress and Simulated Driving Performance. *In* G.B. Graysen (ed.), *Behavioral Research In Road Safety* 8, 20–32. Crawthorn, Berks: Transportation Research Laboratory.

Mattila, A.S., & Wirtz, J. (2001). Congruency of Scent and Music as a Driver of In-Store Evaluations and Behavior. *Journal of Retailing* 77, 273–289.

Maxwell, S. (2001). Negotiations of Car Use in Everyday Life. *In* D. Miller (ed.), *Car Cultures,* 203–222. Oxford: Oxford University Press.

Mayfield, C., & Moss, S. (1989). Effect of Music Tempo on Task Performance. *Psychological Reports* 65, 1283–1290.

McCown, W., Keiser, R., Mulhearn, S., & Williamson, D. (1997). The Role of Personality and Gender in Preference for Exaggerated Bass in Music. *Personality and Individual Differences* 23(4), 543–547.

McCraty, R., Barrios-Choplin, B., Atkinson, M., & Tamasino, D. (1998). The Effects of Different Types of Music on Mood, Tension, and Mental Clarity. *Alternative Therapies* 4(1), 75–84.

McElrea, H., & Standing, L. (1992). Fast Music Causes Fast Drinking. *Perceptual and Motor Skills* 75(2), 362.

McEvoy, S.P., Stevenson, M.R., & Woodward, M. (2006). The Impact of Driver Distraction on Road Safety: Results from a Representative Survey in Two Australian States. *Injury Prevention* 12, 242–247.

McEvoy, S.P., Stevenson, M.R., & Woodward, M. (2007). The Contribution of Passengers Versus Mobile Phone Use to Motor Vehicle Crashes Resulting in Hospital Attendance by the Driver. *Accident Analysis and Prevention* 39, 1170–1176.

McGee, D. (2013). Which Music Fans Are Most Likely to Get Speeding Tickets? *The Globe and Mail (Canada), The List.* www.theglobeandmail.com/globe-drive/car-life/which-music-fans-are-most-likely-to-get-speeding-tickets/article13369645/.

McKnight, A.J., & McKnight, A.S. (1993). The Effect of Cellular Phone Use upon Driving Attention. *Accident Analysis and Prevention* 25(3), 259–265.

McKnight, A.J., & McKnight, A.S. (2003). Young Novice Drivers: Careless or Clueless? *Accident Analysis and Prevention* 35, 921–925.

McNair, D., Lorr, M., & Droppelman, L. (1971). *EITS Manual Fort the Profile of Mood States (POMS)*. San Diego, CA: EITS.

Mesken, J., Hagenzieker, M.P., Rothengatter, T., & de Ward, D. (2007). Frequency, Determinants, and Consequences of Different Driver Emotions: An On-the-Road Study Using Self-Reports (Observed) Behaviour, and Physiology. *Transportation Research Part F* 10(4), 458–475.

MetroNews. (2013). What Does Your Music Say about Your Driving? *Metro News (Canada).* http://metronews.ca/drive/746957/what-does-your-music-say-about-your-driving/.

Meyer, L.B. (1956a). Theory. *In* Meyer, *Emotion and Meaning in Music*, 1–42. Chicago: University of Chicago Press.

Meyer, L.B. (1956b). Note on Image Processes, Connotations, and Moods. *In* Meyer, *Emotion and Meaning in Music* (pp. 250–272). Chicago: University of Chicago Press.

Meyer, L.B. (1994). Emotion and Meaning in Music. *In* R. Aiello & J. A. Sloboda (eds), *Musical Perceptions* (pp. 3–39). New York: Oxford University Press.

Micheyl, C., Carbonnel, O., & Collet, L. (1995). Medial Olivocochlear System and Loudness Adaption: Differences Between Musicians and Non-Musicians. *Brain and Cognition* 29, 127–136.

Miller, D. (2001). Driven Societies. *In* D. Miller (ed.), *Car Cultures* (pp. 1–33). Oxford: Oxford University Press.

Milliman, R.E. (1982). Using Background Music to Affect the Behavior of Supermarket Shoppers. *Journal of Marketing* 46, 86–91.

Milliman, R.E. (1986). The Influence of Background Music on the Behavior of Restaurant Patrons. *Journal of Consumer Research* 13, 286–289.

Milne, S. (2009). Hip-hop Flop: Rap Most Dangerous Driving Music. *AutoTrader Magazine UK.* www.autotrader.co.uk/EDITORIAL/CARS/news/AUTOTRADER/OTHER/hip_hop_flop_rap_most_dangerous_driving_music.html.

Montlick & Associates. (2011). Driver Distracted by Loud Music Hit by Freight Train. *Montlick and Associates Blog.* www.montlick.com/montlick-blog/montlick-law-blog/313-driver-distracted-by-loud-music-hit-by-freight-train.

Morrison, M., Gan, S., Dubelaar, C., & Oppewal, H. (2011). In-Store Music and Aroma Influences on Shopper Behavior and Satisfaction. *Journal of Business Research* 64, 558–564.

Motavalli, J. (2012). Driven to Distraction by Rock and Rap. *Mother Nature Network.* www.mnn.com/green-tech/transportation/blogs/driven-to-distraction-by-rock-and-rap.

Motorola. (n.d.). 1930: The First Motorola Brand Car Radio. *Motorola Solutions: Sound in Motion.* www.motorolasolutions.com/US-EN/About/Company+Overview/History/Explore+Motorola+Heritage/Sound+in+Motion.

Mottram, E. (2002). Blood on the Nash Ambassador: Cars in American Films. *In* P. Wollen & J. Keer (eds), *Autopia: Cars and Culture* (pp. 95–114). London: Reakton Books.

Mouloua, M., Jaramillo, D., Smither, J., Alberti, P., & Brill, J.C. (2011). The Effects of iPod Use on Driver Distraction. *Proceedings of the 55th Annual Meeting of the Human Factors and Ergonomics Society* (pp. 1581–1585).

Mozes, A. (2013). Music Doesn't Hurt Driving Performance, Study Says. *Health Day News.* http://consumer.healthday.com/mental-health-information-25/behavior-health-news-56/listening-to-music-while-driving-not-harmful-may-even-help-study-677535.html.

Mulholland, A. (2013). Do Motorists' Musical Soundtracks Affect Driving Habits? Survey Says: Yes. *CTV News.* www.ctvnews.ca/autos/do-motorists-musical-soundtracks-affect-driving-habits-survey-says-yes-1.1382891.

Nater, U.M., Abbruzzese, E., Krebs, M., & Ehlert, U. (2006). Sex Differences in Emotional and Psychophysiological Responses to Music Stimuli. *International Journal of Psychophysiology* 62, 300–308.

Nater, U.M., Krebs, M., & Ehlert, U. (2005). Sensation Seeking, Music Preference, and Psychophysiological Reactivity to Music. *Musicae Scientiae* 9(2), 239–254.

Nationwide. (2013). Passengers, Phones, and Music. *Nationwide Insurance.* www.nationwide.com/teen-distracted-driving.jsp.

NBLaw Group. (2012). Music Could Add to Distracted Driving, Study Shows. *Nagelbery Barnard Law Group: Personal Injury News.* www.nblawgroup.com/music-could-add-to-distracted-driving-study-shows/.

Neale, V.L. (2011). 100-Car Naturalistic Driving Study. *FOT-Net WIKI.* wiki.fot-net.eu/index.php?title=100-Car_naturalistic_driving_study.

Neale, V.L., Dingus, T.A., Klauer, S.G., Sudweeks, J., & Goodman, M. (2005). *An Overview of the 100-Car Naturalistic Study and Findings* (05–0400): Virginia Tech Transportation Institute and NHTSA. www.nhtsa.gov/DOT/NHTSA/NRD/Multimedia/PDFs/Crash%2520Avoidance/2005/100Car_ESV05summary.pdf.

Nelson, T.M., & Nilsson, T.H. (1990). Comparing Headphone and Speaker Effects on Simulated Driving. *Accident Analyses and Prevention* 22(6), 523–529.

Newcomb, D. (2008). Exploring the World of the Car Audio. *Car Audio for Dummies* (pp. 7–16). Hoboken, NJ: John Wiley and Sons.

NewsHome. (2013). Shakespeare Disi-Rui Configuration Parsing B-Class Car Market New Security Paradigm. July 5. *News Home, Automotive.* www.newshome.us/news-4827568-Shakespeare-Disi-Rui-configuration-parsing-B-class-car-market-new-security-paradigm.html.

NHTSA. (2010). *Traffic Safety Facts, Research Note: Distracted Driving 2009.* Washington, DC: National Highway Safety Administration.

Nicholson, J. (2011). Distracted Driving Concern: Loud Music. *Davis Law Groups, P.S; Distracted Driving Accidents.* www.injurytriallawyer.com/blog/distracted-driving-concern-loud-music.cfm.

Nittono, H., Tsuda, A., Akai, S., & Nakajima, Y. (2000). Tempo of Background Sound and Performance Speed. *Perception and Motor Skills, 90*, 1122.

NME. (2011). Beastie Boys, Prodigy, Kanye West Named Most Dangerous Acts to Listen to While Driving: But Adele, Jack Johnson and Coldplay Are the Safest, According to New Halfords Survey. *New Musical Express (NME)*. www.nme.com/news/the-prodigy/56661.

Norman, W. (n.d.). History of the Car Stereo. *eHow.* Retrieved 9.2.2013, from www.ehow.com/about_6299810_history-car-sterio.html.

North, A.C., & Hargreaves, D.J. (1995). Subjective Complexity, Familiarity, and Liking for Popular Music. *Psychomusicology* 14(Spring/Fall), 77–93.

North, A.C., & Hargreaves, D.J. (1996a). Responses to Music in a Dinning Area. *Journal of Applied Social Psychology* 26(6), 491–501.

North, A.C., & Hargreaves, D.J. (1996b). The Effects of Music on Responses to a Dining Area. *Journal of Environmental Psychology* 16, 55–64.

North, A.C., & Hargreaves, D.J. (1997a). The Musical Milieu: Studies of Listening in Everyday Life. *The Psychologist* 10(7), 309–312.

North, A.C., & Hargreaves, D.J. (1997b). Liking, Arousal Potential, and the Emotions Expressed by Music. *Scandinavian Journal of Psychology* 38, 45–53.

North, A.C., & Hargreaves, D.J. (1997c). Music in Consumer Behaviour. *In* D. Hargreaves, J. & A. North, C. (eds), *The Social Psychology of Music* (pp. 268–289). Oxford: Oxford University Press.

North, A.C., & Hargreaves, D.J. (1998). The Effect of Music on Atmosphere and Purchase in Intentions in a Cafeteria. *Journal of Applied Social Psychology* 28(24), 2254–2227.

North, A.C., & Hargreaves, D.J. (1999a). Can Music Move People? The Effects of Musical Complexity and Silence on Waiting Time. *Environment and Behavior* 31(1), 136–149.

North, A.C., & Hargreaves, D.J. (1999b). Music and Driving Game Performance. *Scandinavian Journal of Psychology* 40(4), 285–292.

North, A.C., & Hargreaves, D.J. (2000). Musical Preferences during and after Relaxation and Exercises. *American Journal of Psychology* 113(1), 43–67.

North, A.C., & Hargreaves, D.J. (2005). Musical Communication in Commercial Contexts. *In* D. Miell, R. MacDonald, & D.J. Hargreaves (eds), *Musical Communication* (pp. 405–422). Oxford: Oxford University Press.

North, A.C., & Hargreaves, D.J. (2006). Problem Music and Self-Harming. *Suicide and Life-Threatening Behavior* 36(5), 582.

North, A.C., & Hargreaves, D.J. (2007a). Lifestyle Correlates of Music Performance: 1. Relationships, Living Arrangements, Beliefs, and Crime. *Psychology of Music* 35(1), 58–87.

North, A.C., & Hargreaves, D.J. (2007b). Lifestyle Correlates of Musical Preference: 2. Media, Leisure Time and Music. *Psychology Of Music* 35(2), 179–200.

North, A.C., & Hargreaves, D.J. (2007c). Lifestyle Correlates of Musical Performance: 3. Travel, Money, Education, Employment, and Health. *Psychology of Music* 35(3), 473–498.

North, A.C., & Hargreaves, D.J. (2010). Music and Marketing. *In* P.N. Juslin & J.A. Sloboda (eds), *Handbook of Music and Emotion: Theory, Research, Applications* (pp. 909–932). Oxford: Oxford University Press.

North, A.C., Hargreaves, D.J., & Hargreaves, J.J. (2004). Uses of Music in Everyday Life. *Music Perception* 22(1), 41–77.

North, A.C., Hargreaves, D.J., & Heath, S.J. (1998). Musical Tempo and Time Perception in a Gymnasium. *Psychology of Music* 26(1), 78–88.

North, A., Shilcock, A., & Hargreaves, D.J. (2003). The Effect of Musical Style on Restaurant Customer's Spending. *Journal of Environment and Behavior* 35(5), 712–718.

North, A.C., Hargreaves, D.J., & McKendrick, J. (1999a). Music and On-Hold Waiting Time. *British Journal of Psychology* 90(1), 161–164.

North, A.C., Hargreaves, D.J., & McKendrick, J. (1999b). The Influence of In-Store Music on Wine Selections. *Journal of Applied Psychology* 84(2), 271–276.

North, A.C., Hargreaves, D.J., & McKendrick, J. (2000). The Effects of Music on Atmosphere in a Bank and a Bar. *Journal of Applied Social Psychology* 30(7), 1504–1522.

North, A.C., Tarrant, M., & Hargreaves, D.J. (2004). The Effects of Music on Helping Behavior: A Field Study. *Environment and Behavior* 36(2), 266–275.

Noseworthy, T.J., & Finlay, K. (2009). A Comparison of Ambient Casino Sound and Music: Effects on Disassociation and on Perceptions of Elapsed Time While Playing Slot Machines. *Journal of Gambling Studies* 25, 331–342.

Noyes, M. (2009). Does Hip-Hop Music Cause Car Accidents and Road Rage? New Study Says Yes. *Perenich, Caulfield, Avril, Noyes, Injury Attorneys*. www.matthewnoyes.com/AutoAccidentsArticle.cfm?ID=483.

NSC. (2010). *Understanding the Distracted Brain: Why Driving While Using Hands-Free Cell Phones Is Risky Behavior* (White Paper): National Safety Council, http://www.fnal.gov/pub/traffic_safety/files/NSC%20White%20Paper%20-%20Distracted%20Driving%203–10.pdf.

NSC. (2012). *Understanding the Distracted Brain: Why Driving While Using Hands-Free Cell Phones Is Risky Behavior* (White Paper): National Safety Council. www.nsc.org/safety_road/Distracted_Driving/Documents/CognitiveDistractionWhitePaper.pdf.

Oakes, S. (2003). Musical Tempo and Waiting Perceptions. *Psychology and Marketing* 20(8), 685–705.

Oakes, S., & North, A.C. (2006). The Impact of Background Musical Tempo and Timbre Congruity upon Ad Content Recall and Affective Response. *Applied Cognitive Psychology, 20*, 505–520.

Oblad, C. (1996). *Music in the Car: Actions and Preferences – On the Car As a Concert Hall* [Musik i bilen. Handling och preferenser. Om bilen som konsertlokal] (KFB 1996:12) *Department of Musicology*. Stockholm: The

Swedish Transport and Communications Research Board (Research Report) and University of Gothenburg (MA Thesis).

Oblad, C. (1997). Music in the Car. *In* A. Gabrielsson (ed.), *Proceedings of the Third Triennial ESCOM Conference, Uppsala University, 7–12 June 1997* (pp. 639–643). Uppsala University, Sweden: Department of Psychology.

Oblad, C. (2000a). *Using Music – On the Car as a Concert Hall [Att använda musik – om bilen som konsertlokal].* University of Gothenburg, Sweden, Stockholm.

Oblad, C. (2000c). On Using Music – About the Car as a Concert Hall. *In* C. Woods, G. Luck, R. Brochard, F. Seddon, & J. A. Sloboda (eds), *Proceedings of the Sixth International Conference on Music Perception and Cognition, Keele University, August 2000.* Staffordshire, UK: Department of Psychology, Keele University.

Oblad, C., & Foward, S. (2000b). *The Impact of Music on Driving* (KFB&VTI 37). Sweden: Swedish Road and Transportation Institute (VTI), SE-581 95 Linkoping, Sweden.

Oda, J., Macpherson, A., Middaugh-Bonney, T., Brussoni, M., Piedt, S., & Pike, I. (2010). Prevalence of Driving Distractions among High School Student Drivers in Three Canadian Cities. *Injury Prevention* 16(Suppl 1), A66.

Oldham, G., Cummings, A., Mischel, L.J., Schmidtke, J.M., & Zhou, J. (1995). Listen While You Work? Quasi-Experimental Relations between Personal-Stereo Headset Use and Employee Work Responses. *Journal of Applied Psychology* 80(5), 547–564.

Oron-Gilad, T. (2003). *Alertness Maintaining Tasks While Driving. PhD thesis.* Ben-Gurion University of the Negev, Beer-Sheva, Israel.

Oron-Gilad, T., Ronen, A., & Shinar, D. (2008). Alertness Maintaining Tasks (AMTs) While Driving. *Accident Analysis and Prevention* 40, 851–860.

Ortmann, O. (1927). Types of Listeners: Genetic Considerations. *In* M. Schoen (ed.), *The Effects of Music* (pp. 38–77.). New York: Harcourt Brace.

Otte, A., Juengling, F.D., & Kassubek, J. (2001). Exceptional Brain Function in Musicians and the Neural Basis of Music Processing. *European Journal of Nuclear Medicine* 28(1), 130–131.

Ouis, D. (2001). Annoyance from Road Traffic Noise: A Review. *Journal of Environmental Psychology* 21, 101–120.

Panksepp, J., & Bernatzky, G. (2002). Emotional Sounds and the Brain: The Neuro-Affective Foundations of Musical Appreciation. *Behavioural Processes* 60(2), 133–155.

Parncutt, R., & Marin, M.M. (2006). Emotions and Associations Evoked by Unfamiliar Music. *Proceedings of the 2006 International Association of Empirical Aesthetics, Avignon France, August 29–September 1, 2006.*

Pascoe, D. (2002). Vanishing Points. *In* P. Wollen & J. Keer (eds), *Autopi: Cars and Culture* (pp. 75–82). London: Reakton Books.

Patel, J., Ball, D.J., & Jones, H. (2008). Factors Influencing Subjective Ranking of Driver Distraction. *Accident Analysis and Prevention* 40, 392–395.

Paterson, M. (2007). *Automobile Politics: Ecology and Cultural Political Economy*. Cambridge: Cambridge University Press.
Pecher, C., Lemercier, C., & Cellier, J.-M. (2009). Emotions Drive Attention: Effects on Driver's Behavior. *Safety Science* 47, 1254–1259.
Peek-Asa, C., Britton, C., Young, T., Pawlovich, M., & Falb, S. (2010). Teenage Driver Crash Incidence and Factors Influencing Crash Injury by Ruralty. *Journal of Safety Research* 41, 487–492.
Philipson, A. (2013). Want to Drive Safely? Listen to Elton John, Aerosmith or S Club 7. *Telegraph*.
Plato. (380BC/1930). *The Republic* (P. Shory, trans.). London: W. Heinemann.
Power, J. (2009). The Effects of Music on Driving: Can Listening to the Radio Affect How We Drive? *Yahoo Voices*. www.associatedcontent.com/article/1870137/the_effects_of_music_on_driving.html?cat=27.
Presta, R. (2013). How Listening to Music while Driving Impacts Your Driving Anxiety. *The Driving Fear Program – Overcome Your Driving Anxiety*. http://drivingfearhelp.com/how-listening-to-music-while-driving-impacts-your-driving-anxiety/.
Price-Styles, A. (2012). 'Born to Roll': An Examination of Jeep Culture in the Music of Masta Ace. *In* C. Hart, M. Duffett, & B. Peter (eds), *Proceedings of the Popular Music and Automobile Culture Conference*. University of Chester, England: www.hartchester.blogspot.com.
PRNewswire. (2010). AAA, *Seventeen* and DOT Recognize National Two-Second Turnoff Day. www.prnewswire.com/news-releases/aaa-seventeen-and-dot-recognize-national-two-second-turnoff-day-103109524.html.
PRNewswire. (2012). Music Driving Us to Distraction – New Research from Allianz You Cover Lifts the Lid on the Effects of the Music We Listen to in Our Cars. *PR Newswire*. www.prnewswire.com/news-releases/music-driving-us-to-distraction--new-research-from-allianz-your-cover-lifts-the-lid-on-the-effects-of-the-music-we-listen-to-in-our-cars-167969396.html.
Quick, D. (2011). Smart Steering Wheel Gives a Health Check While Driving. *GizMag*. www.gizmag.com/smart-steering-wheel/20467/.
Quicken-Insurance. (2000). American Redefine Reckless Driving Habits: Quicken Insurance Survey Finds Loud, Fast Drivers Have Rubber Necks. Mountain View California: Intuit. www.intuti.com/about_intuit/press_releases/2000/11–14a.html.
Quotemehappy. (2011). Driven to Distraction: How Music Impacts Mood and Manners on Britain's roads. *Quote Me Happy*. www.quotemehappy.com/media-centre/Driven-to-distraction and http://blog.quotemehappy.com/news/how-the-music-we-choose-can-change-how-we-drive-98155.html.
Qureshi, M. (2013). What Does Your Music Say about Your Driving? *680 News: All News Radio*. www.680news.com/2013/07/25/what-does-your-music-say-about-your-driving/.

RAC. (2004). 'Grooving While Cruising?'. *RAC Foundation – Press Release.* www.racfoundation.org/assets/rac_foundation/content/downloadables/grooving_while_cruising.pdf.

RAC. (2009). 'Driven to Distraction': Annual Report on Motoring Identifies the Worrying Extent of In-Car Distractions. *Press Releases, 16 September 2009.* www.rac.co.uk/press-centre/press-releases/post/2009/9/driven-to-distraction/.

Radocy, R.E., & Boyle, J.D. (1979). *Psychological Foundations of Musical Behavior*. Springfield: Charles C. Thomas.

Ramos, L.-V. (1993). The Effects of On-Hold Telephone Music on the Number of Premature Disconnections to a Statewide Protective Services Abuse Hot Line. *Journal of Music Therapy* 30(2), 119–129.

Ramsey, K.L., & Simmons, F.B. (1993). High-Powered Automobile Stereos. *Otolaryngology Head and Neck Surgery* 103, 108–110.

Ransdell, S.E., & Gilroy, L. (2001). The Effects of Background Music on Word Processing Writing. *Computers in Human Behavior* 17, 141–148.

Rao, M. (2013). 'Safest' Music to Drive To? Coldplay, According to New Study (Video). *Huffpost Arts and Culture*, the *Huffington Post.* www.huffingtonpost.com/2013/01/17/safest-driving-music-coldplay-study_n_2488652.html.

Read, R. (2013). Rock On, Gold Dust Woman: Listening to Music Isn't a Distraction in the Car. *The Car Connection.* www.thecarconnection.com/news/1084959_rock-on-gold-dust-woman-listening-to-music-isnt-a-distraction-in-the-car.

Recarte, M.A., & Nunes, L.M. (2000). Effects of Verbal and Spatial-Imagery Tasks on Eye Fixations While Driving. *Journal of Experimental Psychology: Applied* 6(1), 31–43.

Recarte, M.A., & Nunes, L.M. (2009). Driver Distraction. *In* C. Castro (ed.), *Human Factors of Visual and Cognitive Performance in Driving* (pp. 65–88). Boca Raton, FL: CRC Press, Taylor and Francis.

Reed, N., & Diels, C. (2011). Emotion in Motion: Comparing the Effects of a Range of Everyday Life Driving Conditions on Driver Behaviour (Paper #76). *Proceedings of the 2nd International Conference on Driver Distractions and Inattention (DDI), Gothenberg, Sweden, 5–7 September 2011*. Gothenburg, Sweden: Chalmers. www.chalmers.se/safer/ddi2011-en/program/papers-presentations.

Rees, A.L. (2002). Moving Spaces. *In* P. Wollen & J. Keer (eds), *Autopia: Cars and Culture* (pp. 83–94). London: Reakton Books.

Reeves, J., & Stevens, A. (1996). A Practical Method for Comparing Driver Distraction Associated with In-Vehicle Equipment. *In* A.A. Gale, S.P. Taylor, & C.M. Haslegrove (eds), *Vision in Vehicles, V: Proceedings of the Glasgow Conference (9–11 September 1993*), pp. 171–178). Amsterdam: Elsevier.

Regan, M.A., Young, K.L., & Lee, J.D. (2009a). Introduction. *In* M. Regan, J.D. Lee, & K.L. Young (eds), *Driver Distraction: Theory, Effects, and Mitigation* (pp. 3–10). Boca Raton, FL: CRCPress, Taylor and Francis.

Regan, M.A., Young, K.L., Lee, J.D., & Gordon, C.P. (2009b). Sources of Driver Distraction. *In* M. Regan, J.D. Lee & K.L. Young (eds), *Driver Distraction: Theory, Effects, and Mitigation* (pp. 249–280). Boca Raton, FL: CRCPress, Taylor and Francis.

Rentfrow, P.J., & Gosling, S.D. (2003). The Do Re Mi's of Everyday Life: The Structure and Personality Correlates of Music Preferences. *Journal of Personality and Social Psychology* 84(6), 1236–1256.

Rentfrow, P.J., & Gosling, S.D. (2006). Message in a Ballad: The Role of Music Preferences in Interpersonal Perception. *Psychological Science* 17(3), 236–242.

Rentfrow, P.J., & Gosling, S.D. (2007). The Content and Validity of Music-Genre Stereotypes among College Students. *Psychology of Music* 35(2), 306–326.

Rentfrow, P.J., & McDonald, J.A. (2010). Preference, Personality, and Emotions. *In* P.N. Juslin & J.A. Sloboda (eds), *Handbook of Music and Emotion: Theory, Research, Applications* (pp. 669–698). Oxford: Oxford University Press.

Rentfrow, P.J., McDonald, J.A., & Oldmeadow, J.A. (2009). You Are What You Listen To: Young People's Stereotypes about Music Fans. *Group Processes and Intergroup Relations* 12(3), 329–344.

Reyner, L.A., & Horne, J.A. (1998). Evaluation of 'In-Car' Countermeasures to Sleepiness: Cold Air and Radio. *Sleep* 21(1), 46–51.

Roballey, T.C., McGreevy, C., Rongo, R.R., Schwantes, M.L., Steger, P.J., Wininger, M.A., & Gardner, E.B. (1985). The Effect of Music on Eating Behavior. *Bulletin of the Psychonomic Society* 23(3), 221–222.

Roberts, K.R., Dimsdale, J., East, P., & Friedman, L. (1998). Adolescent Emotional Response to Music and Its Relationship to Tisk-Taking Behaviors. *Journal of Adolescent Health* 23(1), 49–54.

Robson, C. (1993). *Real World Research: A Resource for Social Scientists and Practitioner- Researchers*. Oxford: Blackwell.

Rodrigues, R., Green, J., Virshup, L., Tsao, J., Brooks, R., Law, A., Powers, C., & Ross, S. (2013). *Radio Today 2013: How America Listens to Radio: Arbitron, Inc.* www.arbitron/com/radiotoday.htm and www.arbitron.com/study/grt.asp.

Rose, B., & Webster, T. (2012). *The Infinite Dial 2012: Navigating Digital Platforms: Arbitron/Edison*. Report found at: www.edisonresearch.com/wp-content/uploads/2012/04/2012_infirnate_dial_companiaon_report.pdf. Slides found at: www.edisonresearch.com/wp-content/uploads/2012/04/Edison_Arbitron_Infinate_Dial_2012.pdf.

RoSPA (2007). *Road Safety Information: Driver Distraction*. Birmingham, UK: Royal Society for the Prevention of Accidents (RoSPA).

Royal, D. (2003). *National Survey of Distracted and Drowsy Attitudes and Behavior: 2002 Vol. I: Findings* (DOT HS 809 566). Washington, DC: National Highway Traffic Safety Adminsitration (NHTSA).

RSC. (2006). *Parliament of Victoria Report of the Road Safety Committee on the Inquiry into Driver Distraction* (No. 209, Session 2003–2006). Victoria, AU.: Parliament of Victoria. www.parliament.vic.gov.au/images/stories/committees/rsc/driver_distraction/Distraction_Final_Report1.pdf.

Russel, P.A. (1997). Musical Tastes and Society. *In* D. Hargreaves, J. & A.C. North (eds), *The Social Psychology of Music* (pp. 141–160). Oxford: Oxford University Press.

Rustad, R.A., Small, J.E., Jobes, D.A., Safer, M.S., & Peterson, R.J. (2003). The Impact of Rock Videos and Music with Suicidal Content on Thoughts and Attitudes about Suicide. *Suicide and Life Threatening Behavior* 33(2), 120–131.

Sagberg, F. (2001). Accident Risk of Car Drivers During Mobile Telephone Use. *International Journal of Vehicle Design* 26(1), 57–69.

Salvucci, D.D., Markley, D., Zuber, M., & Brumby, D.P. (2007). iPod Distraction: Effects of Portable Music-Player Use on Driver Performance.

Sanchez, K. (2013). Turn Up the Volume: Norah Jones, Coldplay Make Your Driving Safer, Study Suggests. *Motor Trend.* http://wot.motortrend.com/turn-up-the-volume-norah-jones-coldplay-make-you-drive-safer-study-suggests-313213.html#axzz2Vc1DGXuX.

Santos, J., Merat, N., Maouta, S., Brookhuis, K., & de Waard, D. (2005). The Interaction between Driving and In-Vehicle Information Systems: Comparison of Results from Laboratory, Simulator, and Real-World Studies. *Transportation Research, Part F: Traffic Psychology and Behaviour* 8, 135–146.

Saperston, B. (1995). The Effect of Consistent Tempi and Physiologically Interactive Tempi on Heart Rate and EMG Responses. *In* T. Wigram, B. Saperston, & R. West (eds), *The Art and Science of Music Therapy: A Handbook* (pp. 58–82). Chur, Swizerland: Harwood Academic Publishers GmbH.

Scheufele, P.M. (2000). Effects of Progressive Relaxation and Classical Music on Measurements of Attention, Relaxation, and Stress Response. *Journal of Behavioral Medicine* 23, 207–226.

Schmithorst, V.J., & Wilke, M. (2002). Differences in White Matter Architecture between Musicians and Non-Musicians: A Diffusion Tensor Imaging Study. *Neuroscience Letters* 321, 57–60.

Schroer, W.J. (n.d.). Generations X, Y, Z and the Others. *The Social Librarian: Bringing the Power of Social Marketing to Library Professionals.* Retrieved 08.2013 from www.socialmarketing.org/newsletter/features/generation3.htm.

Schubert, E. (2007). The Influence of Emotion, Locus of Emotion and Familiarity upon Preference in Music. *Psychology of Music 35*(3), 499–516.

Schwartz, K.D., & Fouts, G.T. (2003). Music Preferences, Personality Style, and Developmental Issues of Adolescents. *Journal of Youth and Adolescence* 32(3), 205–213.

ScienceDaily. (2013). Listening to Music While Driving Has Very Little Effect on Driving Performance, Study Says. *Science Daily.* www.sciencedaily.com/releases/2013/06/130606101550.htm.

Sergent, J. (1993). Mapping the Musician Brain. *Human Brain Mapping* 1, 20–38.

Sethumadhaven, A. (2011). In-vehicle Technologies and Driver Distraction. *Ergonomics in Design: The Quarterly of Human Factors Applications* 19(4), 29–30.

Shawcross, V. (2008). *Booming and Fuming: Noise Nuisance from Car Stereos and Mini-Motorbikes (Report #5)*. London: Greater London Authority, London Assembly. www.london.gov.uk/assembly/reports and http://legacy.london.gov.uk/assembly/reports/... fuming-and-booming.rtf and http://london.gov.uk/sites/default/files/archives/assembly-report-environment-fumming-and-booming.pdf and http://london.gov.uk/sites/default/files/archives/assembly-report-environment-fumming-and-booming.rtf.

Sheller, M. (2004). Automotive Emotions: Feeling the Car. *Theory, Culture and Society* 21(4/5), 221–242.

Sheridan, D. (1998 [2000]). Mass-Observation Revived: The Thatcher Years and After. *In* S.D.D. Bloome & B. Street (eds), *Writing Ourselves: Literacy Practices and the Mass-Observation Project*. Cresskill, NJ: Hampton Press.

Shiloah, A. (1978). *The Epistle on Music of the Ikhwan Al-Safa (Basra Muslim Brotherhood)*. Tel Aviv: Tel Aviv University.

Shinar, D. (1978). *Psychology on the Road: The Human Factor in Traffic Safety*. New York: Wiley.

Shinar, D. (1998). Aggressive Driving: The Contribution of the Drivers and the Situation. *Transportation Research Part F* 1, 137–160.

Shinar, D. (2007). Distraction and Inattention. *In* D. Shinar (ed.), *Traffic Safety and Human Behavior* (pp. 517–564). Oxford: Elsevier.

Simons-Morton, B.G., Ouimet, M.C., Zhang, Z., Klauer, S.E., Lee, S.E., Wang, J., Chen, R., Albert, P., & Dingus, T.A. (2011). The Effect of Passengers and Risk-Taking Friends on Risky Driving and Crashes/Near Crashes among Novice Teenagers. *Journal of Adolescent Health* 49, 587–593.

Sinclare, R. (2012). Do You Become Distracted by Music While Driving? *Autobulbs Direct.* www.autobulbsdirect.co.uk/blog/801442419/do-you-become-distracted-by-music-while-driving/.

Slawinski, E.B., & MacNeil, J.F. (2002). Age, Music, and Driving Performance: Detection of External Warning Sounds in Vehicles. *Psychomusicology* 18, 123–131.

Sloboda, J.A. (1991). Music Structure and Emotional Response: Some Empirical Findings. *Psychology of Music* 19, 110–120.

Sloboda, J.A. (1999). Everyday Uses of Music Listening: A Preliminary Study. *In* S.W.Yi (ed.), *Music, Mind, and Science* (pp. 354–369). Seoul, Korea: Seoul National University Press.

Sloboda, J.A. (ed.). (2005). *Exploring the Musical Mind.* Oxford: Oxford University Press.

Sloboda, J.A., O'Neil, S.A., & Ivaldi, A. (2000). Everyday Experience of Music: An Experience-Sampling Study. *In* C. Woods, G. Luck, R. Brochard, F. Seddon, & J.A. Sloboda (eds), *Proceedings of the Sixth International Conference on Music Perception and Cognition, Keele University, August 2000*. Keele, UK: Department of Psychology, Keele University.

Sloboda, J.A., O'Neil, S.A., & Ivaldi, A. (2001). Functions of Music in Everyday Life: An Exploratory Study Using the Experience Sampling Method. *Musicae Scientiae* 5(1), 9–32.

Sloboda, J.S., Lamont, A., & Greasley, A. (2009). Choosing to Hear Music: Motivation, Process, and Effects. *In* S. Hallam, I. Cross, & M. Thaut (eds), *The Oxford Handbook of Music Psychology* (pp. 431–440). Oxford: Oxford University Press.

Smith, E. (2006). How Listening to Music Affects Driving. *The Legal Examiner.* http://sacramento.legalexaminer.com/automobile-accidents/how-listening-to-music-affects-driving/.

Smith, P.C., & Curnow, R. (1966). 'Arousal Hypothesis' and the Effects of Music on Purchasing Behavior. *Journal of Applied Psychology* 50(3), 255–256.

Spence, C. (2012, September 2012). Drive Safely with Neuroergonomics. *Psychologist* 25, 664–667.

Spence, C., & Ho, C. (2009). Crossmodal Information Processing in Driving. In C. Castro (ed.), *Human Factors of Visual and Cognitive Performance in Driving* (pp. 187–200). Boca Raton, FL: CRC Press, Taylor and Francis.

Spence, C., & Read, L. (2003). Speech Shadowing While Driving: On the Difficulty of Splitting Attention Between the Eye and Ear. *Psychological Science* 14(3), 251–256.

Spinney, L. (1997). Pump Down the Volume. *New Scientist* 155, 22.

Stack, S., & Gunndlach, J. (1992). The Effects of Country Music on Suicide. *Social Forces* 71(1), 211–218.

Standley, J.M. (1986). Music Research in Medical/ Dental Treatment: Meta-Analysis and Clinical Applications. *The Journal of Music Therapy* 23(2), 56–122.

Standley, J.M. (1992). Meta-Analysis of Research in Music and Medical Treatment: Effect Size as a Basis for Comparison across Multiple Dependent and Independent Variables. *In* R. Spintge & R. Droh (eds), *MusicMedicine* (pp. 364–378). St. Louis: MMB Music.

Standley, J.M. (1995). Music as a Therapeutic Intervention in Medical and Dental Treatment: Research and Clinical Applications. *In* T. Wigram, B. Saperston, & R. West (eds), *The Art and Science of Music Therapy: A Handbook* (pp. 3–22). Chur, Swizerland: Harwood Academic Publishers GmbH.

Staum, M.J., & Brotons, M. (2000). The Effect of Music Amplitude on the Relaxation Response. *Journal of Music Therapy* 37, 22–39.

Steg, L. (2005). Car Use: Lust and Must: Instrumental, Symbolic and Affective Motive for Car Use. *Transportation Research: Part A* 39, 147–162.

Steg, L., & Gifford, R. (2005). Sustainable Transportation and Quality of Life. *Journal of Transport Geography* 13, 50–69.

Steg, L., Vlek, C., & Slotegraaf, G. (2001). Instrumental-Reasoned and Symbolic-Affective Motives for Using a Motor Car. *Transportation Research, Part F* 4(3), 151–169.

Stevens, A., & Minton, R. (2001). In-Vehicle Distraction and Fatal Accidents in England and Wales. *Accident Analysis and Prevention* 33, 539–545.

Stevens, A., & Rai, G. (2000). Development of Safety Principles for In-Vehicle Information and Communication System. London: Transportation Research Laboratory. www.esafetysupport.org/download/working_groups/30.pdf.

Stomp, W. (2011). Toyota to Integrate ECG Sensors into Steering Wheel. *medGadget.* www.medgadget.com/2011/07/toyota-to-integrate-ecg-sensors-into-steering-wheels.html.

Stratton, V.N., & Zalanowski, A. (1984). The Effect of Background Music on Verbal Interaction in Groups. *Journal of Music Therapy* 21(1), 16–26.

Strayer, D.L., & Drews, F.A. (2004). Profiles in Driver Distraction: Effects of Cell Phone Conversations on Younger and Older Drivers. *Human Factors* 46(4), 640–649.

Strayer, D.L., Drews, F.A., & Johnston, W.A. (2003). Inattention-Blindness behind the Wheel. *Journal of Vision* 3(9), 157.

Strayer, D.L., & Johnston, W.A. (2001). Driven to Distraction: Dual-Task Studies of Simulated Driving and Conversing on a Cellular Telephone. *Psychological Science* 12(6), 462–466.

Stutts, J.C. (2001). Cell Phones and Driving (Presentation Slides). Washington, DC: AAA Foundation. http://www.aaafoundation.org/multimedia/presentations/CellPhones.ppt.

Stutts, J., Feaganes, J., Reinfurt, D., Rodman, E., Hamlett, C., Gish, K., & Staplin, L. (2005). Driver's Exposure to Distractions in Their Natural Driving Environment. *Accident Analysis and Prevention* 37, 1093–1101.

Stutts, J., Feaganes, J., Rodgman, E., Hamlett, C., Meadows, T., Reinfurt, D., Gish, K., Mercadante, M., & Staplin, L. (2003). *Distractions in Everyday Driving*. Washington, DC: AAA Foundation for Traffic Safety. https://www.aaafoundation.org/research/completed-projects (For Year 2003) and www.aaafoundation.org./pdf/DistarctionsInEverydayDriving.pdf.

Stutts, J.C., Reinfurt, D.W., & Rodgman, E.A. (2001a). The Role of Driver Distraction in Crashes: An Analysis of 1995–1999 Crashworthiness Data System Data. *The Annual Proceedings of the Association for the Advancement of Automotive Medicine* 45, 287–301.

Stutts, J.C., Reinfurt, D.W., Staplin, L., & Rodgman, E.A. (2001b). *The Role of Driver Distraction in Traffic Crashes.* Washington, DC: AAA Foundation for Traffic Safety. https://www.aaafoundation.org/research/completed-projects (For Year 2001) and www.aaafoundartion.org/pdf/distraction.pdf.

Stutts, J.C., Wilkins, J.W., & Vaughn, B.V. (1999). *Why Do People Have Drowsy Driving Crashes? Input from Drivers Who Just Did.* Washington, DC: AAA Foundation for Traffic Safety. https://www.aaafoundation.org/why-do-people-have-drowsy-driving-crashes-input-drivers-who-just-do and https://www.aaafoundation.org/sites/default/files/sleep.pdf.

Styles, W. (2013). Listening to Music Has Little Effect on Your Driving – So Crank It Up. *103 Albany's Rock Station.* http://q103albany.com/listening-to-music-has-little-effect-on-your-driving-so-crank-it/.

Summers, T. (2012). Music in Racing Video Games. *In* C. Hart, M. Duffett, & B. Peter (eds), *Proceedings of the Popular Music and Automobile Culture Conference*. University of Chester, England: www.hartchester.blogspot.com.

Swade. (2008). Mamma Mia – Volvo's Summer Road Trip Songs! www.trollhattansaab.net/archives/2008/07/mamma-mia-volvos-summer-road-trip-songs.html.

Syal, R. (2004). Booming Car Stereos Aren't Just Infuriating – They're Dangerous. *Telegraph.* www.telegraph.co.uk/news/uknews/1457925/Booming-car-stereos-arent-just-infuriating-theyre-dangerous.html.

Szabo, A., Small, A., & Leigh, M. (1999). The Effects of Slow- and Fast-Rhythm Classical Music on Progressive Cycling to Voluntary Physical Exhaustion. *Journal of Sports Medicine and Physcial Fitness* 39, 220–225.

Tarrant, M., North, A.C., & Hargreaves, D.J. (2000). English and American Adolescents' Reasons for Listening to Music. *Psychology of Music* 28(2), 166–173.

Terwogt, M.M., & van Grinsven, F. (1991). Musical Expression of Moodstates. *Psychology of Music* 19, 99–109.

TheMayFirm. (2013). Does Music Cause Car Accidents. *The May Firm, Injury Lawyers, Central Coast, Blog.* www.mayfirm.com/blog/music-car-accidents/.

TheTelegraph. (2009). Rap Fans Most At Risk of Car Accidents. *Road and Rail Transport, in the Telegraph (UK Online).* www.telegraph.co.uk/news/uknews/road-and-rail-transport/6221801/Rap-fans-most-at-risk-of-car-accidents.html.

Thompson, W.F., Schellenberg, E.G., & Letnic, A.K. (2012). Fast and Loud Background Music Disrupts Reading Comprehension. *Psychology of Music* 40(6), 700–708.

Thrasher, M., van der Zwaag, M.D., Bianchi-Berthouze, N., & Westerink, J.H.D.M. (2011). Mood Recognition Based on Upper Body Posture and Movement Features. *In* S. D'Mello, A. Graesser, B. Schuller, & J.-C. Martin (eds), *Proceedings of the 4th International Conference of Affective Computing and Intelligent Interaction, 9–12 October 2011, Memphis, TN (USA)* (pp. 377–386). Berlin: Springer-Verlag.

Thrift, N. (2004). Driving in the City. *Theory, Culture, and Society* 21(4/5), 41–59.

Tijerina, L. (2000). *Issues in Evaluation of Driver Distraction Associated with In-Vehicle Information and Telecommunications Systems*. USA: Transportation Research Center. www.nrd.nhtsa.dot.gov/departments/Human Factors/driver-distraction/PDF/3.pdf and www.skylinewaterways.org/Appeal/%2523Directory-Traffic_Safety/Issues%2520in%2520the%2520Evaluation%2520of%2520Driver%2520Distraction%2520Associated%2520with%2520In-Vehicle%2520Information%2520and%2520Tele%2520System.pdf andsa=Xandscisig=AAGBfm3RpKE54mHf0AxLYRA04oeQ9KJsKwandoi=scholarrandei=93IlUp3fKIjF7AbMvIDoBQandved=0CCkQgAMoADAA.

Titchener, K., White, M., & Kaye, S. (2009). In-Vehicle Driver Distractions: Characteristics Underlying Drivers' Risk Perceptions. Proceedings of the *Road Safety Conference, New South Wales, 10–12 November 2009*. New South Wales, Sydney, Australia: Queensland University of Technology.

Trainor, L.J., & Schmidt, L.A. (2003). Processing Emotions Induced by Music. *In* I. Peretz & R.J. Zatorre (eds), *The Cognitive Neurosciences of Music* (pp. 310–324). Oxford: Oxford University Press.

Tripp, J. (2005). The Car Radio: An Interesting Story; Car Radio History. *Jim's Antique Radio Museum.* www.antiqueradiomuseum.org/thecarradio.htm.

Turley, L.W., & Milliman, R.E. (2000). Atmospheric Effects on Shopping Behavior: A Review of the Experimental Evidence. *Journal of Business Research* 49(2), 193–211.

Turner, M., L., Fernandez, J.E., & Nelson, K. (1996). The Effect of Music Amplitude on the Reaction to Unexpected Visual Events. *The Journal of General Psychology* 123(1), 51–62.

Unal, A.B. (2013b). *Please Don't Stop the Music... The Influence of Music and Radio on Cognitive Processes, Arousal and Driving Performance.* University of Groningen, The Netherlands.

Unal, A.B., de Waard, D., Epstude, K., & Steg, L. (2013c). Driving with Music: Effects on Arousal and Performance. *Transportation Research Part F* 21, 52–65.

Unal, A.B., Platteel, S., Steg, L., & Epstude, K. (2013a). Blocking-Out Auditory Distractors While Driving: A Cognitive Strategy to Reduce Task-Demand on the Road. *Accident Analysis and Prevention* 590, 434–442.

Unal, A.B., Steg, L., & Epstude, K. (2012). The Influence of Music on Mental Effort and Driving Performance. *Accident Analysis and Prevention* 48, 271–278.

Underwood, G., Chapman, P., & Crundall, D. (2009). Experience and Visual Attention in Driving. In C. Castro (ed.), *Human Factors of Visual and Cognitive Performance in Driving* (pp. 89–117). Boca Raton, FL: CRC Press, Taylor and Francis.

Urken, R.K. (2011). Music and Cars Go Together for All Times. www.autos.aol.com/article/best-car-songs/.

Urry, J. (1999). Automobility, Car Culture and Weightless Travel: A Discussion Paper. Lancaster, UK: Department of Sociology, Lancaster University. http://www.lancs.ac.uk/fass/sociology/papers/urry-automobility.pdf.

Urry, J. (2004). The 'System' of Automobility. *Theory, Culture, and Society* 21(4/5), 25–39.

Urry, J. (2006). Inhabiting the Car. *Sociological Review* 54(1), 17–31.

USAToday. (2004). Wagner Tops List of Music Not to Play While Driving. *USA Today Lifestyle.* http://usatoday30.usatoday.com/life/lifestyle/2004–04–14-music-and-driving_x.htm.

van der Zwaag, M., Dijksterhuis, C., de Waard, D., Mulder, B.L.J.M., Westerink, J.H.D.M., & Brookhuis, K.A. (2012). The Influence of Music on Mood and Performance While Driving. *Ergonomics* 55(1), 12–22.

van der Zwaag, M.D., Fairclough, S., Spiridon, E., & Westerink, J.H.D.M. (2011). The Impact of Music on Affect during Anger Inducing Drives. *In* S. D'Mello, A. Graesser, B. Schuller, & J.-C. Martin (eds), *Proceedings of the 4th International Conference of Affective Computing and Intelligent Interaction, 9–12 October 2011, Memphis, TN (USA)* (pp. 407–416). Berlin: Springer-Verlag.

van der Zwaag, M.D., Janssen, J.H., Nass, C., Westerink, J.H.D.M., Chowdhury, S., & de Waard, D. (2013a). Using Music to Change Mood while Driving. *Ergonomics* 56(10), 1504–1514.

van der Zwaag, M.D., Janssen, J.H., & Westerink, J.H.D.M. (2013b). Directing Physiology and Mood through Music: Validation of an Affective Music Player. *IEEE Transactions on Affective Computing* 4(1), 57–68.

Vanderbilt, T. (2008). *Traffic: Why We Drive the Way We Do (And What It Says about Us)*. New York: Vintage Books, Random House.

Vanlaar, W., Simpson, H., Mayhew, D., & Robertson, R. (2008). Fatigues and Drowsy Driving: A Survey of Attitudes, Opinions, and Behaviors. *Journal of Safety Research* 39, 303–309.

Victor, T.W., Engström, J., & Harbluck, J. L. (2009). Distraction Assessment Methods Based on Visual Behavior and Event Detection. *In* M. Regan, J.D. Lee, & K.L. Young (eds), *Driver Distraction: Theory, Effects, and Mitigation* (pp. 135–169). Boca Raton, FL: CRCPress, Taylor and Francis Group.

Victor, T.W., Harbluck, J.L., & Engström, J.A. (2005). Sensitivity of Eye-Movement Measures to In-Vehicle Task Difficulty. *Transportation Research Part F* 8, 167–190.

Volvo. (2008). Volvo Songs and Sounds. *SAC Volvo Club.* www.sacvolvoclub.org/Home/songs.

Walter, H., Vetter, S.C., Grothe, J., Wunderlich, A.P., Hahn, S., & Spitzer, M. (2001). The Neural Correlates of Driving. *NeuroReport,* 12(8), 1763–1767.

Webster, G.D., & Weir, C.G. (2005). Emotional Responses to Music: Interactive Effects of Mode, Texture, and Tempo. *Motivation and Emotion,* 29(1), 19–39.

Webster, T. (2012). The Infinite Dial 2012: Navigating Digital Platforms. *Edison Research: Perspectives, News and Opinions from the Researchers at Edison.* www.edisonresearch.com/home/archives/2012/04/the-infinite-dial-2012-navigating-digital-platforms.php.

Weingroff, R.F. (1996). Road Movies. *Federal Highway Administration, U.S. Department of Transportation, Celebrating the 50th Anniversary of the Eisenhower Interstate Highway System.* www.fhwa.dot.gov/interstate/roadmovies.htm.

Westlake, E.J., & Boyle, L.N. (2012). Perception of Driver Distractions Among Teenage Drivers. *Transportation Research, Part F* 15, 644–653.

White, M.P., Eiser, J.R., & Harris, P.R. (2004). Risk Perceptions of Mobile Phone Use While Driving. *Risk Analysis* 24(2), 323–334.

Widmar, E.L. (2002). Crossroads: The Automobile, Rock and Roll, and Democracy. *In* P. Wollen & J. Keer (eds), *Autopia: Cars and Culture* (pp. 65–74). London: Reakton Books.

Wiese, E.E., & Lee, J.D. (2004). Auditory Alerts for In-Vehicle Information Systems: The Effects of Temporal Conflict and Sound Parameters on Driver Attitudes and Performance. *Ergonomics* 47(9), 965–986.

Wiesenthal, D., Hennessy, D.A., & Totten, B. (2000). The Influence of Music on Driver Stress. *Journal of Applied Social Psychology* 30(8), 1709–1719.

Wiesenthal, D., Hennessy, D.A., & Totten, B. (2003). The Influence of Music on Mild Driver Aggression. *Transportation Research Part F* 6, 125–134.

Wikman, A.-S., Nieminen, T., & Summala, H. (1998). Driving Experiences and Time-Sharing During In-Car Tasks on Roads of Different Width. *Ergonomics* 41(3), 358–372.

Will, J. (2012). Is Loud Music a Driving Hazard? *Globe And Mail.* www.theglobeandmail.com/globe-drive/car-life/is-loud-music-a-driving-hazard/article4171340/.

Williams, A.F. (2003). Teenage Drivers: Patterns of Risk. *Journal of Safety Research* 34, 5–15.

Williams, A.F., Ferguson, S.A., & McCartt, A.T. (2007). Passenger Effects on Teenage Driving and Opportunities for Reducing the Risks of Such Travel. *Journal of Safety Research* 38, 381–390.

Williams, D. (2009). The Arbitron National In-Car Study (2009 Edition). New York: Arbitron, Inc.

Williams, G. (2008). A Brief History of Car Audio. *Calgary Herald.* www.gregwilliams.ca/?p=461.

Williams, J.A. (2010). 'You Never Been on a Ride Like This Befo': Los Angeles, Automotive Listening, and Dr. Dre's 'G-Funk'. *Popular Music History* 4(2), 160–176.

Williams, J.A. (2014). 'Cars with the Boom': Music, Automobility, and Hip-Hop 'Sub' Cultures. *In* S. Gopinath & J. Stanyek (eds), *The Oxford Handbook of Mobile Music, Vol. 2.* (pp. 109–145). Oxford: Oxford University Press.

Williamson, V. (2011a). Music and Driving. *Music Psychology with Dr Victoria Williamson.* http://musicpsychology.co.uk/music-and-driving/.

Williamson, V. (2011b). Music and Driving. *Music Psychology.* http://musicpsychology.co.uk/music-and-driving/.

Wilson, S. (2003). The Effect of Music on Perceived Atmosphere and Purchase Intention in a Restaurant. *Psychology of Music* 31(1), 93–112.

Wollen, P. (2002). Cars and Culture. *In* P. Wollen & J. Keer (eds), *Autopia; Cars and Culture* (pp. 10–20). London: Reakton Books.

Wright, C., & Curtis, B. (2005). Reshaping the Motor Car. *Transportation Policy* 12, 11–22.

Yalch, R., & Spangenberg, E. (1990). Effects of Store Music on Shopping Behavior. *The Journal of Consumer Marketing* 7(2), 55–63.

Yannis, G. (2012). The Effects of Different Types of Driver Distractions: Findings from the EU. *Proceedings of the Canadian Automobile Association, Traffic Injury Research Foundation, International Conference on Distracted Driving, 21 March 2012*. Toronto, Canada. www.nrso.ntua.gr/geyannis.

Yeoh, J.P.S., & North, A.C. (2010a). The Effects of Musical Fit on Choice between Two Competing Foods. *Musicae Scientiae* 14(1), 165–180.

Yeoh, J.P.S., & North, A.C. (2010b). The Effect of Musical Fit on Consumers' Memory. *Psychology of Music* 38(3), 368–378.

Yeoh, J.P.S., & North, A.C. (2012). The Effects of Musical Fit on Consumers' Preferences between Competing Alternate Petrols. *Psychology of Music* 40(6), 709–719.

Young, K.L., & Lenne, M.G. (2010). Diver Engagement in Distracting Activities and the Strategies Used to Minimize Risk. *Safety Science* 48, 326–332.

Young, K.L., Mitsopoulos-Rubins, E., Rudin-Brown, C.M., & Lenne, M.G. (2011). Driver Behaviour and Task-Sharing Strategies When Using a Portable Music Player (Paper #18). *Proceedings of the 2nd International Conference on Driver Distraction and Inattention (DDI), Gothenberg, Sweden, 5–7 September 2011*. Gothenburg, Sweden: Chalmers. www.chalmers.se/safer/ddi2011-en/program/papers-presentations.

Young, K.L., & Salmon, P. (2012). Examining the Relationship between Driver Distraction and Driving Errors: A Discussion of Theory, Studies, and Methods. *Safety Science* 50, 165–174.

Zakay, D. (1989). Subjective Time Attentional Resource Allocation: An Integrated Model of Time Estimation. In I. Levin & D. Zakay (eds), *Time and Human Cognition: A Life-Span Perspective* (pp. 365–397). North Holland: Elsevier Science.

Ziv, N., Hoftman, M., & Geyer, M. (2012). Music and Moral Judgment: The Effects of Background Music on the Ads Promoting Unethical Behavior. *Psychology of Music* 40(6), 738–760.

Zuckerman, M. (1994). *Behavioral Expressions and Biosocial Bases of Sensation Seeking*.

Index